Body Systems Review I: Hematopoietic/Lymphoreticular, Respiratory, Cardiovascular

Body Systems Review I: Hematopoietic/Lymphoreticular, Respiratory, Cardiovascular

 Board Simulator

DEVELOPED By NATIONAL MEDICAL SCHOOL REVIEW

EDITOR

Edward F. Goljan, M.D.

*Associate Professor and Chairman
Department of Pathology
Oklahoma State University
College of Osteopathic Medicine
Director of NMSR's Step 2 Programs
Director of NMSR's Department of Pathology
NMSR Professor of Pathology*

Williams & Wilkins
A WAVERLY COMPANY

BALTIMORE • PHILADELPHIA • LONDON • PARIS • BANGKOK
BUENOS AIRES • HONG KONG • MUNICH • SYDNEY • TOKYO • WROCLAW

1996

Editor: Elizabeth A. Nieginski
Managing Editors: Amy G. Dinkel, Alethea H. Elkins
Development Editors: Melanie Cann, Beth Goldner, Rebecca Krumm, Carol Loyd
Editorial Assistants: Lisa Kiesel, Jackie Jenks
Production Coordinator: Danielle Santucci
Designer: Cotter Visual Communications
Artist: Matthew C. Chansky
Typesetter: Port City Press
Printer: Port City Press
Binder: Port City Press

Copyright © 1996 Williams & Wilkins

351 West Camden Street
Baltimore, Maryland 21201-2436 USA

Rose Tree Corporate Center
1400 North Providence Road
Building II, Suite 5025
Media, Pennsylvania 19063-2043 USA

All rights reserved. This book is protected by copyright. No part of this book may be reproduced in any form or by any means, including photocopying, or utilized by any information storage and retrieval system without written permission from the copyright owner.

Accurate indications, adverse reactions, and dosage schedules for drugs are provided in this book, but it is possible that they may change. The reader is urged to review the package information data of the manufacturers of the medications mentioned.

Printed in the United States of America

First Edition,

Library of Congress Cataloging-in-Publication Data
Body systems review I : hematopoietic/lymphoreticular, respiratory,
    cardiovascular / developed by National Medical School Review — 1st
    ed.
        p.   cm. — (Board simulator)
    Contains simulations of the United States medical licensing
examination (USMLE) Step 1.
    ISBN 0-683-06329-4
    1. Cardiopulmonary system—Pathophysiology—Examinations,
questions, etc.   2. Hematopoietic system—Pathophysiology—
Examinations, questions, etc.   3. Cardiopulmonary system—
Physiology—Examinations, questions, etc.   4. Hematopoietic system—
Physiology—Examinations, questions, etc.   I. National Medical
School Review (Firm)   II. Series.
    [DNLM: 1. Hematopoietic System—examination questions.
2. Respiratory System—examination questions.   3. Cardiovascular
System—examination questions.   WH 18.2 B668 1996]
RC702.B63   1996
616'.0076—dc20
DNLM/DLC
for Library of Congress   96-24459
                                                                                                  CIP

The publishers have made every effort to trace the copyright holders for borrowed material. If they have inadvertently overlooked any, they will be pleased to make the necessary arrangements at the first opportunity.

To purchase additional copies of this book, call our customer service department at (800) 638-0672 or fax orders to (800) 447-8438. For other book services, including chapter reprints and large-quantity sales, ask for the Special Sales department.

Canadian customers should call (800) 268-4178, or fax (905) 470-6780. For all other calls originating outside of the United States, please call (410) 528-4223 or fax us at (410) 528-8550.

Visit Williams & Wilkins on the Internet: http://www.wwilkins.com or contact our customer service department at custserv@wwilkins.com. Williams & Wilkins customer service representatives are available from 8:30 am to 6:00 pm EST, Monday through Friday, for telephone access.

96 97 98 99
1 2 3 4 5 6 7 8 9 10

Contents

Contributors — vii
Preface — ix
Acknowledgment — xi
Guide to Using this Book — xiii
Exam Preparation Guide — xix

Test 1 — 1
- Questions — 3
- Answer Key — 37
- Answers and Explanations — 39

Test 2 — 79
- Questions — 81
- Answer Key — 129
- Answers and Explanations — 131

Test 3 — 175
- Questions — 177
- Answer Key — 215
- Answers and Explanations — 217

Test 4 — 251
- Questions — 253
- Answer Key — 287
- Answers and Explanations — 289

CONTRIBUTORS

GERALD D. BARRY, Ph.D.
Professor of Physiology and Director of the MA/MD
    Biomedical Program
Touro College School of Health Sciences

GRACE BINGHAM, Ed.D.
President and Educational Consultant
Bingham Associates, Inc.
Toms River, NJ
Coordinator of Cognitive Skills
National Medical School Review

GEORGE M. BRENNER, Ph.D.
Professor and Chairman
Department of Pharmacology
Oklahoma State University
College of Osteopathic Medicine

BARBARA FADEM, Ph.D.
Professor, Department of Psychiatry
University of Medicine and Dentistry
New Jersey Medical School

DAILA S. GRIDLEY, Ph.D.
Professor
Department of Microbiology and Molecular Genetics
Department of Radiation Medicine
Loma Linda University
School of Medicine

VICTOR N. GRUBER, M.D.
Founder and Executive Director
National Medical School Review
Newport Beach, California

KENNETH H. IBSEN, Ph.D.
Professor, Emeritus
Department of Biochemistry
University of California at Irvine
Director of Academic Development
National Medical School Review

KIRBY L. JAROLIM, Ph.D.
Professor and Chairman
Department of Anatomy
Oklahoma State University
College of Osteopathic Medicine

KATHLEEN KEEF, Ph.D.
Professor
Department of Physiology and Cell Biology
University of Nevada
School of Medicine

JAMES KETTERING, Ph.D.
Professor and Assistant Chairman
Department of Microbiology and Molecular Genetics
Loma Linda University
School of Medicine

RICHARD M. KRIEBEL, Ph.D.
Professor
Department of Anatomy
Philadelphia College of Osteopathic Medicine

WILLIAM D. MEEK, Ph.D.
Professor
Department of Anatomy
Oklahoma State University
College of Osteopathic Medicine

STANLEY PASSO, Ph.D.
Associate Professor
Department of Physiology
New York Medical College

JAMES P. PORTER, Ph.D.
Associate Professor
Department of Physiology and Biophysics
University of Louisville
School of Medicine

VERNON REICHENBECHER, Ph.D.
Associate Professor
Department of Biochemistry and Molecular Biology
Marshall University
School of Medicine

DAVID SEIDEN, Ph.D.
Professor
Department of Neuroscience and Cell Biology
University of Medicine and Dentistry
Robert Wood Johnson Medical School

PREFACE

Since its establishment in 1988, the goal of National Medical School Review (NMSR) has been to provide medical students and physicians with the information they need to pass their national licensing examinations. During this period, NMSR has established a national reputation for high-quality programs delivered by the best teaching faculty available in United States and Canadian schools. Nearly 10,000 participants in NMSR programs have had access to outstanding faculty lectures as well as to diagnostic and practice examinations and high-yield notes that can be kept as learning tools, and they have achieved an impressive level of success on the United States Medical Licensing Examination (USMLE) Steps 1, 2, and 3. The publication of the *Board Simulator* series inaugurates a truly innovative new educational contribution by which NMSR will help medical students to prepare for and pass Step 1 of the USMLE.

This five-volume series (Vol. I: *General Principles;* Vol. II: *Normal and Abnormal Processes;* Vols. III, IV, and V: *Body Systems I, II, III*) is unique in that it was developed using the USMLE Step 1 Guidelines published by the National Board of Medical Examiners (NBME), rather than being organized strictly by isolated basic science topics (for example, individual volumes covering anatomy, biochemistry, or cell biology). Therefore, many questions are preceded by a clinical vignette with the content structured in the integrated fashion currently favored by the NBME, and questions often require a multistep reasoning process to arrive at a correct response. Thus, not only must students recall a particular fact or principle, but they must also manipulate and apply that information to come up with the correct answer. It is NMSR's belief that judicious use of this series can provide most second-year medical students attending a United States medical school the essential information and test-taking experience required to pass Step 1. This series can also provide valuable assistance to those students who feel they could also benefit from attending a review program.

The *Board Simulator* series provides far more than simulated Step 1 examinations, because each question is answered with a full and detailed explanation. A student who uses these explanations to clarify why the right answers are correct choices and the wrong answers are incorrect choices will have performed a comprehensive, up-to-date content review of the basic biomedical sciences covered in the Step 1 examination. With this goal in mind, NMSR believes it has developed an educational tool that can help maximize a student's opportunity to make reviewing for the USMLE Step 1 a successful and rewarding experience.

Victor Gruber, M.D.

Acknowledgment

The authors thank the Williams & Wilkins editorial staff for their expertise in organizing and editing this book.

Guide to Using this Book

During the past 10 years, a number of changes in curriculum organization have occurred in many U.S. and Canadian medical schools, particularly within the first 2 years of education. A number of meaningful innovations have been implemented, such as more self-directed learning, de-emphasis of lectures as the dominant instructional mode, earlier introduction of clinical experiences, and increases in the proportion of problem-based learning.

Today, regardless of which curriculum a medical school adopts, one trend exists even in those schools that have maintained a traditional stance: a loosening of the boundaries that organized basic science material into large territorial "subject" courses. The move is toward synthesizing domains of medical knowledge into more flexible cross-disciplinary patterns believed to approximate interactions that characterize medical practice today.

From a student's perspective, all the attention given to the number and variety of medical school curriculum reforms may have highlighted only the differences among the curricula and neglected to underscore the important similarities remaining. Regardless of the specific curriculum followed at any school, medical students are still expected to:

1. Read and understand large quantities of material. Whether the access mode is texts, handouts, specialized print materials, or computer modules, the reading demands remain significant.

2. Organize information in meaningful ways. No matter how well written a concept may appear in a text, or how well explained in a lecture, the students should *generate* the pattern of meaning that makes the most sense to them.

3. Develop relationships among experiences. All medical students are expected to engage in higher order thinking processes. New information about a topic will need to be *encoded* and *synthesized* with prior knowledge; *compared* with a lab experiment; *evaluated* through discussion with a colleague; or *solved* as a problem.

4. Store, retrieve, and remember information dependably. Historically, the demands on memory for medical personnel have been exceptional. The rapidity with which new technological advances in diagnosis and increases in treatment options are becoming available makes it more difficult than ever to learn and stay up to date in the field. Distinguishing between what needs to be mastered and recalled readily and what does not is a professional decision with which medical practitioners will struggle for the rest of their careers.

5. Demonstrate achievement through various forms of evaluation. Regardless of their curriculum model, all medical schools still require high levels of performance of their students. Evaluators can choose from a large range of methods to assess students, from the traditional instructor-designed examinations to oral evaluations, on-site observations, behavioral checklists, and product evaluations (e.g., written reports, research papers, problem solutions). Whatever the method, students will need to give evidence of competency.

6. Demonstrate competency on national standardized examinations. To qualify for licensure, students need to be successful on Steps 1, 2, and 3 of the USMLE. After further graduate training, students need to demonstrate success on specialty examinations to qualify for Board certification.

Series Design

Some of the more positive changes that characterize medical learning today are already reflected in the design of current instructional materials. One such change is the design of the five books in this series. The organization differs from the traditional subject-oriented subdivisions of basic science in that it conforms more closely to the content outline of USMLE Step 1 as presented in the *General Instructions* booklet. In Step 1, basic science material is organized along two dimensions: system (consisting of general principles and individual organ systems) and process, which divide each organ system into normal development, abnormal processes, therapeutics, and psychosocial and other considerations.

Book I: General Principles/Quantitative Methods: Normal Development and Normal Processes

Book II: General Principles: Normal and Abnormal Processes

Book III: Hematopoietic and Lymphoreticular Systems, Respiratory System, Cardiovascular System: Normal and Abnormal Processes

Book IV: Gastrointestinal System, Renal/Urinary System, Reproductive System, Endocrine System: Normal and Abnormal Processes

Book V: Nervous System/Special Senses, Skin and Connective Tissue, Musculoskeletal System: Normal and Abnormal Processes

Each book contains four examinations consisting of approximately 160 questions each, for a total of 650 questions per book. The distribution of questions within each test in books III, IV, and V approximates subcategory percentages as follows: normal development, 10%–20%; normal processes, 20%–30%; abnormal processes, 30%–40%; principles of therapeutics, 10%–20%; psychosocial, cultural, and environmental considerations, 10%–20%.

Questions in each test are presented using the two multiple-choice formats that appear in USMLE Step 1: One Best Answer (including negatively phrased items) and Matching Sets, with the largest number of questions of the One Best Answer type.

Who Can Use These Questions?

The group likely to use the questions in this series of books most frequently is students preparing for USMLE Step 1. However, other student groups who can benefit from these questions are medical students in years I, II, III, and IV. These students may turn to individual books in the series or use the entire set of books as a supplemental self-testing resource.

Students Preparing for USMLE Step 1

Students preparing for the Step 1 exam will use these books as a **diagnostic tool,** as a **guide to focus further study,** and as a **self-evaluation device.**

Specific instructions about how to use the question sets for each purpose will be described in the Examination Preparation Guide that follows.

Students in Years I and II

Students in years I and II will use these books for **periodic self-testing.** Students in the first 2 years of medical school take a large number of examinations that evaluate their performance on material covered in their courses or other learning experiences. Those tests are usually compiled by the instructors and reflect the intructors' choice of emphasis. The questions in the five books in this series provide a sampling of the wide range of material typically taught in the first 2 years of medical school. For students who want to practice with questions that go beyond the scope of their specific school's course, these question sets provide another level of testing, one that approximates more closely the expectations of the Step 1 exam. These questions also offer opportunities for practice to those students in medical school courses that use Board shelf exams as one of their required evaluations. The section How to Use These Practice Exams provides guidance on their use.

Students in Years III and IV

Students in years III and IV will use these books for **reactivating and assessing prior learning.** During their third and fourth years, students may find during clinical situations that they have forgotten some material they learned during the first 2 years. One way to stimulate, reactivate, and supplement that knowledge is by responding to questions. The content and organization of the questions in each book of the series make it possible for students in the clinical years to select and use specific segments for review.

How to Use These Practice Exams

The four sets of practice questions in this book assess knowledge of both normal and abnormal processes associated with each of the systems addressed. Whereas the specific content of questions in each system differs, certain categories of knowledge organization recur. Questions in this book relate to:

— Normal processes
  Development and structure
  Metabolic, physiologic, and regulatory processes
  Repair and regeneration
— Abnormal processes
  Disorders arising from a variety of origins:
    Traumatic
    Mechanical
    Metabolic, physiologic, or regulatory
    Neoplastic
    Idiopathic
    Vascular
— Therapeutic processes
  Uses of drugs to treat disorders, and their mechanisms of action
  Adverse effects of drug treatment
— Psychosocial, cultural, and environmental factors affecting disease processes

These question sets may be used in a number of ways: (1) as a diagnostic tool (pretest), (2) as a guide and focus for further study, and (3) for self-evaluation. The least effective use of these questions is to "study" them by reading them one at at time, and then looking at the correct response. Although the questions have been compiled to be representative of the domains of information found in USMLE Step 1, simply knowing the answers to these particular 650 questions does not ensure a passing grade on the exam. The questions are intended to be an integral part of a well-planned review,

These question sets may be used in a number of ways: (1) as a diagnostic tool (pretest), (2) to guide and focus further study, and (3) for self-evaluation. The least effective use of these questions is to "study" them by reading them one at at time, and then looking at the correct response. Although the questions have been compiled to be representative of the domains of information found in USMLE Step 1, simply knowing the answers to these particular 650 questions does not ensure a passing grade on the exam. The questions are intended to be an integral part of a well-planned review, rather than an isolated resource. If used appropriately, the four sets can provide self-assessment information beyond a numeric score.

As a diagnostic tool. It is possible to use each set of questions as a screening device to gather diagnostic information about relative performance across the 10 large topics presented in this book. For those who have been away from basic science study for awhile and have no other recent performance data, using a practice exam in this manner provides a form of feedback before beginning review. This method also allows students to respond to Board-type questions similar to those on the examination so they can experience the structure and complexity of such questions and acquire a sense of what the questions "feel" like.

1. Select any one of the four tests in this book. It does not matter which one you choose, since they are all approximately equal in terms of topics represented and question difficulty.

2. Allow yourself the same amount of time as will be allowed on the Board exam (approximately 60 seconds per question).

3. Use a separate sheet of paper for your answers (instead of writing in the book). This will make it easier for you to score, analyze, and interpret the results.

4. Score your responses (but do not read the correct answer to the question or record the correct response next to your incorrect one). Compute an accuracy level by counting the number of correct responses and dividing by the total number of questions to get the percent correct. Note your score, but be careful not to overreact to this initial score. Remember that this type of sampling provides only a rough indication of how familiar or remote this basic science material seems to you before review. Not reading the correct answer to these questions may seem a bit strange at first, but by not doing so now, you will be able to use these questions again later in your review as a posttest to check progress.

5. Know your distribution of errors across the topics. To find out, categorize each error (e.g., biochemistry, cell biology, genetics; tissue biology, pharmacokinetics, multisystem processes). If in doubt about how to categorize a particular question, check the reference listed in the answer and use that to make your decision.

6. Arrange topics in a hierarchy from relatively strong (few errors) to relatively weak (many errors). Did you do well in those topics you thought you would do well in, and vice versa, or were there some unexpected highs or lows?

To guide further study. After reviewing the material of the major areas noted previously and giving the information a complete "first pass," it is time to test yourself using another question set in this book. Your purpose is to check your estimates of which topics and subtopics have been learned well, which are still shaky, and which are quite weak. To do that:

1. Follow the first five steps described previously.

2. Analyze errors using the guidelines described in the section "Monitoring Functions for Consolidating Information."

3. Focus your follow-up study on the content areas or specific subtopics noted to still be weak. Pay particular attention to whether a pattern of errors has emerged (e.g., questions requiring understanding of genetic principles; questions requiring knowledge of repair and regenerative processes; or questions requiring knowledge of metabolic pathways and associated diseases).

There is another possible use for these questions. If you already know from experience the two or three major topics in this book that cause you the greatest concern, you can:

1. Select from *two* question sets only those questions that deal with those specific topics. (You can identify them easily because each answer is topically keyed.)

2. If the number of questions is large, you may want to divide the number in half and reserve one half for a later test.

3. Follow steps 2 to 4 from the diagnostic testing section.

4. In conducting error analysis, try to pin down more specifically your within-topic errors, so that in follow-up study you can concentrate on strengthening weaknesses that remain.

5. When you feel you have firmed up your information base, test yourself with the other half of the questions and note your progress, as well as any remaining subtopics for follow-up study. The questions not used in the first two sets of questions, as well as the two full sets, can be used as your review progresses.

For self-evaluation. As the last few weeks before the exam approach, some students begin to experience feelings of "approach/avoidance"; they would like to know if they are close to, or even beyond, the minimum needed to pass, but they also fear that if they find a large discrepancy, it may deplete their efforts during the final phase of their review. This situation is less likely to occur with students who have engaged in self-testing throughout their preparation. These students have been collecting and analyzing test data all along and adjusting their study agenda accordingly. The last level of evaluation is not likely to give them any surprises about strengths and weaknesses, but will identify areas in which they can continue to fine-tune.

There are a few different ways to handle the last round of self-testing. Some students feel less anxious if they do the final round of self-evaluation in the first few days of the last week and reserve the rest of the time for last-minute follow-up study. Other students prefer to start the week with a composite test, continue with further study, and then take another practice test 2 or 3 days before the exam.

1. If you have used one question set as a pretest, it would now be informative to start with those questions as a posttest. Follow the steps described previously and compare performance (both total score and the score across each large basic science topic).

2. The three remaining tests can be used as individual question sets, or they can be combined into one large composite set of approximately 480 questions.

3. If none of the tests has been used for pretesting, you might take every other question from all four tests (325 questions) and follow up later with the remaining 325 questions.

4. If you are using other books in this series, you can select a set from each of the other four books to form a comprehensive final evaluation.

Score Interpretation

Keeping in mind that the percentage of items needed to pass the USMLE Step 1 is between 55% and 65% should help you interpret accuracy levels from your self-testing. The practice test samples suggested here (usually 160 questions) provide useful feedback to chart your progress. On your tests, percentages between 55% and 60% are minimal, but encouraging. Percentages between 60% and 75% show you are moving beyond the bare minimum needed for passing. Scores of 75% and above are indicators of substantial strength.

Exam Preparation Guide

USMLE Step 1: What to Expect

USMLE Step 1 is the first examination of the three-step sequence required for medical licensure, so it is not surprising that its approach engenders apprehension in many students. Successful performance is particularly consequential in those medical schools that require successful passage of Step 1 before permitting students to proceed to third year. The "new" Step 1 has been in effect since 1991, and although most people are now familiar with its general contours, some "myths" still circulate.

The sources that contain the most complete and specific information about the examination are those distributed to students when they register to take Step 1: *Bulletin of Information* and *USMLE Step 1— General Instructions, Content Outline, and Sample Items*. **Both books should be read in their entirety before taking the examination.** What follows is a brief summary of what can be found in much more detail in those materials.

Description. The purpose of the Step 1 exam is to assess students' understanding and application of important concepts in the basic biomedical sciences: anatomy, behavioral science, biochemistry, microbiology, pathology, pharmacology, and physiology. Emphasis is placed on **principles and mechanisms underlying health, disease, and modes of therapy.**

A "blueprint" in the Guidelines booklet shows how basic science material is organized for the examination. Two dimensions are used: system and process. The first dimension includes a section on General Principles and ten Individual Organ Systems. The second dimension is divided into normal development; normal processes; abnormal processes; principles of therapeutics; and psychosocial, cultural, and environmental considerations. Also shown are the percentages of questions across categories of the two dimensions. The percentages are rather close between General Principles, 40%–50%, and Individual Organ Systems, 50%–60%. Of the categories in dimension 2, abnormal processes has the largest percentage (30%–40%). A more detailed breakdown of content can be found in the *Step 1 Content Outline*, but not all the topics listed are included in each test administration.

Students are expected to respond to some questions that require straightforward basic science knowledge, but the majority of questions require application of basic science principles to clinical situations. There are also questions that require interpretation of graphic and tabular data and identification of gross and microscopic specimens. There seems to be more coverage of content typically taught in the second year, but interdisciplinary topics such as Immunology, which is usually taught in the first year, receive quite a bit of attention.

Format. The 2-day examination consists of four books with approximately 180 items in each book (total, 720 questions). Two books are given on each of the days. Three hours are allowed to complete each book, or approximately 60 seconds per question.

Question types. Two types of questions are on the exam: single best answer and matching sets, which begin with a list of a certain number of response options used for all items in the set.

Scores. Passing is based on the total score. Raw scores are converted to a standard score scale with a mean of 200 and a standard deviation of 20. A score of 176, or 1.2 standard deviations below the mean, is needed to pass.

Examinees will receive a total test score, a pass/fail designation, and a graphic performance "profile" depicting strengths and weaknesses by discipline and organ system. No individual subscores are reported. A two-digit score is also reported, in which a score of 75 corresponds to the minimum passing score and 82 is equivalent to the mean of 200.

A Framework for Successful Preparation

By the time you reach medical school, you have been a student for most of your life. You have learned in a variety of settings and have achieved a number of personal goals. There is probably little that you have not observed about your own learning. Despite this, you may still approach medical studies with some degree of apprehension and have questions about the effectiveness of your study strategies, specific skills, and attitudes.

After experiencing medical courses during their first year or two, most students accommodate well and, if necessary, make whatever adjustments in their study patterns seem warranted. But, even the most competent student, given the pressure of frequent and demanding examinations, will have occasional doubts regarding the efficiency of a particular study method. For those planning to take USMLE Step 1, many questions occur about how best to proceed. "How much time is adequate for review?" "What materials should I use?" "What should I study and in what order?" In discussions with other students, you will hear about approaches they took and what worked for them. But, eventually, you will need to make important decisions for yourself about how to *initiate* and *sustain* a preparation plan that results in success on the exam.

This preparatory guide selects and summarizes, from many different areas of cognitive and educational psychology, those findings that have most applicability to a medical learning context. Strategies, skills, and functions are organized according to their potential utility for students as they move progressively from initial encounter with new learning at stage I, acquiring information, to stage II, consolidating information, and finally to the goal of self-confident achievement, stage III, reaching mastery. In the sections that follow, the conceptual framework shown in Figure 1 will be used to discuss specific suggestions and activities.

FIGURE 1. Medical learning framework.

Cognitive Learning Strategies

Stage I. Acquiring Information

Stage II. Consolidating Information

Stage III. Reaching Mastery

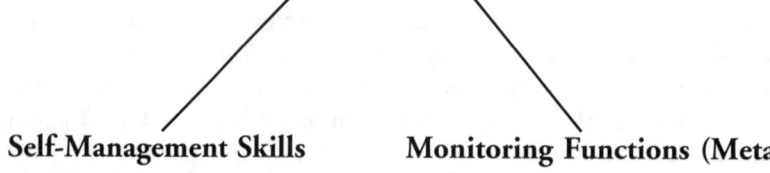

Self-Management Skills   **Monitoring Functions (Metacognitive)**

Time allocation   Study progress

Effort expenditure   Feelings/stress

Study resources   Self-evaluation

Three main subdivisions are represented in this Medical Learning Framework: cognitive learning strategies, self-management skills, and monitoring functions.

Cognitive learning strategies. These strategies can be used to acquire, retain, and master a massive amount of information in the basic sciences. The strategies will be arranged according to which ones are appropriate at each stage in the learning sequence.

Self-management skills. At each stage of learning noted previously, there are skills that can help students allocate time efficiently, expend effort productively, and use study resources effectively.

Monitoring functions. In addition to the cognitive dimensions, medical learning requires metacognitive functions—the ongoing self-regulation that helps students track their progress and decide whether they need to modify or fine-tune any behaviors. Students also need to monitor and try to control potentially interfering negative feelings and stress.

Cognitive Learning Strategies

Any learning experience a student engages in, whether listening to a lecture, reading a text, observing a demonstration, or viewing a video presentation can be said to move through three stages as the learner proceeds from initial encounter to eventual mastery. Many factors influence the progression from one stage to another, among them the characteristics of the student, such as ability, motivation, attitudes, and interests. Also influential are the characteristics of the material, its conceptual difficulty, its organization, and its relationship to the learner's prior knowledge. The specific study activities the student uses also will have an effect. Whether you are trying to learn medical material for the first time, or reviewing information you learned before and need to reactivate and strengthen, the three-stage concept of how learning takes place offers a handy scheme for deciding which study strategies to use when.

Strategies for Acquiring Information

In this first stage, as you read or listen to a lecture, the main task is consciously and intentionally to generate as much *meaning* (understanding) as you can. Because studies have shown that strong initial *encoding* influences to a large extent what will be stored in long-term memory, there is a payoff for being *active* at this stage. The ongoing task is to decide if what you are reading or hearing is unfamiliar information, somewhat familiar, or already part of your fund of knowledge. Rarely will you encounter something that is completely new, but some topics will seem more remote than others if your previous experience with them has been limited. As you move through the information, do so at as brisk a pace as you can without sacrificing meaning. Following are some productive strategies that can be used at this stage:

1. Preview. Before starting to read, notice how a topic or other chunk of material has been organized. Use external arrangements such as titles and subheadings to get an idea of how the topic has been segmented. One technique is to convert these subdivisions into questions. Study any pictorial material such as figures and diagrams. Read the introduction, summary, and questions, if available. Read anything that is printed in different type, such as italic, or highlighted. Notice unfamiliar terms and look them up. Remember that the purpose of previewing is to give you a preliminary cognitive "map" that should help you extract more meaning from your subsequent reading.

2. Read actively. When reading a text, handouts, and notes, some parts will trigger recollection from your previous learning. When you encounter familiarity, try *prompting*: Pause and look away from the page, anticipate what will be coming, and try to bring forth from your memory whatever you can recall about that topic. Also, try to read as if on a *search*. Having looked at the subheading of a section and raised questions in your mind about what to expect, read to see if you can find responses to your questions.

3. Link information. Many medical students acknowledge that this is an important and useful strategy for enhancing understanding, yet few actually implement it. As you read, stop periodically and (a) summarize in your own words, (b) draw relationships to other knowledge by comparing and contrasting, (c) make an educated guess (inference), and (d) raise questions (What would happen if. . .?). If you are wondering whether you have time to think about the material given the usual pressures, remind yourself that these are the very thinking processes that are built into the questions of the Step 1 exam.

4. Construct notes. You probably have been taking notes in class since your earliest school days, and you may have developed a system for reducing and compacting lecture information that has served you well in the past. If so, continue using it. If, however, you are still trying to listen and write as much as you can, and as fast as you can, then perhaps you want to try a different method. When an instructor has provided a handout or other type of script before a lecture, preview it ahead and "cue" the sections that are obscure and need more elaboration. Then, you can limit note taking to what is essential to make sense of that script. Use whatever symbols you wish as cues (e.g., stars, circles, triangles) and assign a particular meaning to each. When you return to that handout after the lecture, you can translate your cues into further study activities (e.g., rewrite a particular section, supplement from a text, memorize a procedure).

One activity you might find helpful to institute fairly early is a last-minute study list consisting of those topics, mechanisms, procedures, and details that you find particularly problematic. Record either a brief explanation or the page and reference source where the information can be found. This list is particularly useful toward the end of your exam preparation sequence when you will want to make the most effective use of whatever time remains.

Self-Management Skills

At this early stage there are certain activities related to time, effort, and resources that are appropriate to carry out.

Time

Form a realistic study plan. Before plunging in, give some thought to how you want to organize your plan of study and which factors you need to consider. What is the amount of time you can reasonably allocate to preparation for the exam? If you are preparing for Step 1, it might be 3 or 4 months. If your experience with basic science material goes back a number of years, you will be doing more than simply activating former learning. There will be chunks of recent scientific knowledge that will require more intensive processing and more study time to reach a level of familiarity.

There are a few principles worth observing regardless of the total time actually allocated: (a) Use whatever diagnostic information you have (data-based, if possible) to assign time on the basis of relative strengths and weaknesses. Your review should be comprehensive, but some topics should be given more time than others. (b) Draft a long-range, tentative plan across the time you have available and estimate approximately how much you want to assign to each segment of content. Even a rough plan written down will reduce concern about whether you can fit everything in. You will be able to observe whether you underestimated the number of hours needed and increase them as you implement your plan. (c) Leave the last 2 weeks unscheduled so that you can return to areas that need a second pass. (d) At the end of each week, look at your plan and make changes based on your experiences during that week.

Effort

Get started. Perhaps what takes the most effort at this stage is just "lifting off" and getting into some type of study routine. You may find yourself putting off the actual start until you can finish other "essential" things, but you are probably procrastinating. It may help if you begin by studying something that you are strong in because a feeling of success will encourage you to continue. Gradually, shift to a topic that is less familiar and requires a little more intentional effort.

Select conditions conducive to study. Find a place where you can sustain a study block with few or no distractions. Put yourself in an active study posture, sitting upright, not lying on a couch or bed. Make yourself go to your study place as part of your routine. Staying in your apartment may be convenient, but it also makes it tempting to give in to other distractions.

Establish a reasonable, steady pace. If you are highly motivated, you may be tempted to work for exceptionally lengthy stretches, particularly during the early days of your review. Try, instead, to establish a reasonable routine that allows you to get a return from each study block. Know what your peak work periods are and do your most difficult studying at those times. Pay attention to whether you are getting fatigued and losing your ability to concentrate. Build in breaks that will reenergize you and help you feel refreshed when you return to studying.

Resources
Select effective study materials. Whether you are studying for a class exam or USMLE Step 1, finding just the right study material often can prove frustrating. Although quite a number of study resources are available in bookstores, each differs in purpose, format, depth, and comprehensiveness of coverage. For review, your own notes, charts, and handouts are good sources if you still have them available. They are familiar and have personal associations helpful for recalling information. To initiate review, look for publications that summarize or "compact" information, are not excessively wordy, but still provide enough narrative for you to make sense of the topic. The purpose of such books (e.g., Williams & Wilkins' *Board Review Series*) is to stimulate recall of material learned previously. Finally, have a reliable text available in each of the basic sciences so that you can use them as a supplemental resource, if needed.

Monitoring Functions

Since you are just beginning to get into your study routine, this is the time to:

Initiate self-observations. These are the "informal" impressions, thoughts, and reactions that you form as you experience certain learning activities. For example, as you listen to a lecture, everything is making sense and fitting in with what you already know. Or, you feel some discomfort because the lecture is moving at too rapid a pace for you to process material meaningfully. Your reactions may be telling you that all is going well, and you should continue without change, or they may be signaling the need for some attention and possible adjustment in your study strategies.

Monitor emerging negative thoughts. If in reviewing you are reactivating without difficulty material you studied previously, you will feel productive and have a sense of accomplishment. But, there will be times when the proportion of understanding will seem relatively meager, and some discouragement will be felt. Try to confine your discouragement to the specific event that prompted the feeling without letting it generalize to *all* study activities.

Strategies for Consolidating Information

After you have listened to lectures or have read sections of material, you probably have acquired a reasonable percentage of the meaning. But, you also know that to *retain* what you understood, you will need to engage in other study activities. Of the multitude of activities from which you could choose, the following have been found to be effective to *maintain*, *consolidate*, *integrate*, and *synthesize* your knowledge.

1. Fill in gaps in your understanding. As soon after a lecture as is practical, follow up any of the "cued" sections in your notes or handouts by filling in what was unclear or incomplete. You can use another reference book, discuss the lecture with a peer, or ask for clarification from the instructor. Whatever action you take will make your learning stronger and move the information to long-term memory.

2. Reorganize for recall. Most students are familiar with the devices that can be used to reorganize information for better retrieval and recall: outlines, charts, index cards, concept maps, tree diagrams, and so forth. Following are guidelines for whether you should bother restructuring information and, if so, when it should be done.

If the material being used for study is already well organized, little if any restructuring may be needed. Sometimes, however, a different schematic format may make even well-organized material easier to recall.

If you reorganize, arrange the information so that *meaning* is emphasized. Note prototypes such as the most common and least common disease for a category, and the most frequent and least frequent treatment. In a set of diseases sharing similar symptoms, note particularly the differentiating feature(s).

If you decide to use one or more of the preceding devices, remember to do so during this stage, rather than close to the exam deadline, so you will have sufficient time to incorporate what you have restructured into your memory.

3. Synthesize from multiple sources. Avoid studying the same topic in three or four different sources. Use one substantive source as your "road map" and check other sources if you think yours is not comprehensive enough. Notice what needs to be added to make yours more complete, but end up with one dependable "script" that you can use for any subsequent study.

4. Rehearse to strengthen recall. Many students read things over and over. Rereading alone is not likely to be effective. The following habits could lead to more durable learning because they involve more active processing:

Use visual imaging. Visualize what you are trying to learn by "seeing" it in the form you will use when you want to retrieve it later (e.g., an anatomic structure as you saw it in lab, or as a schematic representation from the text, or as the instructor detailed it on a transparency).

Form analogies. Wherever possible, try to associate a new concept to a similar and simpler one that is already familiar to you.

Elaborate verbally. Talk about what you want to remember. Say it either to yourself or to others, but in your own words. "Stretch" beyond the script in the book or handout and develop inferences (make reasonable guesses about other relationships or applications).

Use mnemonics. These mental cues can be used to associate a wide range of medical information. Many can be found in student resources, or you can construct your own. Although you can be creative and even bizarre, avoid complexities that make the mnemonic harder to remember than the material itself. Using the first letter of each word to **form acronyms** is a common mnemonic device. For example, the causes of coma are AEIOU TIPS, which means *a*lcoholism, *e*ncephalopathy, *i*nsulin excess or deficiency, *o*piates, *u*remia, *t*rauma, *i*nfection, *p*sychosis, *s*yncope. **Method of loci** is one of the oldest mnemonic devices. You "place" mentally what you want to remember in certain familiar locations, such as rooms in your house, or locations within a room.

5. Establish patterns of practice. Certain "essentials" may need to be memorized and recalled almost verbatim. For such learning, **distribute the practice** so that you rehearse for a number of short periods, with breaks and other activities interspersed, rather than trying to sustain one lengthy period. Try **cumulative practice** by learning a few "chunks" at one session; then at a subsequent session, review those and add a few more. Continue the same pattern until all you want to memorize has been incorporated.

6. Study with others. This can be an effective study activity if used properly. An initial exploration of material by each student in the group will make group discussion more valuable. Discussion can then focus on clarifying material and confirming and extending understanding. Studying with others works best if the group is small so everyone participates, and if some ground rules are established about how the sessions will be conducted.

7. Self-test periodically. The purpose of self-testing at this stage is **to guide further study.** Self-testing can help you decide which topics need more intense study, which are fairly close to being learned, and which have been learned well. Resources to use for this purpose and the sequence to follow will be described in later sections.

Applying Effective Testing Skills

Following are suggestions that will increase the likelihood that what you have learned and can recall from your study will translate into correct responses on multiple-choice examinations.

General test-taking skills

Read carefully for comprehension, not speed and respond to questions in sequence. Mark every item on your answer sheet as you go along, even if you are not completely sure of your choice. Cue the questions to which you want to return if there is time.

Be positive. Suppose the first question you see as you open the test book is a particularly difficult one, and you can feel yourself getting anxious. After giving it a try, go to the second question and respond to that one, which in all likelihood will feel more manageable.

Avoid mechanical errors. At the end of each page of questions, before going to the next page, check to make sure the **number of the question** you just finished **matches the number on your answer sheet.**

Be alert to key terms in the question stem such as "most," "least," "primarily," "frequently," "most often," and "most likely." Notice transition words that signal a change in meaning, such as "but," "although," and "however."

Let your original response stand unless you have thought of additional information.

Pace yourself. You will have approximately 60 seconds per question. Avoid dwelling on any one question, or rushing to finish. Set up checkpoints in your test booklet of where you want to be at the end of the first hour, second hour, and so on. You will know before you get close to the end whether you need to adjust your pace.

Analyzing questions

When you first read a question and look at the options, the answer may not be immediately apparent. Although you may be uncertain, don't just pick an answer arbitrarily. You can apply systematic skills of logic and deduction to narrow the five options to two or three possibilities.

Search for key information. As you read the question stem, notice key information (e.g., age, symptoms, lab results, chronic or acute condition, history). Highlight the key information by underlining or circling. **Notice particularly the request of the question** in phrases such as "the most likely diagnosis is," "the most appropriate initial step in management is," "which initial diagnostic evaluation is most appropriate." Take a quick look at the last line of the stem before reading the specific information in the remainder of the stem, especially if it is lengthy.

Analyze options. As you read each option, **try to eliminate** those that are inconsistent with the information you highlighted in the stem. For example, if a question concerns a 65-year-old woman, you would eliminate a procedure that you know applies only to children.

Cue each option. As you consider each option, mark down your initial reaction. In a "one best answer" question there are four "false" options and one "true" answer. For negative one best answer questions, the reverse applies. As you read each option, cue those you are sure of with a symbol such as "F," "N," or a minus sign. Cue true responses with a "T," "Y," or a plus sign. Cue those options you are uncertain about with the symbol you are using and a question mark.

Analyze structural clues in words. Pay attention to the meaning of prefixes, suffixes, and root words, which can sometimes help you decide whether to eliminate an option.

Approaching questions strategically. The following examples show how you might approach questions found in the books of this series.

Example: Which bronchial condition is characterized by normal IgE levels and late adulthood onset?
  A. Bronchiectasis
  B. Extrinsic bronchial asthma
  C. Chronic bronchitis

xxv

D. Intrinsic bronchial asthma
E. Bronchiolitis obliterans—organizing pneumonia

Analysis: I know that some of these conditions share similar features, such as cough and bronchospasms, but which one is distinguishable by the two features in the stem? I will eliminate A, since other, more pertinent identifying characterstcs would probably have been used (e.g., irreversible dilatation of the bronchi, destruction of bronchial tissue). For the same reason, I will eliminate E since that condition, too, has other, more clinically meaningful characteristics that would have been noted (e.g., narrowed bronchioles, dyspnea). Given the choices remaining, I am inclined to say the answer is either B or D. Now, what do I remember about the two types of asthma? Extrinsic asthma is triggered by various allergens, and the IgE levels are elevated. I'm not sure about the time of onset, but I think it starts earlier than late adulthood. I know that intrinsic bronchial asthma does not have any particular triggering antigens, so IgE levels are probably normal. it Intrinsic bronchial asthma does start in later adulthood; therefore, my choice is D.

Comment: In dealing with diseases such as the ones in this question, it is often difficult for students to recall all the features that characterize each one, especially if they have some features in common. It is important to identify the one or two *key* differentiating characteristics that are associated with each disease. It is particularly important that students be familiar with the diagnostic features of those diseases likely to be encountered in a physician's daily practice (e.g., asthma).

Example: A middle-aged woman who complained of an uncomfortable feeling in her chest presents with a fast and irregular radial pulse. The patient's electrocardiogram results show normal QRS complexes with no distinct P waves. The patient most likely has
A. first-degree heart block
B. second-degree heart block
C. third-degree heart block
D. atrial fibrillation
E. ventricular fibrillation

Analysis: This is a cardioarrhythmia problem, and I have to think about where the disordered rhythm is originating that could produce normal QRS complexes and no distinct P waves. I can eliminate E because in ventricular fibrillation, the electrocardiogram tracing would have an irregular appearance. hat would need to be happening in order to produce no P waves? This has to do with electrical activity in the atria. If impulses were disordered and going in all directions, then that could account for the absence of P waves. Since heart blocks occur at the AV node, SA node, or at the bundle branches, there are P waves in all three heart blocks, so I can eliminate A, B, and C. Option D fits the description best because, in addition to the absence of P waves, the fast but irregular pulse is another feature of atrial fibrillation.

Comment: This question tests students' understanding of the basic physiologic mechanisms that characterize various arrhythmias. In addition, students need to know the representations of arrhythmias as electrocardiogram tracings and need to be able to integrate those with presenting clinical symptoms to differentiate one arrhythmia from another. While medical students eventually become quite facile at performing these tasks, at first there are numerous possibilitles for confusion or misinterpretation.

Relying on test question cues

The ability to use the characteristics or the formats of the test itself to increase your score is sometimes referred to as "test wiseness." It is possible to make use of idiosyncrasies in the way the questions are constructed to decide on the correct choice. This technique should be used only if you are unable to answer the question based on direct knowledge or reasoning. The following are examples of the principles of test wiseness, but you may have little opportunity to use them on USMLE Step 1, because the experts who construct the questions eliminate these cues.

Length of an option. If an option is much longer or much shorter than the others, it is more likely to be correct.

Grammatical consistency. Options that are not grammatically aligned with the stem are probably false.

Specific determiners. Options that contain words such as "all," "always," and "never" overqualify an option and are likely to be false.

Overuse of the same words or expressions. Some test makers have a tendency to repeat words or phrases in the options. If you are unsure of an answer, select from the options with the repeated words or phrases. Another variation of this principle is to select an option in which a key word from the stem is repeated.

Numeric midrange. When all options can be listed in numeric order (e.g., percentages), the correct choice will most often be one of the two middle values.

Guessing

The following are "last resort" strategies, but you should be aware of them.

1. If you have eliminated one or two options, but have no idea about the remaining ones, choose the first in the list.

2. If you are unable to eliminate any options, choose A, B, or C.

3. If you have a number of questions left to do and time is running out, **do not leave blanks.** Choose A, B, or C, and fill in the same letter for all remaining questions.

Self-Management Skills

Time

Study in blocks. Assuming you plan to study 4 to 5 hours each night, you might consider dividing those hours into two study blocks. This allows you to study two areas, one that is weaker and therefore requires more time, and another that is relatively strong and can be allocated less time. The advantage of studying two sciences concurrently is that you will move through your strong science with ease and feel a sense of accomplishment, even if the weaker science does not reach the same level of confidence.

Set goals for each study block. Begin by identifying a few goals you think can be accomplished within that block of time. The goals need not be elaborately stated. Identifying what you think is important to study increases the chances that you will study *actively* (with heightened awareness) since you are controlling the purpose, direction, and rate of the studying.

Set realistic deadlines. Although your accuracy in estimating how long it takes you to complete a study agenda will vary, observe whether you habitually overestimate. Arbitrary deadlines are self-defeating if you have little or no chance of meeting them. Set more realistic targets and attempt to meet them most of the time.

Use record-keeping devices. Calendars and appointment books will help you schedule your study agenda and permit you to look ahead and adjust plans to meet deadlines.

Control distractions. There are many kinds of distractions, some of which are self-imposed. Others, such as telephone calls, can interrupt concentration and make it harder to get back to work. If a call is not urgent, decide on a response within the first 30 seconds (e.g., "I'll call you back later. I'm in the middle of something important."). When you do call later, use it as a reward for having worked well, and enjoy It. Also, learn to say "No" to requests that take time and distract you from your schedule.

Effort

Avoid activities that dissipate effort. Be aware of whether there are things you do each day that reduce your total energy, and particularly the energy you want to give to studying. Think about which of the "nonessential" tasks you can delegate to other family members, or to friends who want to be helpful. Give them some direction about what kind of help you would appreciate most.

Try to anticipate crises. There are disruptive life events that happen to all of us that we cannot anticipate but must deal with as best we can. But, there are other events of a less traumatic nature that, if they occur, can interrrupt the flow of a study plan and throw you off course (e.g., the car breaking down and needing immediate repair, a relative who wants to come and stay with you, or a friend who needs your advice on a troublesome problem). Anticipate crises that may happen during the span of your preparation and have alternative plans ready that permit you to be a part of what is going on but do not derail you completely.

Reward yourself for good effort. After having sustained a stretch of "heavy duty" studying, reward yourself by doing something that for you is pleasurable. A phone call to or from a friend that might be a distraction if it happens when you are trying to study can be a source of pleasure if you can defer it until you have completed your agenda for that day.

Resources

The following testing materials are appropriate for self-testing to guide further study. Their use is described in the next section.

Instructor content tests. These tests consist of questions prepared by the instructor who taught a particular segment of content. Some medical schools retain former course exams on file for student practice. Although the questions may not be structured as they appear on the Step 1 exam, they are good for pinpointing specific gaps or confusions in your knowledge base. Keep records of each practice test result and note relative performance across sciences and across topics in each science. For example, in pathology, note if one system (e.g., respiratory, cardiovascular, endocrine) is notably weaker than another. Cue topics that will need more sustained study. Take advantage of the instructor's presence to seek help, if needed.

Published books of practice questions. The books in this series arrange basic science content into "principles" (two books) and "systems" (three books).

Monitoring Functions

Monitor study progress

During this phase, when you are strengthening your learning, you will want to get **data-based feedback** using numeric scores to chart your progress.

Use questions to monitor progress

The pattern that works best is study, test, follow-up. Although there are times when it is appropriate to use questions before study to stimulate motivation or trigger recall, at this stage the best use of questions is after preliminary study. When you believe you have learned a segment of material, try a batch of questions. If time permits, you might want to test yourself on each major topic after completing its study, and before testing yourself on a mixed batch of topics in a science. However, if time is limited, select those topics about which you feel the most uncertainty, and use the feedback to guide additional study.

Select a representative sample of questions and complete them using the same time limits as will be used on your class exam or Board exam (approximately 60 seconds per question). Record your answers on a sheet of paper rather than in the question book or on the class practice exam. A separate sheet will allow you to do error analysis (described later) and keeps the book "clean" for future question retakes.

Do not do questions one at a time and then read the answer. The purpose is not to learn a particular question, but to find out which topics require follow-up.

Score your responses, but do not read answers immediately since you may want to give some questions a second try. After performing error analysis, decide which topics need further study and the type of study needed.

Compute an accuracy percentage by dividing the number correct by the total number of questions. Keep a record of your scores and note whether your accuracy level is approaching the percentage required for passing (for Board exams, between 55% and 65%). For class exams, the percentage may be higher.

Analyze for errors. It is important to analyze more specific aspects of your study and test-taking behavior to direct further study and make it more focused and productive.

1. Were patterns of errors noted (e.g., questions related to DNA principles, or questions about immune responses, or questions regarding quantatitive methods)?
2. Did you misread or misinterpret the question?
3. Were questions missed because, although you understood the concept, you forgot important details?
4. Did you note errors in addressing the *decision* required by the question? For example, although you knew much about the disease process described, you could not differentiate a likely diagnosis, or you were unable to form a judgment about a mechanism involved, draw an inference about the appropriate next step in management, or make a prediction about which drug would cause an adverse effect. In other words, you were unable to transform your conceptual and factual knowledge to meet the request of the question.

Monitor test anxiety

One aspect of self-testing that you should be monitoring is whether you are experiencing *inordinate* anxiety when dealing with test questions. It is not unusual to feel some elevation in anxiety when facing a comprehensive and consequential examination such as Step 1. But, if the amount of worry and the physiologic aspects (rapid breathing, sweaty palms, increased heartbeat) become so preoccupying that they interfere with productive studying, then some professional attention may be needed. If, however, test anxiety is of reasonable proportion, then remember what many studies have found. The best defense against test anxiety is a combination of strong review of subject matter, practice on tests similar to the target test, and positive self-reinforcement throughout the preparation process.

Combat negative self-statements

Part of your monitoring should include awareness of your moods and general state of being. Be sensitive to when you are about to give yourself a negative self-evaluation and combat it with an accurate, but positive one. "What if I don't. . ." statements will intrude periodically and, if permitted, can change your mood and distract you from your study. Start practicing self-talk by having a positive statement ready to use to redirect yourself back to your agenda ("I've been studying well and my scores show I'm making progress. . . I just need to keep going!" or "I can't afford the time to worry now; maybe late tonight; back to the topic now").

Strategies for Reaching Mastery

By the time you reach this stage you should feel more confident that your knowledge is firmer and that you can retrieve information dependably. The tasks of this stage deal with refining, or fine-tuning, for increased accuracy.

1. Focus on follow-up study. Your study agenda at this stage should be based on findings from your error analysis of questions you practiced, as well as any behavioral observations you noted from monitoring your performance.

If there are topics you need to reinforce, check your resources to note if those explanations are adequate, or if confusions still remain that may need to be clarifed through use of another text or discussion with a peer.

If details are eluding you, engage in some of the memory strengthening activities noted in the previous stage, particularly use of mnemonics, and cumulative practice.

If you misread information in questions, remember to highlight pertinent cues as you read, to focus on comprehension and avoid regressions, and to vary your rate to emphasize meaning.

If one of the patterns you noted is that errors were made on questions with very long stems, practice by first reading the "request" at the end of the question stem. You may then be able to interpret the direction and relevance of the information in the question more quickly and accurately.

If you found that you were unable to translate your knowledge to the specific *thinking* requirement of the question (form a judgment, integrate information to form a conclusion, draw an inference), first check to make sure you know the principles or mechanisms the question assumes you know (e.g., biosynthesis and degradation, dose–effect relationships, alterations in immunologic function). Then, analyze the question through a "think aloud" procedure, with a peer if possible, and try to identify why your thinking is inaccurate.

2. Engage in comprehensive self-evaluation. If you have practiced questions topically, and through a systems approach, as in this series, and followed up with focused review, you should be ready to test yourself with comprehensive question sets. These sets will contain questions that sample most of the domains of information represented on the target exam. The procedure in using those questions, scoring them, and analyzing errors is the same as described previously. Since the question sets are likely to be longer (approximately 150 questions), schedule them during the last 2 or 3 weeks, with time between each to benefit from the feedback. Some students end self-evaluation before the last week because further testing too close to the exam date heightens their anxiety.

3. Deal with interfering test-taking behaviors and attitudes. If your self-observations have noted any test-taking behaviors that need improvement, this is the time to correct them.

Impulsive responding. Do you find yourself getting annoyed if the answer to a question is not immediately apparent, and simply choose an option impulsively? Try to curb your impatience, and remind yourself that some questions are designed to engage you in an internal dialogue before deciding on a response.

Inability to move on. Are you unable to disengage from a particularly troublesome question and move on to the next one? This is especially bothersome when you have that tip-of-the-tongue feeling that the answer is something you know, but seems just a little out of reach. Difficult as it seems, try not to allow yourself to become irritated and get "stuck" to that question. Choose an answer and move on. It is likely that if you return to it later on, something may trigger recall.

Carrying over previous unsuccessful testing experiences. If comprehensive multiple-choice exams have been problematic for you in the past, and particularly if a recent attempt has not been successful, you may be tempted to see yourself as a "poor test-taker" and allow a defeatist attitude to permeate your self-testing activities. It would be better to start by asking yourself, "Why do I not do as well as I would like on multiple-choice exams?" Then, through your self-observations and data-based assessments, note any interfering behaviors you would like to change and implement activities that are more effective and can lead to success on such exams.

Self-Management Skills

Time

Set and maintain study priorities. One of the biggest problems experienced by some medical students at any level of training is approaching an exam deadline realizing that there is still so much to learn that they will not reach the stage of "mastery." After some last minute cramming they may even "pass," but the feeling of personal accomplishment eludes them. Although this has happened to all of us at one time or another, if it occurs as an ongoing pattern, then some change is needed. Make a list at the beginning of the week of all the study activities you want to accomplish and rank them in order of *importance* and *urgency*. At the beginning of the next week look at any low-priority items left undone and decide where to arrange them in that week's list. Make a record of each time you procrastinated or gave in to other distractions. Also note how often you kept to your schedule—it can motivate and encourage you to stay with it.

Schedule time for self-testing. Avoid deferring your first self-testing until just before an exam deadline. Build it into your schedule as part of your ongoing study activities and benefit from the feedback. You can then do "last minute" testing to aim for greater accuracy.

Effort

Avoid excessive fatigue. It is expected that you will work hard to be ready for an exam, but allowing yourself to get excessively tired and sleep deprived sabotages your goal. Respond to your body's need for rest instead of pushing for another hour's study with little to show for it. Try to pace yourself so that you have energy left to think clearly when you take the exam.

Keep motivation high. One of the possible pitfalls toward the end of your review is a reduction in your level of attention and concentration, because of either fatigue or emerging apprehension about the imminence of the exam. If you have done some record keeping during the preparation sequence, it is now helpful to look back and acknowledge how far you have come from the point where you started. Reward yourself for progress by planning a pleasurable activity following a block of concentrated study, and enjoy it without guilt.

Resources

Comprehensive question sets. For USMLE Step 1, materials such as Williams & Wilkins' *Review for USMLE Step 1* (NMS series), which provides five practice exams with approximately 200 questions in each exam, will be useful for comprehensive evaluation.

Monitoring Functions

As the exam deadline approaches, you may find that you are experiencing frequent mood shifts. When things are going well, your spirits may be high, but after a disappointing day you may feel blue, gloomy, or even angry. Recent research has found that some techniques work better than others to escape from a bad mood:

1. Take some action. If possible, do something to solve the problem that is causing the bad mood.
2. Spend time with other people, particularly to shake sadness. Focus on something other than what is getting you down.
3. Exercise. The biggest boost comes to people who are usually sedentary, rather than the already aerobically fit.
4. Pick a sensual pleasure, such as taking a hot bath or listening to a favorite piece of music. Be careful of using eating for this purpose; it may work in the short run, but may backfire, leaving you feeling guilty. Drinking and drugs are to be avoided for obvious reasons.
5. Try a mental maneuver such as reminding yourself of previous successes to help bolster your self-esteem.
6. Take a walk. Cool down before confronting whatever gave rise to your negative feelings.
7. Try to see the situation from the other person's point of view—why someone might have done whatever provoked your anger.
8. Lend a helping hand to someone in need. If you are studying, offer to help someone understand a science topic in which you feel very competent.
9. Use stress reduction techniques. Among the most effective are progressive relaxation, which uses tension and tension release in the body's muscle groups; mental imagery, which is putting yourself mentally in a location that evokes feelings of calm and peacefulness; and meditation, which aims for a state of relaxed alertness.

Grace Bingham, Ed.D.

Test 1

QUESTIONS

DIRECTIONS:

Each of the numbered items or incomplete statements in this section is followed by answers or by completions of the statement. Select the ONE lettered answer or completion that is BEST in each case.

1. In reviewing a lateral projection of a barium swallow of a patient, a physician notes that the anterior wall of the esophagus is compressed by an enlarged structure immediately anterior to it. The most likely structure compressing the esophagus is the

(A) right ventricle
(B) left ventricle
(C) right auricular appendage
(D) left auricular appendage
(E) left atrium

2. Which of the following statements about the nasal cavity is true?

(A) The nasolacrimal duct opens into the middle meatus
(B) The ostium of the maxillary sinus opens into the inferior meatus
(C) There is an opening in the sphenoethmoidal recess that communicates with the posterior ethmoidal sinus
(D) There is an opening in the superior meatus that communicates with the sphenoid sinus
(E) The nasal cavity receives blood supply from branches of the ophthalmic, maxillary, and facial arteries

3. In the pathophysiology of right-sided heart failure that is not associated with left-sided failure, which of the following would most likely occur?

(A) An increase in the left ventricular end-diastolic pressure (LVEDP)
(B) An increase in renal reabsorption of sodium
(C) An increase in effective arterial blood volume
(D) An inhibition of aldosterone release
(E) An increase in pulmonary capillary wedge pressure when measured by a Swan-Ganz catheter

4. Patients with von Willebrand disease (vWD) have a decreased incidence of coronary artery disease. The most likely explanation for this is

(A) decreased platelet adhesion to the endothelium
(B) decreased factor VIII coagulation concentration
(C) decreased factor VIII antigen concentration
(D) patients die of vWD at an early age
(E) patients with vWD have lower levels of low-density lipoprotein (LDL) than the general population

Questions 5-9

A 52-year-old black man presents to the emergency room with a 1-year history of intermittent chest pain. He states that the pain has occurred once every 2 weeks while jogging and resolved after 3 to 5 minutes of rest. The pain was substernal and aching and remained localized. Two months ago, he noticed that the chest pain occurred at rest, lasting 5 to 10 minutes before resolving, and occurred at least once every 3 to 4 days. This morning, while eating breakfast, he experienced a severe bout of substernal chest pain that took his breath away. The pain radiated to his left jaw and down his left arm. He felt dizzy, sweaty, and nauseous, but he did not faint. The pain was not aggravated by twisting his torso or by taking deep breaths. Three hours after the onset of chest pain, he continues to feel moderate discomfort.

Physical examination shows an afebrile patient with normal respirations and an average blood pressure of 100/60 mm Hg on 3 separate readings. His pulse is 98 beats/min, weak and regular. Eye grounds show decreased caliber of the retinal vessels, arteriovenous nicking, but neither exudates nor flame hemorrhages are present. The optic disc is normal. The patient has no neck vein distention, and his lungs are clear. An S_4 heart sound is present, and $A2$ is accentuated. No murmurs are present, and there is no hepatosplenomegaly. The patient is admitted to the hospital.

On day 3 of hospitalization, the patient has a progressive onset of dyspnea associated with a nonproductive cough. His temperature is normal, and his respiratory rate is 28/min. Neck vein distention is present. Wet inspiratory rales are diffusely scattered throughout both lung fields. The patient does not experience chest pain on inspiration. A grade III pansystolic murmur is present at the apex of the heart, with radiation into the axilla. An S_3 heart sound is present. His urine output has decreased during the previous two days.

5. Based on the patient's past and present history and physical examination in the emergency room, which of the following would NOT be on a list of differential diagnoses?

(A) Prior history of exertional angina
(B) Essential hypertension
(C) History of unstable angina two months ago
(D) Acute myocardial infarction
(E) Left-sided heart failure

6. What is the most likely etiology of the patient's S_4 heart sound?

(A) Left ventricular hypertrophy
(B) Mitral insufficiency
(C) Left-sided heart failure
(D) Pulmonary hypertension
(E) Aortic regurgitation

7. Which of the following findings would most likely be noted 3 hours after onset of chest pain?

(A) Normal serum creatine kinase MB isoenzyme
(B) Pallor of his cardiac muscle
(C) "Bat wing" infiltrate in the lungs on chest radiograph
(D) Lactate dehydrogenase (LDH) isoenzymes with $LDH_1 > LDH_2$ flip
(E) Q waves with S-T segment depression

8. Which of the following findings is LEAST likely present on day 3 of hospitalization?

(A) Increased peritubular capillary oncotic pressure
(B) Secondary aldosteronism
(C) Increased central venous hydrostatic pressure
(D) Increased lung compliance
(E) Increased left ventricular end-diastolic pressure

9. Which of the following findings is most likely present on day 3 of hospitalization?

(A) An $LDH_1 > LDH_2$ flip
(B) Elevated creatine kinase MB isoenzyme
(C) Normal left ventricular ejection fraction
(D) Multiple lobar consolidations on chest radiograph
(E) Arterial blood gases with a normal alveolar-arterial gradient

10. Which of the following is the most likely cause of high-output cardiac failure in a patient with severe anemia?

(A) Hypoxic injury to the cardiac muscle
(B) Decreased viscosity of blood
(C) Increased peripheral resistance
(D) Decreased contractility of cardiac muscle
(E) Myocardial infarction

11. A 55-year-old man with renal disease has intractable heart failure. A renal biopsy shows an accumulation of material that has an apple-green birefringence when specifically stained and polarized. His cardiac disease is most likely caused by

(A) a restrictive cardiomyopathy
(B) a congestive (dilated) cardiomyopathy
(C) a hypertrophic cardiomyopathy
(D) a constrictive pericarditis
(E) ischemic heart disease

12. Which of the following sequences represents developing Eisenmenger syndrome in a patient with a ventricular septal defect?

(A) Left-to-right shunt, right ventricular hypertrophy, pulmonary hypertension, right-to-left shunt
(B) Right-to-left shunt, pulmonary hypertension, right ventricular hypertrophy, left-to-right shunt
(C) Left-to-right shunt, right ventricular hypertrophy, right-to-left shunt, pulmonary hypertension
(D) Left-to-right shunt, pulmonary hypertension, right ventricular hypertrophy, right-to-left shunt
(E) Right-to-left shunt, right ventricular hypertrophy, pulmonary hypertension, left-to-right shunt

13. A 32-year-old man has hypertrophic cardiomyopathy (idiopathic hypertrophic subaortic stenosis). Which of the following would cause the patient's condition to worsen?

(A) Calcium channel blockers
(B) Squatting
(C) Valsalva maneuver
(D) β-Blockers
(E) Lying down

14. An 80-year-old female presents with a severe headache localized to the left temporal area. She also has visual impairment in her left eye. Which of the following is the most cost-effective screening test to exclude or confirm a diagnosis?

(A) Temporal artery biopsy
(B) Erythrocyte sedimentation rate (ESR)
(C) Serum C3 complement
(D) Antineutrophil cytoplasmic antibody
(E) Serum antinuclear antibody

15. An 8-year-old boy who had an upper respiratory infection 1 week ago presents with fever, nonmigratory joint pain, crampy abdominal pain, and palpable purpura limited to his buttocks and lower extremities. A urinalysis reveals hematuria and red blood cell casts. The pathogenesis of his disease is most likely associated with

(A) an infection of endothelial cells
(B) an immunoglobulin E (IgE)-mediated antibody reaction
(C) a cytotoxic antibody reaction
(D) an immune complex reaction
(E) a delayed hypersensitivity reaction

16. A 62-year-old man with a 30-year history of cigarette smoking presents to his physician with substernal chest pain. He had a heart attack 8 years ago. He is normotensive and has diabetes mellitus type II. His father died of a heart attack at 40 years of age. Lipid profiles performed on two separate occasions show the following mean values: serum cholesterol, 330 mg/dl; high-density lipoprotein, 25 mg/dl; triacylglycerol, 450 mg/dl; and low-density lipoprotein, 215 mg/dl. Using the National Cholesterol Education Panel 1993 criteria for coronary artery disease, how many major positive risk factors does this patient have?

(A) 2
(B) 3
(C) 4
(D) 5
(E) 6

17. A 55-year-old man is hospitalized with an acute anterior myocardial infarction. On day 4, he experiences substernal chest pain and suddenly develops an S_3 heart sound and bibasilar rales at both lung bases. A grade II pansystolic murmur is noted at the apex with radiation into the axilla. His neck veins are not distended. Laboratory studies confirm the presence of creatine kinase MB isoenzyme. This patient most likely has

(A) recurrent angina pectoris precipitating left-sided heart failure
(B) reinfarction with subsequent development of left-sided heart failure
(C) a ruptured posteromedial papillary muscle progressing into left-sided heart failure
(D) a pulmonary embolism with right-sided heart failure
(E) ruptured his left ventricle producing cardiac tamponade

18. A lactate dehydrogenase (LDH) isoenzyme study is performed on a patient with a 3-day-old anterior myocardial infarction. The study reveals an $LDH_1 > LDH_2$ flip and an isolated elevation of LDH_5. The creatine kinase MB isoenzyme is absent. The patient most likely

(A) sustained a pulmonary infarction from venous thromboembolism
(B) developed left- and right-sided heart failure
(C) reinfarcted his left ventricle
(D) developed a mural thrombus that has embolized to the spleen
(E) developed a mural thrombus that has embolized to the brain

19. The pulmonary capillary wedge pressure determined by a Swan-Ganz catheter is a measure of left ventricular end-diastolic pressure (LVEDP) [normal < 12 mm Hg]. Which of the following LVEDP readings are most likely in a patient with hypovolemic shock caused by massive blood loss *and* in a patient with acute myocardial infarction involving the left ventricle?

	Hypovolemic shock	Acute myocardial infarction involving the left ventricle
(A)	2 mm Hg	8 mm Hg
(B)	15 mm Hg	5 mm Hg
(C)	2 mm Hg	30 mm Hg
(D)	30 mm Hg	15 mm Hg
(E)	30 mm Hg	30 mm Hg

20. The following diagram shows two cross-sections of a heart. The heart on the left is normal and the heart on the right is abnormal. The abnormal heart most likely came from a patient with

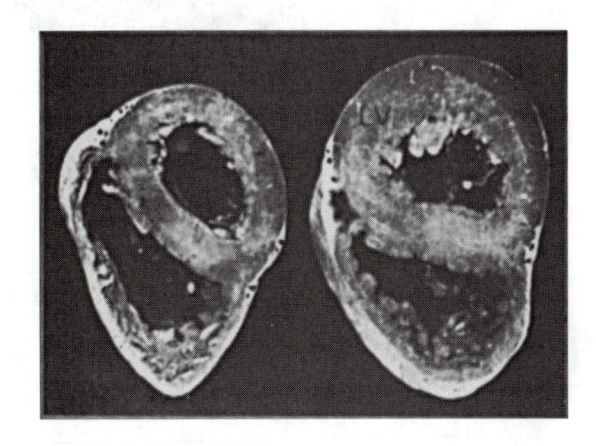

(A) mitral regurgitation
(B) essential hypertension
(C) mitral stenosis
(D) an aortic arch aneurysm
(E) mitral valve prolapse

21. The following autopsy finding is from the heart of a patient with long-standing valvular heart disease. Which of the following is suspected?

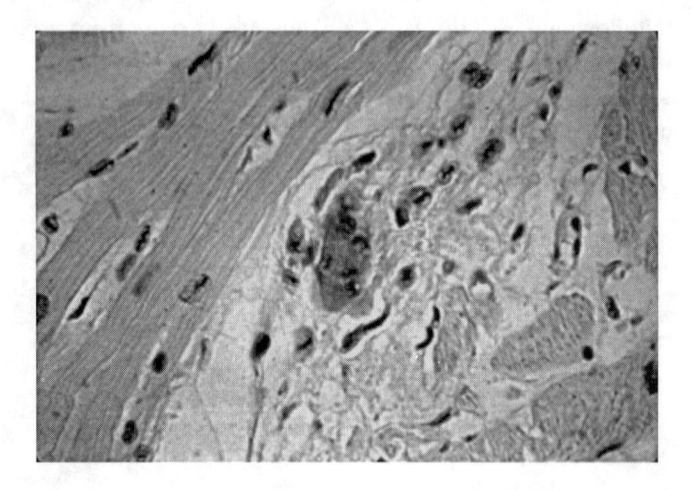

(A) Aortic stenosis
(B) Mitral regurgitation
(C) Aortic regurgitation
(D) Mitral stenosis
(E) Mitral valve prolapse

22. A peripheral smear from a 3-year-old boy with recurrent bacterial infections is shown below. The boy has a normal nitroblue tetrazolium (NBT) test. The mechanism for this patient's disease most likely relates to which one of the following factors?

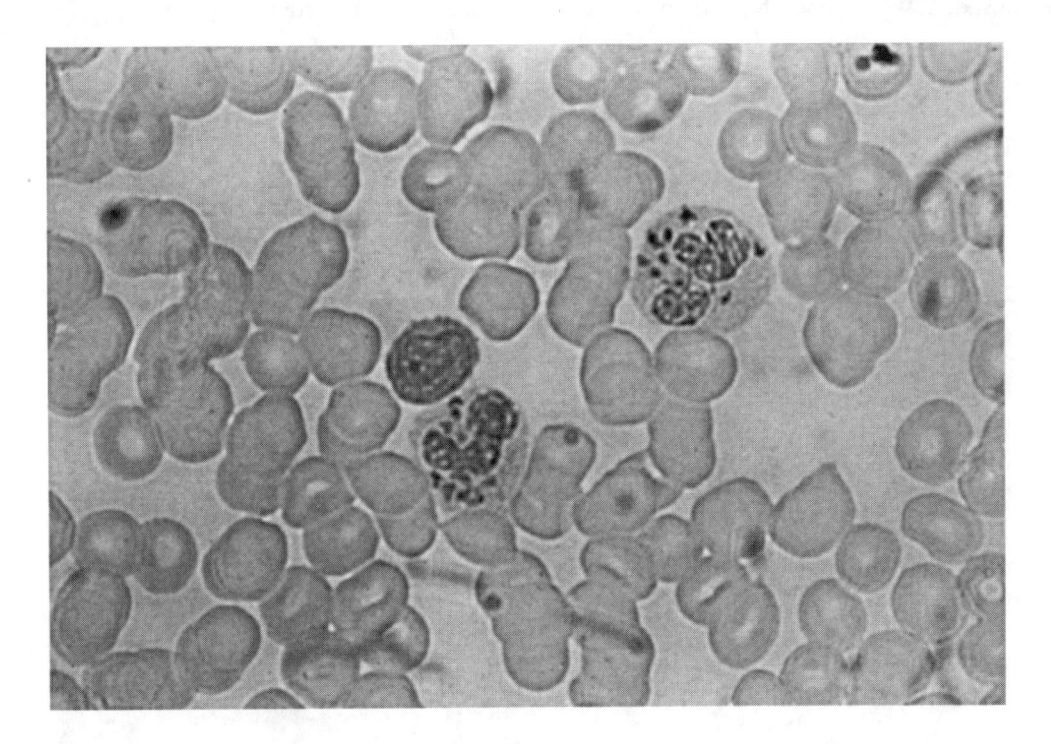

(A) A defect in microtubule polymerization
(B) A sex-linked recessive disease
(C) A deficiency of C3 complement component
(D) A deficiency of reduced nicotinamide-adenine dinucleotide phosphate (NADPH) oxidase
(E) A glucose-6-phosphate dehydrogenase deficiency

23. A peripheral smear from a 72-year-old man with a chief complaint of fatigue is shown below. Physical examination reveals generalized, nontender lymphadenopathy and hepatosplenomegaly. He has a mild normocytic anemia, mild thrombocytopenia, and a leukocyte count of 63,000 cells/μL (98% of which are similar to those noted on the slide). Which one of the following findings would be expected?

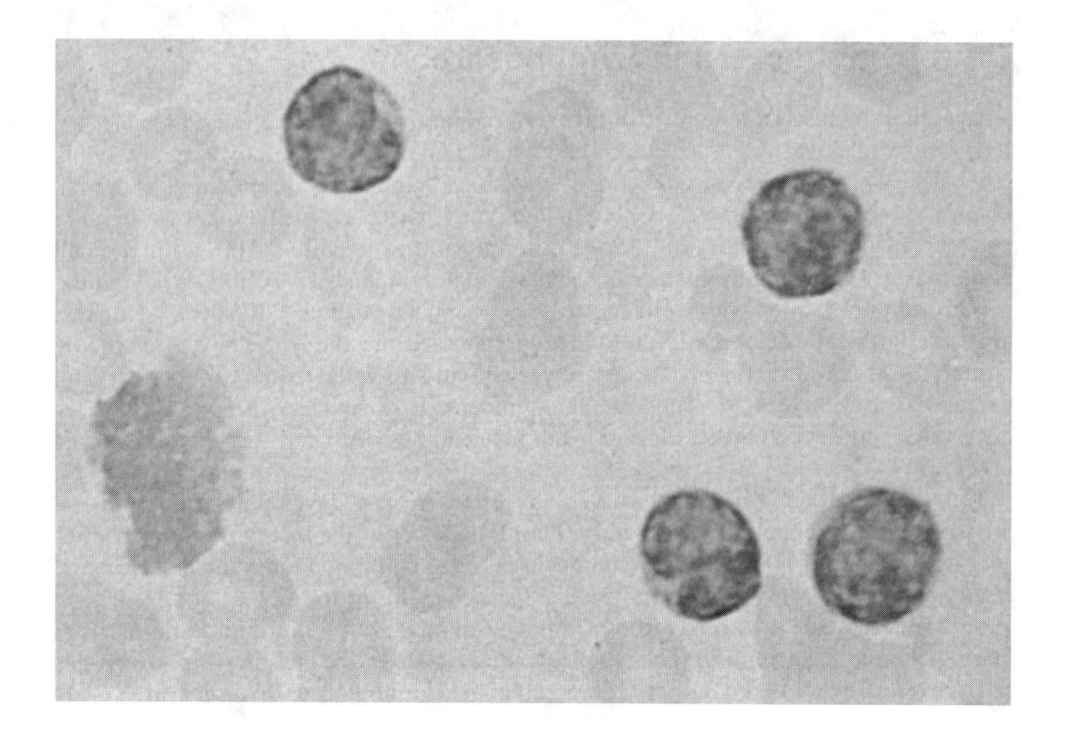

(A) A positive tartrate-resistant acid phosphatase stain on the lymphocytes
(B) Reed-Sternberg cells on a lymph node biopsy
(C) An increased leukocyte alkaline phosphatase score
(D) Positive B-cell marker studies
(E) A polyclonal gammopathy

24. A peripheral smear from a 48-year-old man who presents with fever and weight loss is shown below. He has a normocytic anemia with a corrected reticulocyte count of less than 2% and a white blood cell count of 85,000 cells/μl with 1% myeloblasts. His platelet count is increased. Physical examination reveals hepatosplenomegaly and generalized, nonpainful lymphadenopathy. A bone marrow examination reveals hypercellularity and a leukocyte differential similar to that in the peripheral blood. Which one of the following test combinations would be most useful in arriving at the diagnosis for this patient?

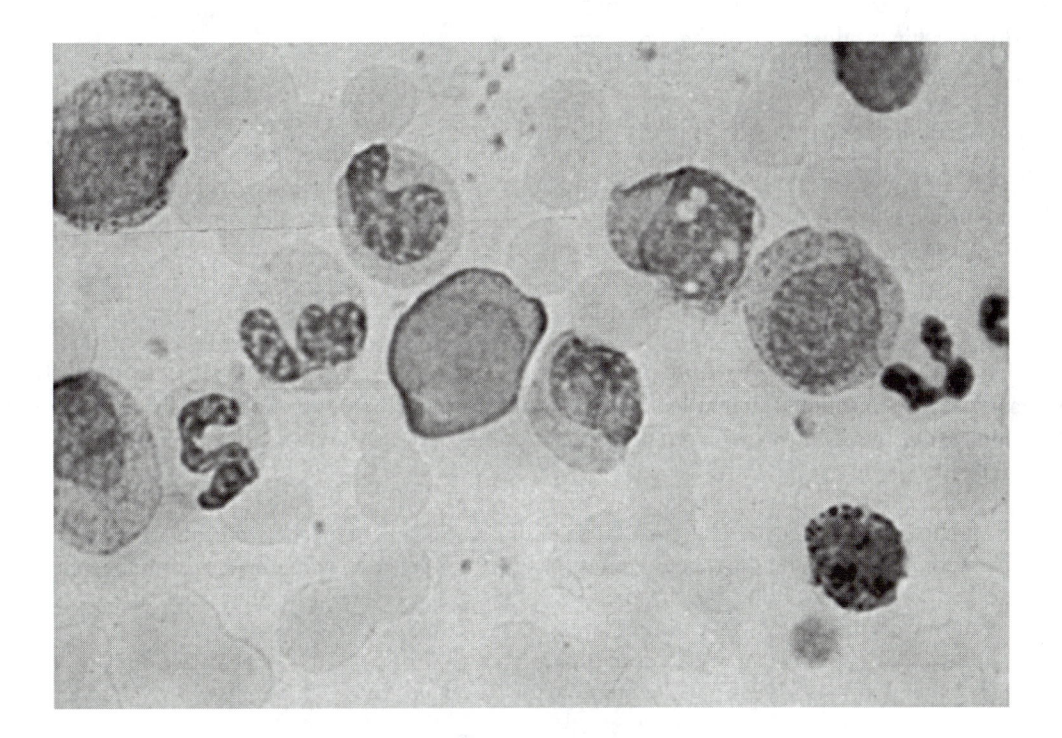

(A) Leukocyte marker studies and a leukocyte alkaline phosphatase score
(B) Arterial blood gas and red blood cell mass/plasma volume studies
(C) Chromosome study and a leukocyte alkaline phosphatase score
(D) Serum protein electrophoresis and computed tomography scan of the abdomen
(E) Lymph node biopsy and chromosome analysis

25. A 22-year-old male college student presents with widespread ecchymoses, hepatosplenomegaly, and bleeding gums. He has a normocytic anemia, thrombocytopenia, and granular-appearing blast cells with numerous auer rods. Coagulation studies exhibit prolonged prothrombin and partial thromboplastin times, low levels of fibrinogen, and increased fibrinogen degradation products (split products). The diagnosis for this patient is

(A) acute monocytic leukemia
(B) acute myelomonocytic leukemia
(C) chronic myelogenous leukemia
(D) acute proganulocytic leukemia
(E) a myelodysplastic syndrome

26. A peripheral smear from a 60-year-old man who presents with fever and hepatosplenomegaly is shown below. He has pancytopenia and scattered leukocytes similar to those noted in the slide. The finding most likely expected in this patient would be

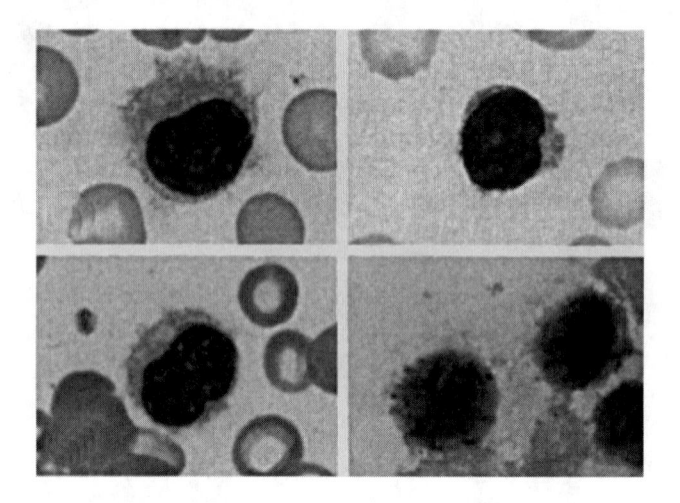

(A) absence of bone marrow involvement
(B) neoplastic cells primarily in the white pulp of the spleen
(C) an unfavorable response to interferon-α therapy
(D) a positive tartrate-resistant acid phosphatase stain on the peripheral blood leukocytes
(E) a terminal lymphoblast crisis in a few years

27. Which one of the following is more likely to be a feature of iron deficiency rather than β-thalassemia minor?

(A) A progression from a normocytic anemia to a microcytic anemia
(B) A normal red blood cell distribution width (RDW)
(C) A low mean corpuscular volume with an increased red blood cell count
(D) An increase in hemoglobin A_2 and F
(E) A defect in globin chain synthesis

28. Which of the following non-Hodgkin lymphomas has the best overall prognosis?

(A) Nodular (follicular) lymphocytic lymphoma
(B) Nodular (follicular) histiocytic lymphoma
(C) Diffuse histiocytic lymphoma
(D) Burkitt lymphoma (small noncleaved lymphocytic lymphoma)
(E) Immunoblastic lymphoma (large cell immunoblastic lymphoma)

29. A 16-year-old girl presents with bowel obstruction. An exploratory laparotomy reveals markedly enlarged para-aortic lymph nodes with entrapment of the small bowel. Sections through the lymph nodes exhibit a diffuse neoplastic infiltrate of small, round lymphocytes with a "starry sky" appearance on low power. The cytoplasm of some of the lymphocytes is vacuolated and fat stains are positive. You would expect the neoplastic cells to demonstrate a

(A) positive non-specific esterase stain of the cells
(B) positive specific esterase stain of the cells
(C) low leukocyte alkaline phosphatase score
(D) t(9;22) translocation
(E) t(8;14) translocation

30. A 65-year-old man presents with multiple plaque-like lesions scattered over his body. A biopsy reveals a dermal infiltrate of atypical-appearing lymphoid cells, some of which occupy spaces within the epidermis. The peripheral smear exhibits lymphocytes with a prominent nuclear cleft. A periodic acid schiff (PAS) stain demonstrates areas of PAS-positive material in the cytoplasm of these cells. The most likely diagnosis for this patient is a

(A) high-grade B-cell malignancy
(B) malignant plasma-cell disorder
(C) malignant T-cell disorder
(D) malignant histiocytosis
(E) benign T-cell disorder

31. Which of the following diseases has a spleen size that is markedly different from the other diseases listed?

(A) Chronic myelogenous leukemia
(B) Gaucher disease
(C) Agnogenic myeloid metaplasia
(D) Amyloidosis
(E) Idiopathic thrombocytopenic purpura

32. The order of best prognosis to worst prognosis followed by the variant of Hodgkin disease with the most to the least number of Reed-Sternberg (RS) cells is

1. lymphocyte depletion
2. nodular sclerosis
3. lymphocyte predominant
4. mixed cellularity

	Best-to-worst prognosis	Most-to-least number of RS cells
(A)	2-4-3-1	3-2-4-1
(B)	3-2-4-1	1-4-2-3
(C)	2-3-1-4	1-2-4-3
(D)	1-4-2-3	3-2-4-1
(E)	3-2-4-1	4-1-3-1

33. A 3-year-old boy presents with polyuria, exophthalmos, and multiple lytic lesions in the skull. The infiltrate is composed of CD1+ cells. This patient most likely has a neoplastic proliferation of which one of the following cell types?

(A) Plasma cells
(B) B lymphocytes
(C) Mast cells
(D) Histiocytes
(E) T lymphocytes

34. Alzheimer disease, restrictive cardiomyopathy, and nephrotic syndrome are features associated with which of the following conditions?

(A) The glycogenoses
(B) Iron overload
(C) Chronic renal disease
(D) Amyloid deposition
(E) Diabetes mellitus

35. A 15-year-old boy presents with epistaxis and multiple, nonpalpable petechiae on his back and chest approximately one week after an upper respiratory infection. He has no lymphadenopathy or hepatosplenomegaly. The complete blood count is unremarkable except for a platelet count of 5000 cells/μL (150,000–400,000 cells/μL). The prothrombin and partial thromboplastin times are normal. A bone marrow aspirate reveals a slightly increased number of megakaryocytes. The mechanism for this patient's thrombocytopenia most closely relates to

(A) a platelet-adhesion defect
(B) immunoglobulin G antibodies against the platelets
(C) hypersplenism
(D) consumption of the platelets in clots
(E) a platelet production problem

36. A 35-year-old female has unusually extensive vaginal bleeding 24 hours after a vaginal hysterectomy for menorrhagia. The excessive bleeding has not responded to two units of packed red blood cells. Laboratory studies reveal normal prothrombin and partial thromboplastin times, a normal platelet count, no fibrinogen degradation products, and a prolonged bleeding time. The patient's diagnosis is most likely

(A) a primary fibrinolysis
(B) disseminated intravascular coagulation
(C) a qualitative platelet defect
(D) von Willebrand disease
(E) a coagulopathy associated with birth control pills

37. Which one of the following tests is abnormal in both classical von Willebrand disease and hemophilia A?

(A) Ristocetin platelet-aggregation study
(B) Bleeding time
(C) Factor VIII antigen levels
(D) Partial thromboplastin time
(E) Prothrombin time

38. Destruction of the trilineage myeloid stem cell in the bone marrow would most likely result in which one of the following conditions?

(A) A pure T-cell deficiency state
(B) Polycythemia rubra vera
(C) Acute lymphocytic leukemia
(D) A pure red blood cell aplasia
(E) Aplastic anemia

39. A patient with a hemoglobin of 10 gm/dl and an uncorrected reticulocyte count of 12% has a corrected reticulocyte count of

(A) 2%
(B) 4%
(C) 6%
(D) 8%
(E) 12%

40. A normal mean corpuscular volume (MCV) would be expected in a patient with which one of the following conditions?

(A) Resection of the terminal ileum
(B) β-thalassemia minor
(C) 2 days post-acute bleed
(D) Chronic bleeding duodenal ulcer
(E) Pregnant and not taking any vitamin supplements

41. A patient with a hemoglobin of 4 gm/dl receives 4 units of packed red blood cells for his anemia. The following day, the patient's hemoglobin is 4 gm/dl. Which one of the following is an unlikely explanation for failure of the hemoglobin to increase after transfusion?

(A) Severe iron-deficiency anemia with depleted iron stores
(B) A hemolytic transfusion reaction with destruction of the transfused cells
(C) An autoimmune hemolytic anemia with destruction of the transfused cells
(D) An active site of bleeding, most likely in the gastrointestinal tract
(E) Splenomegaly with trapping of the transfused cells

42. A 35-year-old woman presents with fatigue and exercise intolerance. Physical examination reveals pallor of the conjunctiva and palmar creases. The patient's hemoglobin is 6 gm/dl. Which one of the following tests is the next most important study to order on this patient?

(A) Serum folate/B_{12} measurement
(B) Coombs test
(C) Bone marrow aspirate
(D) Reticulocyte count
(E) Serum ferritin

43. A poor reticulocyte response would be expected in a patient with which one of the following conditions?

(A) Iron deficiency treated with ferrous sulfate after 1 week of therapy
(B) Autoimmune hemolytic anemia
(C) Congenital spherocytosis
(D) Untreated folate deficiency
(E) Thrombotic thrombocytopenic purpura

44. Which one of the following would more likely result in a folate deficiency rather than a vitamin B_{12} deficiency?

(A) Crohn disease
(B) Pure vegetarian diet
(C) Pregnancy and lactation
(D) Bacterial overgrowth in the small bowel
(E) Chronic pancreatitis

45. Which one of the following indicators is present in both pernicious anemia and folate deficiency?

(A) Decreased vibratory sensation in the lower extremities
(B) Anti-parietal cell antibodies
(C) Megaloblastic bone marrow
(D) Atrophic gastritis of the body and fundus
(E) Schilling test corrected by adding intrinsic factor

46. A 35-year-old man presents with petechia, ecchymoses, fatigue, and a low-grade fever. His physical examination is otherwise unremarkable. A complete blood cell count reveals a moderately severe normocytic anemia, thrombocytopenia, and absolute neutropenia. The corrected reticulocyte count is less than 2%. The serum blood urea nitrogen and creatinine are normal. The next step in the management of this patient is to order which one of the following tests?

(A) Direct Coombs test
(B) Sugar water test
(C) Serum vitamin B_{12}/folate level measurement
(D) Bone marrow aspirate and biopsy
(E) Computed tomography scan of the abdomen

47. A 45-year-old man presents with a severe normocytic anemia and a normal platelet and leukocyte count. His corrected reticulocyte index is less than 2%. A bone marrow aspirate reveals absent red blood cell precursors, but granulocytes and megakaryocytes are normally represented. These findings indicate which one of the following diagnoses?

(A) Iron deficiency
(B) Megaloblastic anemia
(C) Lead poisoning
(D) Anemia of chronic disease
(E) Parvovirus infection

48. A 52-year-old woman with end-stage diabetic nephropathy has a moderately severe normocytic anemia with a corrected reticulocyte count less than 2%. Her anemia will most likely respond to administration of which one of the following substances?

(A) Ferrous sulfate
(B) Vitamin B_{12}
(C) Erythropoietin
(D) Folate
(E) Corticosteroids

49. A bone marrow aspirate with a special stain was performed on a 70-year-old man who has a severe anemia that requires a packed red blood cell transfusion every 3 to 4 months. His peripheral smear consistently shows a dimorphic red blood cell population associated with pancytopenia. His bone marrow is megaloblastic, has 5% myeloblasts, and contains normoblasts similar to the one present in the slide. The most likely diagnosis for this patient is

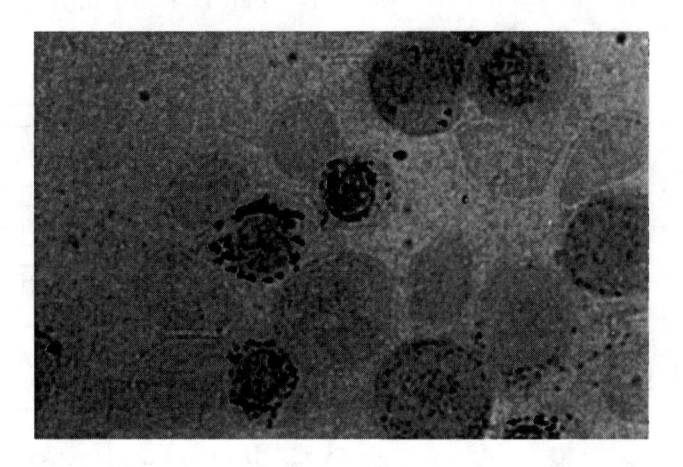

(A) myelodysplastic syndrome
(B) chronic lymphocytic leukemia
(C) chronic myelogenous leukemia
(D) lead poisoning
(E) hairy cell leukemia

50. A 4-year-old boy presents with fever, epistaxis, generalized nontender lymphadenopathy, bone pain, and hepatosplenomegaly. He has a normocytic anemia, thrombocytopenia, and an absolute lymphocytosis with 10% immature lymphoid cells present in the peripheral blood. His bone marrow examination is reported as abnormal. The patient's prognosis is most closely related to

(A) the degree of thrombocytopenia
(B) the degree of anemia
(C) cytogenetic studies of cells taken from a bone marrow aspirate
(D) presence or absence of the common acute lymphoblastic leukemia antigen (CALLA)
(E) whether there is lymph node involvement

51. A positive inotropic agent has which one of the following actions?

(A) It decreases the stroke volume
(B) It decreases the pulse pressure
(C) It decreases the end-systolic volume
(D) It increases the end-diastolic volume
(E) It decreases the end-diastolic reserve volume

Questions 52-53

Blood is flowing through the circuit shown below. The inflow pressure (in the large arteries) is 100 mm Hg, and the outflow pressure (in the capillaries) is 20 mm Hg. The resistance of each of the four branches is 2 mm Hg/ml/min.

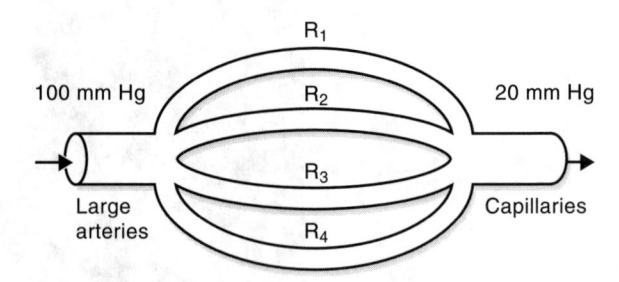

$R_1, R_2, R_3, R_4 = 2$ mm Hg/ml/min

52. If an additional vessel of very high resistance were added to the circuit above, the flow would

(A) become pulsatile
(B) decrease
(C) increase
(D) become turbulent
(E) not change

53. What is the flow across the circuit?

(A) 100 ml/min
(B) 160 ml/min
(C) 12.5 ml/min
(D) 800 ml/min
(E) 500 ml/min

54. In which one of the following sites is turbulent flow most likely to occur?

(A) Ascending aorta
(B) Small arteries
(C) Capillaries
(D) Small veins
(E) Large veins

55. The following information was obtained during catherization of a patient.

Right ventricular pressure	=	25/0 mm Hg
Mean pulmonary arterial pressure	=	15 mm Hg
Pulmonary capillary wedge pressure	=	5 mm Hg
Pulmonary venous pressure	=	10 mm Hg
Cardiac output	=	5 L/min

What is the resistance across the pulmonary circulation?

(A) 1.6 mm Hg/L/min
(B) 5.0 mm Hg/L/min
(C) 2.0 mm Hg/L/min
(D) 3.6 mm Hg/L/min
(E) 1.0 mm Hg/L/min

56. The following data were obtained from a man weighing 176 pounds.

- Aorta oxygen (O_2) content = 20.0 vol%
- Femoral vein O_2 content = 16 vol%
- Coronary sinus O_2 content = 10 vol%
- Pulmonary artery O_2 content = 15 vol%
- Total body O_2 consumption = 400 ml/min

What is the cardiac output of this man?

(A) 10 L/min
(B) 4 L/min
(C) 5 L/min
(D) 6 L/min
(E) 8 L/min

Questions 57-59

The data presented below were obtained from a healthy volunteer before and during moderate exercise.

	Before	During
Stroke volume (ml)	70	100
Heart rate (beats/min)	80	120
Total peripheral resistance (mm Hg/L/min)	16	8
Systolic blood pressure (mm Hg)	120	144
Diastolic blood pressure (mm Hg)	75	72
Arterial P_{O_2} (mm Hg)	100	100
Right atrial pressure (mm Hg)	0	2

57. Based on the information given above, which one of the following takes place during exercise?

(A) Afterload is reduced
(B) Mean arterial blood pressure is increased
(C) Alveolar ventilation does not change
(D) Atrial stretch receptor activity increases vagal stimulation of the heart
(E) Increased sympathetic activity to the heart raises the preload

58. What is the effect of the exercise on pulse pressure?

(A) Pulse pressure is not changed
(B) Pulse pressure is increased
(C) Pulse pressure is decreased
(D) Pulse pressure cannot be estimated

59. The decline in total peripheral resistance is caused by which one of the following?

(A) The muscle pump
(B) Sympathetic vasodilator fiber discharge
(C) Metabolic vasodilation of arterioles
(D) Metabolic vasodilation of venules
(E) Peripheral chemoreceptor activation of the sympathetic nervous system

60. Which one of the following adaptations to endurance training allows the athlete to increase his or her cardiac output more than a normal sedentary individual? The low resting

(A) heart rate
(B) stroke volume
(C) diastolic blood pressure
(D) oxygen extraction by the tissues
(E) cardiac output

61. Pulse pressure widens in normal individuals as they age because of which one of the following factors?

(A) Stroke volume increases
(B) Venous return increases
(C) Heart rate increases
(D) Arterial compliance decreases
(E) Total peripheral resistance increases

62. The rapid infusion of 1 L of plasma into a femoral vein will have which one of the following effects?

(A) Increase the end-diastolic reserve volume
(B) Directly reduce the firing rate of the sino-atrial node
(C) Increase the end-systolic volume
(D) Increase the pulse pressure
(E) Decrease the end-systolic volume

63. Venoconstrictor agents would be expected to have which one of the following effects?

(A) Increase the end-diastolic volume
(B) Decrease the stroke volume
(C) Increase the end-diastolic reserve volume
(D) Decrease the work of the heart
(E) Act primarily on the large veins

64. During a deep inspiration, which one of the following events takes place?

(A) Venous return to the left atrium increases
(B) Cardiac output of the right ventricle increases
(C) Pulmonary blood volume decreases
(D) The pressure gradient between the peripheral veins and the right atrium decreases
(E) Cardiac output from the left ventricle increases

65. A 25-year-old, 70-kg man suffered several broken ribs as a result of a fall from a ladder. His treatment in a nearby hospital included stabilizing his chest with bandages. The bandages were tied in a way that reduced his tidal volume by 50%. To compensate, he doubled his respiratory rate. Two hours later, an arterial blood sample was taken. Which of the following conditions would have been observed?

(A) Increased P_{O_2} and decreased P_{CO_2}
(B) No change in P_{O_2} or P_{CO_2}
(C) Decreased P_{O_2} and increased P_{CO_2}
(D) Increased P_{O_2} and increased P_{CO_2}
(E) No change in P_{O_2} and increased P_{CO_2}

66. Curve A in the diagram below depicts intra-alveolar pressure (IAP) during a normal respiratory cycle. Curve B represents

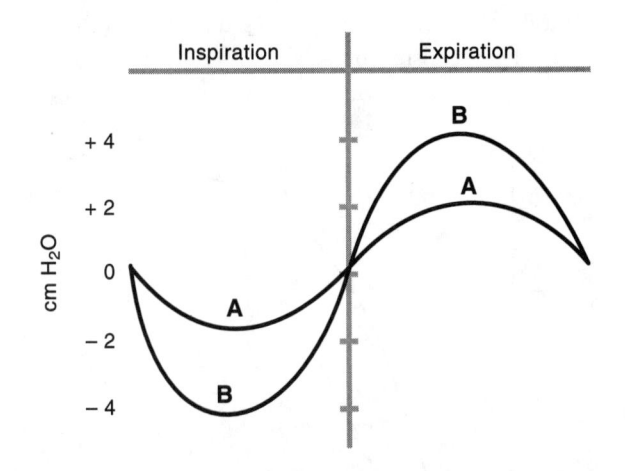

(A) IAP during a forced expiration
(B) IAP during a larger breath
(C) intrapleural pressure (IPP) during the same breath as in Curve A
(D) IPP during a larger breath
(E) IPP during a forced expiration

67. The graph depicts the change in intrapleural pressure during a normal respiratory cycle. The rate of airflow is 0 at which of the following points on the graph?

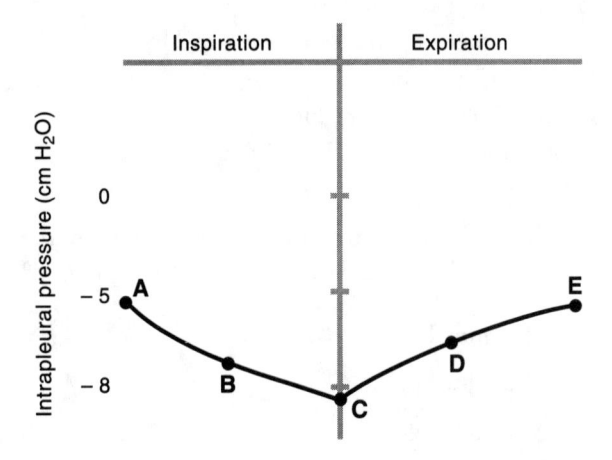

(A) A only
(B) A and E
(C) A, C, and E
(D) B only
(E) E only

Questions 68-69

As part of a helium dilution test, a patient is asked to inhale a gas mixture containing 10% helium, 20% oxygen, and 70% nitrogen. Use the following information to answer the questions.

Total barometric pressure = 760 mm Hg
PH_2O = 47 mm Hg
Alveolar P_{CO_2} = 40 mm Hg
Respiratory exchange ratio = 0.8

68. What is the partial pressure of helium in the anatomic dead space at the end of a normal inspiration?

(A) 71.3 mm Hg
(B) 21.3 mm Hg
(C) 31.3 mm Hg
(D) 142.6 mm Hg
(E) 92.6 mm Hg

69. What is the partial pressure of oxygen in the alveoli?

(A) 100 mm Hg
(B) 102.6 mm Hg
(C) 99 mm Hg
(D) 142.6 mm Hg
(E) 82.6 mm Hg

70. The two oxyhemoglobin dissociation curves and the accompanying data were obtained from blood perfusing skeletal muscle. One curve was obtained at rest and the other during exercise. What is the percent saturation of hemoglobin with O_2 in the venous blood leaving the exercising muscle?

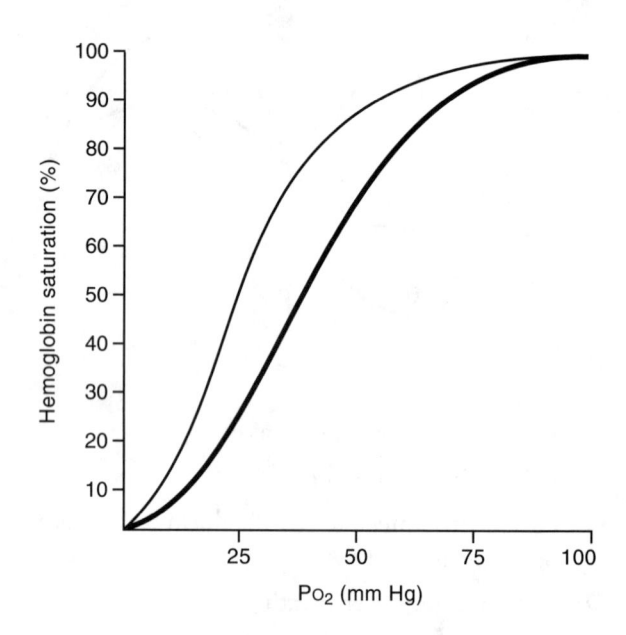

	P_{O_2} (arterial)	P_{O_2} (venous)
Muscle at rest	100 mm Hg	50 mm Hg
Exercising muscle	100 mm Hg	25 mm Hg

(A) 0%
(B) 25%
(C) 50%
(D) 70%
(E) 100%

71. If a patient is breathing room air (21% oxygen), arterial P_{CO_2} is 80 mm Hg, and arterial P_{O_2} is 40 mm Hg, which of the following is the alveolar-arterial (A-a) gradient of the patient?

(A) 10 mm Hg
(B) 20 mm Hg
(C) 30 mm Hg
(D) 40 mm Hg
(E) 50 mm Hg

72. An arterial blood gas sample exposed to room air would be expected to have which of the following alterations in P_{O_2} and P_{CO_2}?

	Arterial P_{O_2}	Arterial P_{CO_2}
(A)	Falsely increased	Falsely increased
(B)	Falsely decreased	Falsely decreased
(C)	Falsely increased	Falsely decreased
(D)	Falsely decreased	Falsely increased
(E)	No change	No change

73. Which of the following clinical conditions would most likely be accompanied by a normal alveolar-arterial (A-a) gradient in the lungs?

(A) Chronic obstructive pulmonary disease
(B) Pulmonary embolus
(C) Paralysis of the diaphragms
(D) Adult respiratory distress syndrome
(E) Coal worker's pneumoconiosis

Questions 74-75

The accompanying diagram is a schematic of the respiratory unit of the lung. The letters A, B, and C represent three subdivisions of the unit.

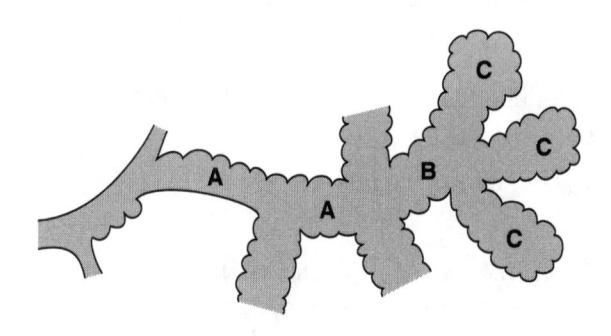

74. A 28-year-old, nonsmoking man presents with chronic lung disease limited to the lower lobes of the lungs. A sweat test is negative, and an abnormality is detected in a serum protein electrophoresis. Which of the following areas of the lung would exhibit evidence of this man's disease?

(A) Area A only
(B) Area B only
(C) Area C only
(D) Areas A, B, and C
(E) Areas B and C

75. A 49-year-old man who worked in coal mines most of his life has severe dyspnea and a productive cough with numerous dust cells present in the sputum. Which of the following areas of the lung would exhibit evidence of this man's disease?

(A) Area A only
(B) Area B only
(C) Area C only
(D) Areas A, B, and C
(E) Areas B and C

76. Twenty-four hours after elective surgery for chronic cholecystitis, a 46-year-old woman develops fever. The pathogenesis of this fever most likely relates to

(A) a pulmonary infection
(B) absorption atelectasis secondary to mucous plugs
(C) a pulmonary embolus with infarction
(D) a wound infection
(E) the normal healing process

77. A 42-year-old man who smokes and is an alcoholic has an abrupt onset of spiking fever, a productive cough with blood-tinged mucoid sputum, and pleuritic chest pain. Physical examination reveals dullness to percussion, crepitant rales, and increased tactile fremitus in the right upper lobe. A chest x-ray reveals a lobar consolidation in the right upper lobe. The Gram stain, which was positive, would be expected to show which of the following?

(A) Gram-positive rods
(B) Gram-negative rods with a capsule
(C) Gram-positive cocci in clumps
(D) Gram-negative diplococci
(E) Gram-positive filamentous bacteria

78. Which of the following disorders decreases both the partial pressure of oxygen (PaO$_2$) and the oxygen saturation (SaO$_2$) of arterial blood?

(A) Anemia
(B) Carbon monoxide poisoning
(C) Respiratory acidosis
(D) Decreased cardiac output
(E) Cyanide poisoning

79. A 69-year-old man with a 60 pack-year history for smoking presents with dizziness and visual disturbances. Physical examination reveals a purplish discoloration of his face, arms and neck; retinal vein engorgement; and visible distention of his neck veins. His complete blood count reveals a normocytic anemia. The pathophysiology of this patient's clinical presentation most likely involves

(A) right heart failure secondary to left heart failure
(B) obstruction of the superior vena cava by a primary tumor arising in the lungs
(C) metastatic disease to the cervical lymph nodes with compression of the jugular veins
(D) polycythemia rubra vera
(E) hyperviscosity syndrome secondary to an underlying plasma cell malignancy

80. A 52-year-old, nonsmoking man complains of pain and paresthesias in his right hand, particularly at night. Physical examination reveals a diminished radial pulse when he abducts his arm when his head is turned to either side. Percussion over the brachial plexus reproduces the pain and paresthesias in his arm. A bruit is audible over the upper right anterior chest, particularly when he pushes his shoulders back and turns his head in either direction. His neurologic examination is unremarkable. You suspect the patient has

(A) a Pancoast tumor
(B) a cervical rib
(C) cervical disc disease
(D) subclavian steal syndrome
(E) superior vena caval syndrome

81. In people with cystic fibrosis, the cause of death is most likely to be related to

(A) acute and chronic pulmonary infections
(B) pulmonary infarction
(C) cor pulmonale
(D) left heart failure
(E) bronchiectasis

82. In a person with situs inversus, recurrent sinopulmonary infections, and sterility, the mechanism most likely responsible for these findings is

(A) a defect in humoral immunity
(B) an immune complex disease
(C) a defect in ciliary structure and function
(D) a defect in cellular immunity
(E) a defect in exocrine secretions

83. The lung depicted in the photograph was removed at autopsy from a 45-year-old individual. Which of the following clinical descriptions would best describe this patient?

(A) Alcoholic with fever, fetid breath, and sputum containing a mixture of gram-positive cocci, gram-negative rods, and fusobacterium
(B) Alcoholic with a history of weight loss, fever, and night sweats; and a positive purified protein derivative (PPD) skin test
(C) Smoker with a history of cough, weight loss, and a sputum cytology showing malignant squamous cells
(D) Smoker with a history of cough, weight loss, hilar mass, and severe hyponatremia
(E) Woman with a history of infiltrating ductal carcinoma of the breast

84. A 25-year-old woman with a history of sinusitis, nasal polyps, and asthma is brought to the emergency department because of severe bronchospasm. Given these findings, one would suspect that the patient

(A) took aspirin for pain
(B) has a history of a positive sweat test
(C) has a strong family history for allergies
(D) has immunoglobulin A deficiency
(E) also has situs inversus

Questions 85-87

A 9-year-old girl is brought by ambulance to the emergency room in respiratory distress. Her mother states that she developed a cold 4 days before and has had progressive difficulty breathing for the last 24 hours. This had happened previously, either after a respiratory infection or during the spring or autumn, but her breathing problems usually improved within a few hours. She is not currently on prophylactic medications.

Physical examination reveals a child in marked respiratory distress. She is afebrile, has a respiratory rate of 30/min, a pulse of 108/min with a regular rhythm, and blood pressure of 100/72 mm Hg. Her nasal mucosa is boggy and pale bilaterally. There is tonsillar enlargement but no exudate. Postnasal discharge is present. She has marked indrawing of her respiratory muscles. The anteroposterior diameter of her chest is increased and there is hyper-resonance to percussion. Generalized inspiratory and expiratory wheezes are present. Her heart sounds are distant. No murmurs or abnormal heart sounds are present.

A complete blood cell count reveals normal hemoglobin and hematocrit, normal white blood cell count, and 13% eosinophils in the white blood cell differential. A chest x-ray exhibits hyperaeration with flattening of the diaphragms. No infiltrates are present.

85. This patient's arterial blood gases at this time would be expected to exhibit which of the following conditions?

(A) Primary metabolic acidosis
(B) Primary respiratory acidosis
(C) Primary metabolic alkalosis
(D) Primary respiratory alkalosis
(E) Mixed primary respiratory alkalosis and primary metabolic acidosis

86. The findings of increased anteroposterior diameter, hyper-resonance to percussion, and flattened diaphragms on chest x-ray are most closely related to

(A) an increase in her tidal volume
(B) an increase in her residual volume
(C) an increase in her vital capacity
(D) an increase in her forced expiratory volume in 1 second (FEV_1)
(E) a component of emphysema in the distal airways

87. The pathogenesis of this patient's pulmonary problems is most closely related to which of the following conditions?

(A) Increased lung compliance
(B) Inflammation in the small-caliber airways
(C) Intrapulmonary shunting
(D) Loss of surfactant
(E) Inflammation in the segmental bronchi

88. Aspirin and ticlopidine share which of the following characteristics?

(A) Both may cause reversible agranulocytosis
(B) Both inhibit adenosine diphosphate (ADP)-induced platelet aggregation
(C) Both inhibit cyclooxygenase and the formation of thromboxane
(D) Both can be used to prevent stroke and myocardial infarction
(E) Both are equally effective in preventing venous and arterial thrombi

89. Which of the following antibiotics should typically be administered to prevent bacterial endocarditis in patients with cardiac valve disease who are undergoing dental surgery?

(A) Ciprofloxacin
(B) Amoxicillin
(C) Ceftriaxone
(D) Doxycycline
(E) Erythromycin

90. Angiotensin-converting enzyme inhibitors such as enalapril also act to increase the activity of which of the following substances?

(A) Vasopressin
(B) Endothelins
(C) Bradykinin
(D) Substance P
(E) Enkephalins

91. Antihistamines associated with an increased risk of cardiac arrhythmias in patients receiving erythromycin or ketoconazole include

(A) diphenhydramine
(B) clemastine
(C) loratadine
(D) terfenadine
(E) chlorpheniramine

92. Propranolol and prazosin share which of the following physiologic effects?

(A) Increase vagal tone
(B) Decrease vascular resistance
(C) Inhibit renin secretion
(D) Decrease cardiac output
(E) Reduce arterial pressure

93. Nifedipine and verapamil share which of the following physiologic effects?

(A) Decrease systemic vascular resistance
(B) Decrease heart rate
(C) Block vascular and cardiac calcium channels equally
(D) Selectively block cardiac calcium channels
(E) Selectively block calcium channels in smooth muscle

Questions 94-95

Plasma levels of the following substances have been measured in a group of patients before and after receiving enalapril therapy:

(A) Aldosterone
(B) Angiotensinogen
(C) Angiotensin I
(D) Angiotensin II
(E) Serum potassium
(F) Renin

94. Levels of which substances would be increased after enalapril treatment?

(A) A, C, and E
(B) D and F
(C) E and F
(D) C, E, and F
(E) A and D

95. Levels of which substances would be decreased after enalapril treatment?

(A) B, C, and D
(B) A, D, and F
(C) D and E
(D) A and E
(E) A and D

DIRECTIONS:

Each of the numbered items or incomplete statements in this section is negatively phrased, as indicated by a capitalized word such as NOT, LEAST, or EXCEPT. Select the ONE lettered answer or completion that is BEST in each case.

96. All of the following statements regarding the lymphatic vascular system are true EXCEPT

(A) the thoracic duct receives lymph from regions above and below the diaphragm
(B) lymph is a fluid that enters lymphatic capillaries from the interstitial space of various tissues
(C) lymph from the right and left bronchomediastinal trunks enters the thoracic duct
(D) lymph enters and leaves the lymph nodes
(E) lymph always enters the venous portion of the cardiovascular system

97. All of the following statements regarding the blood vessels of the heart are true EXCEPT

(A) the right coronary artery typically arises from the right aortic sinus
(B) the left coronary artery passes anterior to the left auricle
(C) the circumflex branch of the left coronary artery typically lies in the coronary sulcus
(D) the great cardiac vein drains into the coronary sinus
(E) the atrioventricular nodal artery is typically a branch of the anterior interventricular artery

98. A patient with asthma is having difficulty breathing. All of the following muscles assist with inspiration EXCEPT

(A) the pectoralis major
(B) the pectoralis minor
(C) the rectus abdominis
(D) the anterior scalene
(E) the middle scalene

99. All of the following statements regarding sympathetic innervation to the heart are true EXCEPT

(A) stimulation causes an increased heart rate
(B) stimulation causes an increased force of contraction
(C) stimulation causes dilatation of the coronary vessels
(D) postganglionic fibers arise from the cervical and upper thoracic ganglia
(E) preganglionic fibers arise from all cervical and upper thoracic spinal cord segments

100. All of the following statements regarding the intercostal vessels are true EXCEPT

(A) the left fifth intercostal vein drains into the hemiazygos vein
(B) the right fifth intercostal vein drains into the azygos vein
(C) the left fifth intercostal artery lies along the upper border of the left sixth rib
(D) the fifth intercostal vein lies superior to the fifth intercostal artery
(E) the fifth intercostal artery communicates with the aorta posteriorly and the internal thoracic artery anteriorly

101. All of the following statements regarding the heart are true EXCEPT

(A) the nerve fibers that conduct the pain associated with ischemia accompany the sympathetic nerve supply to the heart
(B) the sinoatrial node does not receive a nerve supply and thus initiates the heart beat autonomously
(C) the atrioventricular node is located in the interatrial septum
(D) the diaphragmatic surface of the heart is composed primarily of the left ventricular wall
(E) the valve of the inferior vena cava is an elevation of the endocardium between the anterior edge of the inferior vena cava and the limbus of the fossa ovalis

102. A 40-year-old woman who has leukemia is hospitalized in the intensive care unit. When conducting a psychiatric examination, which of the following is a clinician LEAST likely to see?

(A) Agitation
(B) Noncompliance with medical advice
(C) Insomnia
(D) Disorientation
(E) Narcissistic personality disorder

103. Diastolic heart failure is an impairment in the filling of the left ventricle. Which of the following is LEAST likely to occur?

(A) Improvement of the patient's condition with administration of a calcium channel blocker
(B) Decreased compliance of the heart
(C) Increased left atrial pressure
(D) Improvement of patient's condition with administration of a positive inotropic agent
(E) Pulmonary congestion

104. All of the following statements regarding electrocardiographic (ECG) changes in the natural history of an acute myocardial infarction are true EXCEPT

(A) Q waves represent the area of infarct with cell death
(B) elevated ST segment represents the area of subendocardial cell death
(C) flipped T wave represents the area of ischemia
(D) Q waves remain after the acute injury has subsided
(E) Q waves localize the site of involvement in the heart

105. All of the following findings are associated with pure left-sided heart failure EXCEPT

(A) pulmonary edema
(B) hepatojugular reflux
(C) paroxysmal nocturnal dyspnea
(D) S_3 heart sound
(E) prerenal azotemia

106. All of the following associations regarding vasculitis are correct EXCEPT

(A) Kawasaki disease—coronary arteritis
(B) polyarteritis nodosa—hepatitis B antigenemia
(C) Wegener granulomatosis—Raynaud phenomenon
(D) Takayasu arteritis—lack of pulse in the upper extremity
(E) thromboangiitis obliterans—male smoker with gangrenous digits

107. Immune complexes are LEAST likely involved in the pathogenesis of

(A) infective endocarditis
(B) rheumatoid vasculitis
(C) dissecting aortic aneurysm
(D) polyarteritis nodosa
(E) hypersensitivity vasculitis syndromes

108. In the response to injury theory of atherosclerosis, which of the following factors is LEAST likely to have a significant role in the formation of a fibrofatty plaque in a vessel?

(A) Monocytes/macrophages
(B) Platelet-derived growth factor
(C) Free radical injury
(D) Smooth muscle cells
(E) Neutrophils

109. All of the following statements regarding low-density lipoproteins (LDLs) are true EXCEPT

(A) they are the best parameter to follow for diet and drug therapy
(B) they are derived from very-low-density lipoproteins (VLDLs)
(C) they contain the majority of cholesterol circulating in the blood
(D) they are calculated from the cholesterol, high-density lipoprotein (HDL), and triacylglycerol levels
(E) an increase in LDLs is the most important major risk factor for coronary artery disease

110. An 82-year-old man with endotoxic shock is oozing blood from every venipuncture site. Which of the following tests is LEAST indicated in the work-up of this patient?

(A) Partial thromboplastin time
(B) Platelet count
(C) Serial fibrinogen levels
(D) Bone marrow aspirate
(E) Fibrinogen degradation products (split products)

111. In a normal individual, the apparent viscosity of blood is determined by all of the following EXCEPT the

(A) hematocrit
(B) axial streaming of red blood cells
(C) size of the blood vessel
(D) velocity of blood flow
(E) plasma protein concentration

112. A sudden increase in total peripheral resistance has all of the following effects EXCEPT

(A) increase the diastolic blood pressure
(B) reduce the stroke volume
(C) increase the mean arterial blood pressure
(D) increase left ventricular systolic pressure
(E) increase cardiac output

113. Study the table listing characteristics of heparin and warfarin. Which of the lettered sets of characteristics are NOT correct?

	Variable	Warfarin	Heparin
(A)	Crosses the placenta	No	Yes
(B)	Onset of action	Delayed	Rapid
(C)	Active in vitro	No	Yes
(D)	Administration	Oral	Parenteral

114. Which of the following statements characterizing nitric oxide is NOT correct?

(A) It appears to be identical to endothelium-derived relaxing factor
(B) Its formation is stimulated by acetylcholine, bradykinin, and substance P
(C) It is formed from nitroglycerin, nitroprusside, and other nitrovasodilator drugs
(D) It stimulates activation of cyclic AMP-dependent protein kinases
(E) Its physiologic roles include penile erection and inhibition of platelet aggregation

115. All of the following agents relax bronchial smooth muscle in patients with bronchoconstriction EXCEPT

(A) terbutaline
(B) ipratropium
(C) albuterol
(D) cromolyn sodium
(E) epinephrine

116. All of the following agents may be useful in treating acute allergic rhinitis EXCEPT

(A) terfenadine
(B) albuterol
(C) pseudoephedrine
(D) cromolyn sodium
(E) beclomethasone

117. Adverse effects associated with the use of β_2-adrenergic agonist aerosols in the treatment of asthma include all of the following EXCEPT

(A) skeletal muscle tremor
(B) tachycardia
(C) oral candidiasis
(D) central nervous system stimulation
(E) palpitations

118. All of the following drugs are effective antitussive agents EXCEPT

(A) guaifenesin
(B) dextromethorphan
(C) codeine
(D) benzonatate
(E) hydrocodone

Body Systems Review I: Hematopoietic/Lymphoreticular, Respiratory, Cardiovascular

DIRECTIONS:
Each set of matching questions in this section consists of a list of three to twenty-six lettered options (some of which may be in figures) followed by several numbered items. For each numbered item, select the ONE lettered option that is most closely associated with it. To avoid spending too much time on matching sets with a large number of options, it is generally advisable to begin each set by reading the list of options. Then for each item in the set, try to generate the correct answer and locate it in the option list, rather than evaluating each option individually. Each lettered option may be selected once, more than once, or not at all.

Questions 119-120

In the following cross-section of the heart, the lettered areas represent the areas of distribution of the coronary artery blood supply of the left anterior descending, left circumflex, and right coronary arteries. Match the most likely location for an acute myocardial infarction with the associated clinical descriptions.

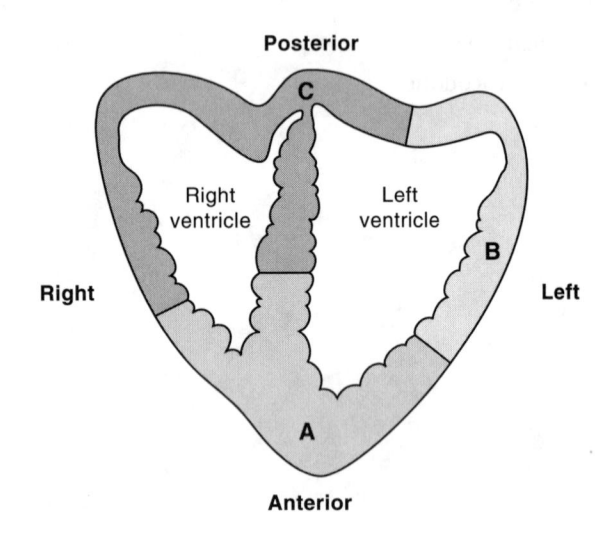

119. Papillary muscle rupture

120. Sinus bradycardia

Questions 121-123

A schematic drawing of a lymph node is shown below. Correlate the following clinical disorders with the area of the lymph node that is primarily involved.

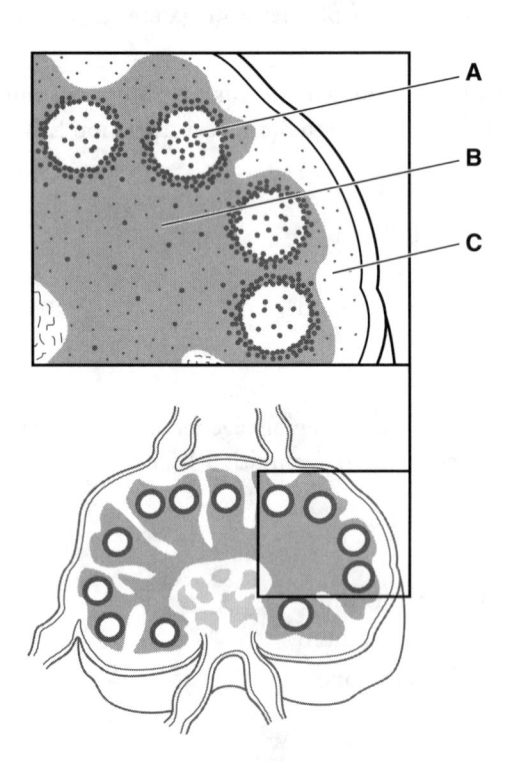

121. Most common site for the origin of malignant lymphomas

122. Most common site for the initial metastasis of a carcinoma

123. Site that would be depleted in a patient with DiGeorge syndrome

Questions 124-127

Match the following terms used in describing hemolytic anemias with the hemolytic anemia it best describes.

(A) Extrinsic hemolytic anemia/intravascular hemolysis
(B) Extrinsic hemolytic anemia/extravascular hemolysis
(C) Intrinsic hemolytic anemia/intravascular hemolysis
(D) Intrinsic hemolytic anemia/extravascular hemolysis

124. Paroxysmal nocturnal hemoglobinuria

125. Warm autoimmune hemolytic anemia (majority of cases)

126. Sickle cell anemia

127. Anemia associated with a prosthetic heart valve

Questions 128-131

Match the following clinical scenarios with the most likely set of laboratory findings.

	RBC mass	Plasma volume	Sao_2	Erythropoietin concentration
(A)	Normal	Decreased	Normal	Normal
(B)	Increased	Increased	Normal	Low
(C)	Increased	Normal	Decreased	Increased
(D)	Increased	Normal	Normal	Increased

RBC = red blood cell; Sao_2 = oxygen saturation.

128. 62-year-old man with a plethoric face, neutrophilic leukocytosis, thrombocytosis, splenomegaly, and an erythrocyte sedimentation rate of 0

129. 72-year-old woman with a 4-cm cystic renal mass, cannon-ball metastases in the lung, and polycythemia

130. 4-year-old boy with uncorrected tetralogy of Fallot

131. 22-year-old female marathon runner who is dehydrated from lack of water supplementation during the race

Questions 132-135

Match each of the following cellular descriptions with the correct muscle in which the action occurs.

(A) Cardiac muscle
(B) Skeletal muscle
(C) Cardiac and skeletal muscle
(D) Neither cardiac nor skeletal muscle

132. Calcium ion (Ca^{2+}) influx from the extracellular fluid occurs during the action potential

133. Increase in cytosolic Ca^{2+} caused by release by the sacroplasmic reticulum

134. The strength of contraction is related to the extracellular fluid Ca^{2+} concentration

135. Pharmacologic blockade of the fast sodium ion channels prevents the development of a normal action potential

Questions 136-137

Match each of the following clinical scenarios to the hematology test that best contributes to a diagnosis.

(A) Heinz body preparation
(B) Hemoglobin electrophoresis
(C) Coombs test
(D) Osmotic fragility test
(E) Sugar water test
(F) Urine test for hemosiderin
(G) Serum haptoglobin test

136. Anemia in a patient taking methyldopa (Aldomet)

137. Acute onset of anemia precipitated by eating fava beans

Questions 138-142

The following graph depicts the relationship between stroke volume and end-diastolic volume. Match each of the following effects to the proper area of the graph.

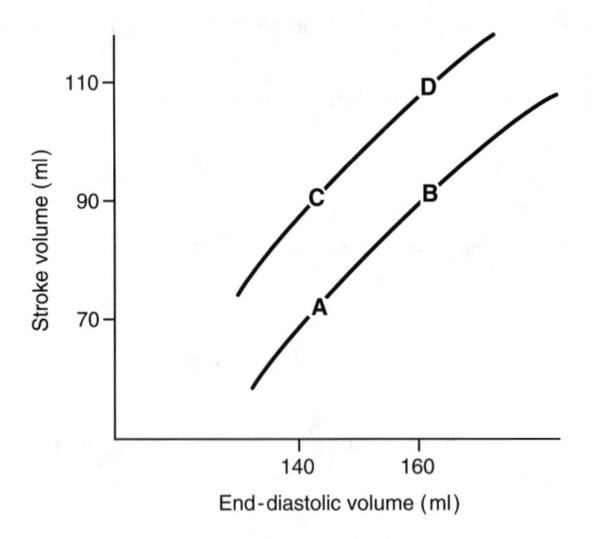

(A) A to B
(B) A to C
(C) A to D
(D) D to C
(E) D to B

138. Indicates an increase in preload

139. Results from the injection of epinephrine

140. Occurs during β-1 blockade

141. Occurs during moderate exercise

142. Results from rapid pacing of the heart

Questions 143-145

The pressure changes described below occurred during a single respiratory cycle. Match each description to the appropriate diagram.

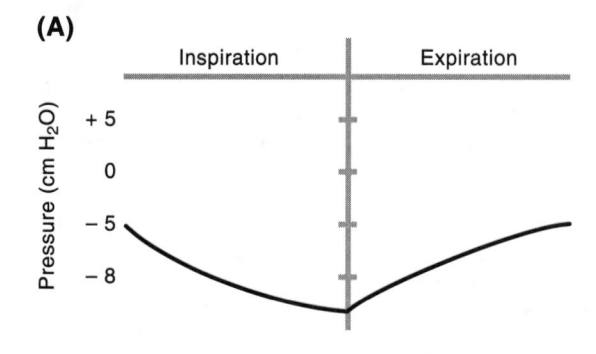

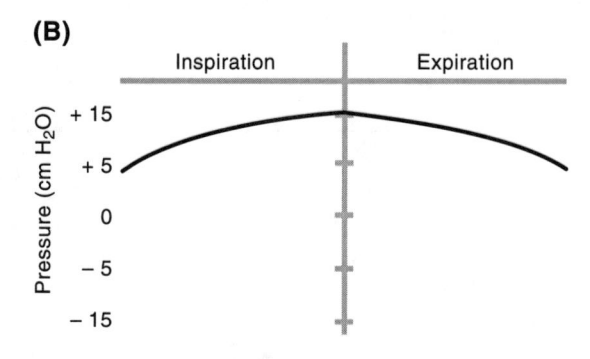

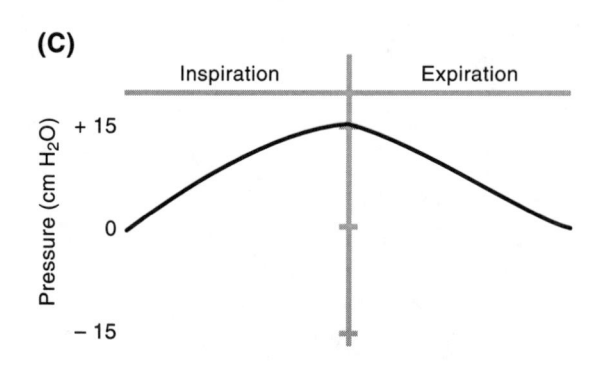

(A)

[Graph showing Pressure (cm H₂O) vs Inspiration/Expiration, curve dipping from 0 to -8 and returning to -5]

(B)

[Graph showing Pressure (cm H₂O) vs Inspiration/Expiration, curve rising from +5 to +15 and returning toward -5]

(C)

[Graph showing Pressure (cm H₂O) vs Inspiration/Expiration, curve rising from 0 to +15 and returning to 0]

143. The change in intra-alveolar pressure during positive pressure breathing

144. The change in intra-alveolar pressure during positive end-expiratory pressure (PEEP) breathing

145. The change in intrapleural pressure during a normal respiratory cycle

Questions 146-155

Match each of the following clinical descriptions with the arterial blood gas results that are most appropriate.

	pH	Paco₂ (mm Hg)	HCO₃⁻ (mEq/L)
(A)	7.22	69	27
(B)	7.50	29	22
(C)	7.26	26	11
(D)	7.51	48	38
(E)	7.40	40	24
(F)	7.42	22	14
(G)	7.60	32	30
(H)	7.33	68	34
(I)	7.37	62	36
(J)	7.00	50	12

146. 26-year-old medical student with an acute anxiety attack and a respiratory rate of 26/min

147. 28-year-old pregnant woman with excessive vomiting (hyperemesis gravidarum)

148. 5-year-old boy with infantile polycystic kidney disease and chronic renal failure

149. 45-year-old woman with chronic renal failure who is on a loop diuretic for heart failure

150. Semi-comatose 27-year-old man with heroin overdose

151. 65-year-old man with cardiorespiratory arrest

152. 55-year-old woman with disabling rheumatoid arthritis who is on toxic doses of salicylates for pain. Her respiratory rate is increased and she is hypotensive.

153. 45-year-old man with pyloric obstruction secondary to chronic duodenal ulcer disease, and is on continuous nasogastric suction

154. 47-year-old female smoker with chronic bronchitis

155. 52-year-old woman with long-term chronic obstructive pulmonary disease who is vomiting secondary to acute gastritis

Questions 156-159

For each numbered profile of cardiovascular effects depicted in the table, select the corresponding drug that best matches it.

(A) Dopamine
(B) Dobutamine
(C) Norepinephrine
(D) Isoproterenol
(E) Epinephrine
(F) Digoxin
(G) Phenylephrine

Cardiovascular Effect

	Heart Rate	Contractility	Vasodilation	Vasoconstriction
156.	0	++ to +++	0	+
157.	++++	++++	++++	0
158.	++	++++	++	0 to ++++
159.	0	0	0	++++

0 = no effect; ++++ = strong effect.

Questions 160-163

For each ergot alkaloid drug listed, select the description that is most accurate.

(A) Is highly selective in stimulating uterine muscle contractions
(B) Its vasoconstrictive effect enables termination of acute migraine headache
(C) A serotonin antagonist/partial agonist used for migraine prophylaxis
(D) A potent dopamine-receptor agonist used for hyperprolactinemia and Parkinson disease
(E) A full agonist at serotonin receptors and α-adrenergic receptors

160. Bromocriptine

161. Ergonovine

162. Ergotamine

163. Methysergide

ANSWER KEY

1. E	29. E	56. E	83. A	110. D	137. A
2. E	30. C	57. B	84. A	111. E	138. A
3. B	31. E	58. B	85. D	112. E	139. B
4. A	32. B	59. C	86. B	113. A	140. E
5. E	33. D	60. A	87. B	114. D	141. C
6. A	34. D	61. D	88. D	115. D	142. D
7. A	35. B	62. D	89. B	116. B	143. C
8. D	36. C	63. A	90. C	117. C	144. B
9. A	37. D	64. B	91. D	118. A	145. A
10. B	38. E	65. C	92. E	119. C	146. B
11. A	39. D	66. B	93. A	120. C	147. G
12. D	40. C	67. C	94. D	121. A	148. C
13. C	41. A	68. A	95. E	122. C	149. E
14. B	42. D	69. C	96. C	123. B	150. A
15. D	43. D	70. B	97. E	124. C	151. J
16. E	44. C	71. A	98. C	125. B	152. F
17. B	45. C	72. C	99. E	126. D	153. D
18. B	46. D	73. C	100. C	127. A	154. H
19. C	47. E	74. D	101. B	128. B	155. I
20. B	48. C	75. A	102. E	129. D	156. F
21. D	49. A	76. B	103. D	130. C	157. D
22. A	50. D	77. B	104. B	131. A	158. A
23. D	51. C	78. C	105. B	132. A	159. G
24. C	52. C	79. B	106. C	133. C	160. D
25. D	53. B	80. B	107. C	134. A	161. A
26. D	54. A	81. A	108. E	135. B	162. B
27. A	55. C	82. C	109. E	136. C	163. C
28. A					

ANSWERS AND EXPLANATIONS

1. The answer is E. *(Gross anatomy; Cardiovascular)* The left atrium is the most posterior chamber of the heart. Immediately posterior to the heart is the esophagus. Therefore, enlargement of the left atrium may compress the anterior wall of the esophagus. Left atrial enlargement may be symptomatic of mitral stenosis or mitral insufficiency. Compression of the esophagus by the left atrium may cause difficulty in swallowing solid food.

2. The answer is E. *(Gross anatomy; Pulmonary/respiratory)* The nasal cavity receives multiple sources of blood supply, which allows for circulation. Gravitational drainage of the sinus occurs from the frontal sinus with the head erect, from the sphenoid sinus with the head tilted forward, and from the maxillary sinus with the head tilted to the contralateral side. The nasolacrimal duct opens into the inferior meatus. The frontal, maxillary, anterior, and middle ethmoidal sinuses open into the middle meatus. The posterior ethmoidal sinus opens into the superior meatus. The sphenoid sinus opens into the sphenoethmoidal recess.

3. The answer is B. *(Pathology; Cardiovascular)* In right-sided heart failure not associated with left-sided failure, there is a reduction, not an increase, in the effective arterial blood volume as the cardiac output decreases. The baroreceptors in the atria and great vessels innervated by the vagus nerve stimulate the release of antidiuretic hormone (ADH) in the hypothalamus. ADH acts on the late distal and collecting tubules by increasing the reabsorption of free water. The peritubular oncotic pressure is increased, because the hydrostatic pressure in the kidneys caused by reduced blood flow (which decreases the glomerular filtration rate) is decreased. An increase in peritubular oncotic pressure causes enhanced reabsorption of sodium and water in isotonic proportions from the proximal tubule. The reduced blood flow to the kidneys causes a release of renin from the juxtaglomerular apparatus, leading to the stimulation, not inhibition, of aldosterone (i.e., secondary aldosteronism). Aldosterone increases the distal tubule reabsorption of sodium and water in isotonic proportions. The presence of ADH and an increased proximal and distal reabsorption of sodium and water result in the reabsorption of a hypotonic fluid to restore the perceived reduction in the effective arterial blood volume. Unfortunately, the presence of right-sided heart failure and increased hydrostatic pressure in the venous system favors the movement of this hypotonic fluid into the interstitial space by Starling forces, resulting in pitting edema. The pulmonary capillary wedge pressure, when measured by a Swan-Ganz catheter, represents the left atrial and left ventricular end-diastolic pressure (LVEDP). Because right-sided heart failure indicates the inability of the right heart to pump blood into the lungs, the LVEDP is decreased, not increased, because less blood is pumped into the left heart from the failed right heart.

Right-sided heart failure is most commonly caused by left-sided failure. In left-sided heart failure, there is vasoconstriction of the pulmonary arteries induced by the increased pulmonary capillary pressure as blood accumulates behind the failed left heart. Other causes of right-sided heart failure include right ventricular infarction and cor pulmonale, which is pulmonary hypertension with associated right ventricular hypertrophy. A large pulmonary embolus is the most common cause of acute right-sided heart failure, because it produces acute strain on the right heart caused by blockage of the pulmonary vessels and generalized pulmonary vasoconstriction caused by hypoxemia. Inability of the right ventricle to pump blood into the lungs results in a "backward failure" into the systemic venous circulation, which markedly increases the hydrostatic pressure (or central venous pressure), leading to jugular neck vein distention, liver congestion (nutmeg liver), splenomegaly, splanchnic congestion (anorexia, nausea, and vomiting), kidney congestion, pitting edema of the lower extremities, and ascites in severe cases. Right-sided heart failure is more often detected by the physician based on signs on physical examination than by the patient complaining about symptoms.

4. The answer is A. *(Pathology; Hematopoietic/lymphoreticular)* Von Willebrand disease (vWD) is the most common hereditary coagulation deficiency. The most common variant has an autosomal dominant inheritance pattern. Laboratory findings in classic vWD include:

- Decreased factor VIII coagulation concentration, which is present in the intrinsic coagulation system
- Decreased factor VIII antigen concentration
- Decreased von Willebrand factor (vWF), which is responsible for platelet adhesion. Platelets have receptors for vWF.

In the response to injury theory of atherosclerosis, platelets play a significant role in development of fatty streaks and fibrofatty plaques. Platelet-derived growth factor is a mitogen that causes the smooth muscle cells in the media of the vessel to proliferate and migrate to a subendothelial location. At this location, the uptake of low-density lipoproteins (LDLs) into these cells produces foam cells. If the platelets cannot adhere to areas of injury on the endothelium, this component of atherogenesis is less prominent, thus contributing to the decreased incidence of coronary artery disease.

5-9. The answers are: 5-E, 6-A, 7-A, 8-D, 9-A. *(Pathology; Cardiovascular)*
The patient has a 1-year history of angina with exercise that resolved with rest. During the past 2 months, the pain has occurred at rest, lasted longer, and occurred more frequently, representing unstable angina. On admission, the pain is substernal, radiating to the left jaw and down the left arm. The patient is diaphoretic (i.e., dizzy, sweaty, and nauseous), and the pain is present 3 hours after onset. These symptoms most likely represent an acute myocardial infarction. Physical examination shows an S_4 heart sound, which is most commonly found in patients who have a stiff, noncompliant ventricle caused by hypertrophy. The presence of an accentuated A2, caused by closure of the aortic valve under increased pressure, indicates hypertension. Essential hypertension is more common in blacks than in whites. In addition, the retinal examination shows reduced caliber of the retinal vessels and arteriovenous nicking, indicating grade II hypertensive retinopathy. The blood pressure is reduced because of his acute myocardial infarction, resulting in left ventricular dysfunction (decreased contractility). On admission, there are no findings in the chest consistent with left-sided heart failure (e.g., wet rales, S_3 heart sound, mitral insufficiency from stretching of the mitral valve ring). There is no accentuation of P_2 to indicate pulmonary hypertension. Because the infarct is only 3 hours old, a lactate dehydrogenase $(LDH)_1$ > LDH_2 flip is not expected. This flip begins to develop 14 hours after a infarct and peaks in 2 to 3 days. The infarcted myocardial tissue would not be pale until 12 hours after an infarct. A chest radiograph would not show a "bat wing" infiltrate in the lungs, because this is a sign of pulmonary edema. Echocardiography has a sensitivity of 70% in an acute myocardial infarction. The classic findings include a Q wave, ST segment elevation, and peaked (i.e., hyperacute) T waves. The total creatine kinase usually increases 4 to 6 hours after an infarct, peaks in 12 to 24 hours, and returns to normal in 3 to 4 days.

On day 3 of hospitalization, the patient develops dyspnea, tachypnea, neck vein distention, rales in the lungs, an S_3 heart sound, and a pansystolic murmur heard at the apex with radiation into the axilla (i.e., murmur of mitral regurgitation). He also has a 2-day history of reduced urine output. These findings indicate left-sided heart failure. Left-sided heart failure increases the left ventricular end-diastolic pressure with a back-up of blood into the lungs, producing pulmonary edema, which decreases lung compliance (i.e., the ability of the lungs to stretch on inspiration). Mitral regurgitation may be caused by dilatation of the mitral valve ring from the increase in left ventricular end-diastolic volume or papillary muscle dysfunction from the myocardial infarction. Blood rushing into the distended ventricle causes an S_3 heart sound. Neck vein distention indicates that he has progressed into right-sided heart failure.

Left- and right-sided heart failure are associated with a decreased cardiac output and a subsequent decrease in the effective arterial blood volume (EABV). Baroreceptors in the atria and great vessels innervated by the vagus nerve stimulate the release of antidiuretic hormone (ADH) in the hypothalamus. ADH acts on the late distal and collecting tubules by increasing the reabsorption of free water. The peritubular oncotic pressure increases, because the hydrostatic pressure in the kidneys resulting from reduced blood flow is decreased. An increase in peritubular oncotic pressure results in enhanced reabsorption of sodium and water in isotonic proportions from the proximal tubule. The reduction in blood flow to the kidneys causes a release of renin from the juxtaglomerular apparatus, leading to the stimulation of aldosterone (i.e., secondary aldosteronism). Aldosterone increases the distal tubule reabsorption of sodium and water in isotonic proportions. The presence of ADH and increased proximal and distal reabsorption of sodium and water result in the reabsorption of a hypotonic fluid to restore the perceived reduction in the EABV. This patient also has right-sided heart failure, so the increased hydrostatic pressure in the venous system (i.e., increased central venous pressure) favors the movement of this hypotonic fluid into the interstitial space by Starling forces. Pitting edema is likely to occur in this setting.

On day 3 of hospitalization, an $LDH_1 > LDH_2$ flip and the absence of creatine kinase MB isoenzyme is likely. Arterial blood gases usually show an increased alveolar-arterial gradient caused by a diffusion abnormality in the lungs that results from the pulmonary edema. Hypoxemia usually occurs, unless the patient is on oxygen. A chest radiograph would most likely show a "bat wing" configuration of pulmonary edema rather than multiple lobar consolidations. The ejection fraction from the left ventricle would be decreased, and the stroke volume would be decreased due to decreased contractility of the damaged left ventricular muscle.

10. The answer is B. *(Pathology; Cardiovascular)*
Severe anemia produces high-output failure by reducing the viscosity of blood, which decreases the peripheral resistance. Cardiac output is equal to the mean arterial blood pressure minus the right atrial pressure, divided by the total peripheral resistance; thus a decrease in total peripheral resistance increases cardiac output. Poiseuille equation involves those factors that change the resistance of blood vessels:

$R = 8nl/\pi r^4$, where
R = resistance,
n = viscosity,
l = the length of the vessel,
and r^4 = the radius of the vessel to the fourth power

Decreasing the viscosity of blood (anemia) or the length of the vessel, or increasing the radius of the vessel (e.g., aorta) reduces the peripheral resistance, which causes an increase in cardiac output.

Hypoxic injury of the muscle, increased peripheral resistance, decreased contractility of cardiac muscle, or a myocardial infarction would decrease cardiac output.

11. The answer is A. *(Pathology; Cardiovascular)*
The patient has amyloid deposition in the heart, which is an example of a restrictive cardiomyopathy. Amyloidosis is a disorder that involves the deposition of a variety of fibrillary proteins into interstitial tissues, resulting in organ dysfunction. An amyloid deposition has a hyaline, eosinophilic appearance on hematoxylin and eosin staining of tissue. Amyloid appears red with Congo red stain and has an apple-green birefringence under polarized light. Electron microscopy shows linear, nonbranching fibrils having a diameter of 7.5 to 10 nm, with hollow cores that are often mistaken for collagen.

Amyloid is derived from various proteins, including light chains, serum-associated amyloid (SAA), prealbumin, β proteins, and peptide hormones (e.g., calcitonin). There are two main types of amyloidosis: primary amyloidosis, in which the amyloid fibrils are composed of light chains, and secondary (reactive) amyloidosis, in which the amyloid fibrils are derived from SAA protein. Common sites of involvement include the spleen, tongue, adrenals, liver, kidneys, and the heart. The pathophysiology of amyloidosis involves its ability to deposit in interstitial tissue to produce compression atrophy, organ dysfunction, and vascular compromise through vessel involvement. In the heart, amyloidosis produces a restrictive cardiomyopathy that is characterized by decreased compliance of the heart with increased resistance to filling, thus reducing cardiac output.

Congestive and hypertrophic cardiomyopathies, constrictive pericarditis, and ischemic heart disease are not associated with amyloid deposition and multisystem disease.

12. The answer is D. *(Pathology; Cardiovascular)*
In congenital heart disease, shunts may be left-to-right (e.g., ventricular septal defect) or right-to-left (e.g., tetralogy of Fallot). In left-to-right shunts, as in a ventricular septal defect, there is volume overload of the right heart. This volume overload causes an increased blood flow through the pulmonary artery, leading to pulmonary hypertension, right ventricular hypertrophy, and reversal of the shunt (right-to-left) when the right-sided pressures are greater than those of the left side. This reversal is called Eisenmenger syndrome, or cyanosis tardive. Right-to-left shunts, in

which unoxygenated blood directly passes into oxygenated blood, produce cyanosis at birth or shortly thereafter.

13. The answer is C. *(Pathology; Cardiovascular)*
A Valsalva maneuver exacerbates the clinical findings associated with hypertrophic cardiomyopathy. The return of blood to the right heart is decreased with forceful holding of the breath, which decreases left ventricular volume and causes further reduction of cardiac output.

Hypertrophic cardiomyopathy, or idiopathic hypertrophic subaortic stenosis, is a cardiomyopathy with disproportionate hypertrophy of the interventricular septum and the free wall of the left ventricle, with concomitant diminution of the left ventricular chamber. Myofiber disarray and abnormalities in the conduction system are frequently seen in autopsy specimens. Hypertrophic cardiomyopathy usually occurs in young adults. In 20% to 30% of cases, it may have an autosomal dominant inheritance pattern. The pathophysiology of outflow obstruction in the left ventricle is associated with the asymmetric hypertrophy of the interventricular septum combined with the apposition of the anterior leaflet of the mitral valve against the interventricular septum, facilitated by the negative vacuum created behind the rapidly ejected blood. The main clinical findings, similar to those of valvular aortic stenosis, are dyspnea and chest pain.

If left ventricular volume (i.e., preload) is increased by decreasing inotropism, increasing venous return of blood to the right heart, or both, the patient's signs and symptoms improve. Maneuvers that increase left ventricular volume include sitting and squatting, which increase venous return. β-Blockers and calcium channel blockers (negative inotropism) also increase left ventricular volume. Initial treatment for hypertrophic cardiomyopathy usually is β-blockers. Verapamil is considered one of the best calcium channel blockers for treatment of this disease.

Reducing the volume of blood in the heart causes the patient's signs and symptoms to worsen. In addition to the Valsalva maneuver, digitalis (positive inotropism), β-adrenergic agents (positive inotropism), venodilators (e.g., amyl nitrite, nitroglycerin, nitroprusside), and exercise decrease left ventricular volume.

14. The answer is B. *(Pathology; Pulmonary/respiratory)*
The patient has temporal arteritis, or giant cell arteritis, which is a multifocal granulomatous vasculitis that primarily involves the temporal artery and branches of the internal carotid artery. It is the most common vasculitis and has a predilection for elderly women. Pathogenesis may involve cell-mediated immunity against arterial antigens in the media smooth muscle cells. Serious complications include blindness and stroke. Polymyalgia rheumatica and pain and stiffness in the neck, shoulders, and hip occurs in 50% of patients.

If the patient has this clinical history in conjunction with an elevated erythrocyte sedimentation rate (ESR), treatment with corticosteroid therapy to prevent blindness, stroke, or both is indicated. A normal ESR essentially excludes this diagnosis. A temporal artery biopsy is negative in 40% of cases; it is not the screening test of choice because it is invasive and has low sensitivity. Serum C3 complement levels, antineutrophil cytoplasmic antibody, and serum antinuclear antibody are not useful for determining this diagnosis.

15. The answer is D. *(Pathology, Immunology; Pulmonary/Respiratory)*
The patient has Henoch-Schönlein purpura, which is an immune vasculitis [immunoglobulin A (IgA)-dominant immune complexes and complement]. It most commonly occurs in children following an upper respiratory infection. Palpable purpura is commonly limited to the lower extremities and buttocks. Polyarthritis, abdominal pain (sometimes accompanied with melena), and acute glomerulonephritis presenting with hematuria is the standard clinical presentation. In children, the disease is usually self-limited and resolves in 1 to 6 weeks.

Immunoglobulin E (IgE)-mediated antibody reaction, cytotoxic antibody reaction, and delayed hypersensitivity reaction are not implicated in Henoch-Schönlein purpura. Infection of endothelial cells is seen in rickettsial diseases and certain viral diseases (e.g., rubella).

16. The answer is E. *(Pathology; Psychosocial, cultural, and environmental influences)*

Using the National Cholesterol Education Panel 1993 criteria, the patient has 6 major risk factors for coronary artery disease. The major positive risk factors include:

1. Age, which is the most important overall risk factor (for men, 45 years of age or older; for women, 55 years of age or older)
2. Family history of premature coronary artery disease, including myocardial infarction or sudden death before 55 years of age in father or other first-degree male relative, or before 65 years of age in mother or other first-degree female relative
3. Current cigarette smoking
4. Hypertension (blood pressure ≥ 140/90 mm Hg), or a patient who is taking antihypertensive medication
5. High-density lipoprotein < 35 mg/dl
6. Diabetes mellitus
7. Low-density lipoprotein ≥ 160 mg/dl

17. The answer is B. *(Pathology; Cardiovascular)*
The patient has a reinfarction of the left ventricle with subsequent backup of blood into the lungs, producing pulmonary edema (bibasilar rales). Normally, the creatine kinase MB isoenzyme fraction begins increasing in 4 to 8 hours, peaks in 24 hours, and disappears in 1.5 to 3 days. The reappearance of creatine kinase MB isoenzyme after 3 days indicates reinfarction. Dead or injured myocardial tissue loses its contractility, resulting in a reduction in stroke volume and an increase in the left ventricular end-diastolic volume. This stretches the mitral valve ring and often produces a mild degree of mitral insufficiency, represented by a pansystolic murmur at the apex. Rapid ventricular filling in this setting causes an S_3 heart sound, which is the first cardiac sign of heart failure.

Recurrent angina pectoris neither causes reappearance of creatine kinase MB isoenzyme nor precipitates an acute onset of left-sided heart failure. A ruptured posteromedial papillary muscle produces a loud pansystolic murmur due to mitral regurgitation. However, most papillary muscle ruptures are caused by right coronary artery thrombosis, because the right coronary artery typically supplies this muscle. The patient has an anterior myocardial infarction, which is most commonly due to thrombosis of the left anterior descending coronary artery. A pulmonary embolism induces right-sided heart failure caused by increased right ventricular strain. In addition, it would not explain the presence of an S_3 heart sound and creatine kinase MB isoenzyme. A ruptured left ventricle producing cardiac tamponade does not produce an increase in creatine kinase MB isoenzyme, and heart sounds are difficult to hear with blood in the pericardial sac. Also, neck vein distention resulting from cardiac tamponade would be present.

18. The answer is B. *(Pathology; Cardiovascular)*
A patient with a 3-day-old anterior myocardial infarction whose lactate dehydrogenase (LDH) isoenzyme study has an $LDH_1 > LDH_2$ flip, an isolated elevation of LDH_5, and an absence of creatine kinase MB isoenzyme most likely has developed right-sided heart failure (secondary to left-sided heart failure) with subsequent development of congestive hepatomegaly (i.e., nutmeg liver). LDH isoenzymes are tetramers of heart (H) and muscle (M) polypeptides that form 5 distinct isoenzymes when separated by electrophoresis. Typically, $LDH_2 > LDH_1$, $LDH_1 > LDH_3$, and $LDH_3 > LDH_4$ and LDH_5, which are roughly equal. The distribution of LDH isoenzymes in tissue is shown in the following table. Organs are listed in descending order of the amount of the LDH isoenzyme.

LDH_1	LDH_2	LDH_3	LDH_4	LDH_5
Heart	Red blood cells	Lung	Lung	Liver
Red blood cells	Heart		Adrenals	Skeletal muscle

In an acute myocardial infarction, LDH_1 increases over LDH_2 (greater amount in LDH_1 fraction) causing an $LDH_1 > LDH_2$ flip. This flip begins to develop in 14 hours, peaks at 48 to 72 hours, and resolves in 7 to 10 days. Because LDH_5 is present in the liver, the LDH_5 fraction increases when liver damage occurs. For example, a backup of blood in the central veins of the liver may occur during right-sided heart failure. The blood compresses the hepatocytes located around the central vein, causing the release of this isoenzyme as well as transaminases (e.g., serum aspartate aminotransferase, alanine aminotransferase).

19. The answer is C. *(Pathology; Cardiovascular)*
The left ventricular end-diastolic pressure (LVEDP) is decreased (< 12 mm Hg) in a patient with hypovolemic shock caused by massive blood loss because the volume of blood is decreased. The LVEDP is increased (> 12 mm Hg) in a patient with an acute myocardial infarction involving the left ventricle because the contractility (and therefore the stroke volume) is reduced in infarcted muscle. This increases the end-diastolic volume and pressure in the left ventricle.

20. The answer is B. *(Pathology; Cardiovascular)*
The abnormal heart shows a concentric hypertrophy of the left ventricle, which is most commonly caused by essential hypertension. Compensation by muscle hypertrophy reduces wall stress, or the afterload the left ventricle must overcome to eject blood. Wall stress is determined by the following equation:

$$\text{Pressure} \times \text{radius} / \text{Wall thickness} \times 2$$

Therefore, an increase in wall thickness decreases wall stress. When the left ventricle fails, left ventricular dilatation occurs. This dilatation increases the radius of the left ventricle and increases wall stress, resulting in left-sided heart failure.

Left ventricular hypertrophy due to an increased peripheral resistance is noted in essential hypertension, aortic stenosis, and coarctation of the aorta. Volume overload produces both hypertrophy and dilatation, because the heart works harder to eject greater volumes of blood by imposing Frank-Starling mechanisms. Volume overload is subclassified into conditions associated with an increased venous return of blood from the right heart or valvular incompetence in the left heart. Increased venous return from the right heart occurs in ventricular and atrial septal defects and patent ductus arteriosus. Valvular incompetence occurs in aortic and mitral regurgitation.

21. The answer is D. *(Pathology; Cardiovascular)*
The histologic section of myocardium reveals an Aschoff body, which is the pathognomonic lesion of acute rheumatic fever. The mitral valve is the most commonly involved valve in rheumatic fever, and mitral stenosis is the most common chronic valvular lesion in long-standing disease. Aschoff bodies contain Anitschkow cells, surrounded by interstitial edema and fibrinoid necrosis (inflammation of immunologic disease). Anitschkow cells are reactive histiocytes, which have a caterpillar-like appearance when cut longitudinally. Fusion of these cells often produces giant cells, one of which is present in the center of the slide. Aschoff bodies are located in the subendocardium and interstitial tissue of the myocardium.

Aortic stenosis is most commonly caused by a congenital bicuspid aortic valve. Mitral and aortic regurgitation are most commonly caused by rheumatic fever, but mitral stenosis is the most common overall lesion. Mitral valve prolapse is not associated with rheumatic fever.

22. The answer is A. *(Pathology; Hematopoietic/lymphoreticular)*
The peripheral smear has segmented neutrophils with numerous large granules in the cytoplasm representing giant lysosomes. The patient has Chediak-Higashi syndrome, which is an autosomal recessive disease with a primary defect in the polymerization of microtubules in leukocytes. Because microtubules are necessary for cell movement and the emptying of lysosomes into phagosomes, this defect causes problems with chemotaxis (directed movement), phagocytosis, and the formation of phagolysosomes that are necessary to kill bacteria. The lysosomes are enlarged because they are unable to empty their enzymes into the vacuoles. The clinical presentation of the patient is that of a child with recurrent infections. A negative nitroblue tetrazolium (NBT) test indicates that the respiratory burst mechanism (generation of superoxide from molecular oxygen) is intact, thus ruling out chronic granulomatous disease of childhood, which is caused by a deficiency of reduced nicotinamide-adenine dinucleotide phosphate (NADPH) oxidase. Glucose 6-phosphate dehydrogenase deficiency is associated with a hemolytic anemia that is often precipitated by infections; however, it is not associated with giant lysosomes.

Complement component C3 deficiency results in severe infections because C3b is required for the opsonization of bacteria. However, it is not associated with giant lysosomes.

23. The answer is D. *(Pathology; Hematopoietic/lymphoreticular)*

The patient has chronic lymphocytic leukemia (CLL), which is the most common leukemia and cause of generalized lymphadenopathy in patients older than 60 years. The peripheral smear exhibits normal-appearing lymphocytes and occasional nuclear remnants representing smudge cells (fragile lymphocytes). CLL is a disorder of virgin B cells, which are long-lived but unable to differentiate into plasma cells, which is the reason for hypogammaglobulinemia in these patients. Approximately 50% of patients have chromosomal abnormalities, the most common being trisomy 12. In addition to generalized lymphadenopathy, patients have hepatosplenomegaly and an increased incidence of autoimmune hemolytic anemia/thrombocytopenia, hypogammaglobulinemia, and secondary malignancies. A monoclonal IgM spike is frequently noted on a serum protein electrophoresis. Peripheral blood findings include a normocytic anemia, thrombocytopenia, and an absolute lymphocytosis with lymphocyte counts ranging from 15,000 to 200,000 cells/μl. The bone marrow is diffusely or focally infiltrated with lymphocytes. Approximately 50% of patients die of infection. The median survival of a patient with CLL is 4 to 6 years. A positive tartrate-resistant acid phosphatase stain of lymphocytes is the characteristic finding in a patient with hairy cell leukemia, the cells of which have hairy projections extending from the cytoplasm. In addition, lymphadenopathy is generally absent. Reed-Sternberg cells are the neoplastic cell of Hodgkin disease. An increased leukocyte alkaline phosphatase score is seen in patients with benign granulocytic proliferations.

24. The answer is C. *(Pathology; Hematopoietic/lymphoreticular)*
The patient has chronic myelogenous leukemia (CML), which is a myeloproliferative disorder. The peripheral smear exhibits a myeloblast in the center of the slide and other cells in the neutrophil series, including myelocytes, band neutrophils, and segmented neutrophils.

Peripheral blood findings also include leukocyte counts ranging between 50,000 to 200,000 cells/μl, normocytic anemia, and thrombocytosis in 40% to 50% of cases. Thrombocytosis is unique to CML, unlike the majority of leukemias in which thrombocytopenia is the rule. Increased numbers of basophils and eosinophils are common, a feature shared by all of the myeloproliferative diseases. There is a low leukocyte alkaline phosphatase (LAP) score, which is extremely useful in separating CML from other disorders with an increased neutrophil count. The low score is caused by the absence of alkaline phosphatase in the specific granules of the neoplastic granulocytes.

CML is a neoplastic clonal expansion of the pluripotential stem cell in the bone marrow, so virtually any lineage may be involved, including red blood cells (RBCs), granulocytes, megakaryocytes, B cells, and T cells. Approximately 95% of patients have a t(9;22) translocation of the *abl* oncogene from chromosome 9 to chromosome 22 (Philadelphia chromosome) with fusion at the break cluster region (*bcr*) to form a fusion gene. This results in continual tyrosine kinase activity, which ultimately increases mitotic activity. The presence of a *bcr:abl* fusion gene has a sensitivity and specificity of 100% for CML. The Philadelphia (Ph) chromosome has a high sensitivity for diagnosing CML, but it is not 100% specific because it occurs in other leukemias.

Key clinical findings include signs of a hypermetabolic state (i.e., fever, weight loss) caused by the increased turnover of large numbers of cells, massive hepatosplenomegaly, and leukemic infiltrates in lymph nodes and other organs. Leukocyte marker studies are only useful if a lymphoid blast crisis is suspected. Arterial blood gas and red blood cell mass/plasma volume studies are useful in the work-up of the polycythemias. A serum protein electrophoresis has no defined role in the work-up of CML. Computed tomography scans of the abdomen may be useful in identifying organ involvement. A lymph node biopsy is not usually necessary in diagnosing CML.

25. The answer is D. *(Pathology; Hematopoietic/lymphoreticular)*
The patient has acute progranulocytic leukemia complicated by disseminated intravascular coagulation (DIC). This leukemia accounts for approximately 5% to 10% of cases of acute nonlymphocytic leukemias. The blasts have a very granular cytoplasm, which contain numerous auer rods (fused lysosomes). Acute progranulocytic leukemia is the most common leukemia associated with DIC, which, in this patient, was manifested by prolonged prothrombin and partial thromboplastin times, a low level of fibrinogen, and increased

fibrinogen degradation products. Additional findings include a characteristic t(15;17) translocation and abnormal retinoic acid metabolism. High-dose vitamin A therapy is useful in inducing remission in some patients. The morphologic features of the blasts and the presence of DIC essentially excludes monocytic, myelomonocytic, and chronic myelogenous leukemia as well as a myelodysplastic syndrome.

26. The answer is D. *(Pathology; Hematopoietic/lymphoreticular)*
The cells in the peripheral blood demonstrate the irregular cytoplasmic projections of hairy cell leukemia (HCL). Frame D in the micrograph represents a positive tartrate-resistant acid phosphatase stain (TRAP), which is the key stain to confirm the diagnosis.

HCL is a B-cell malignancy that is most commonly seen in middle-aged men. Splenomegaly is the most common physical finding (90%). Leukemic cells specifically infiltrate the red pulp, which is very unusual in most leukemias, which favor the white pulp. The spleen is an important site for proliferation of the neoplastic HCL cells, so splenectomy is important in treatment of the disease. Hepatomegaly is infrequently present and lymphadenopathy is conspicuous by its absence. Autoimmune syndromes, including vasculitis, arthritis, lytic bone lesions, and an increased incidence of *Mycobacterium avium-intracellulare* infections round out the clinical picture of a patient with HCL. In the peripheral blood, pancytopenia is a characteristic finding. The bone marrow is always packed with leukemic cells that look like fried eggs. The tartrate-resistant acid phosphatase stain is positive in the majority of cases. Electron microscopy of the leukemic cells reveals characteristic complex lamellae. The treatment involves splenectomy followed by interferon-α therapy and pentostatin, which blocks adenine deaminase (adenine is toxic to lymphocytes). A terminal lymphoblast crisis is not a feature of HCL. Patients have a 50% 5-year survival.

27. The answer is A. *(Pathology; Hematopoietic/lymphoreticular)*
Iron deficiency progresses through various stages before becoming a microcytic anemia, unlike β-thalassemia minor, which is an autosomal recessive disease that is microcytic from the start. The stages of iron deficiency in the order of development are: absent iron stores (no anemia), decreased serum ferritin (no anemia), decreased iron, increased total iron binding capacity, decreased percent saturation (no anemia), normocytic normochromic anemia, microcytic normochromic anemia, then microcytic hypochromic anemia.

β-Thalassemia minor has a normal red blood cell distribution width (RDW), whereas iron deficiency usually has an increased RDW. β-Thalassemia minor characteristically has a low mean corpuscular volume (MCV) in the presence of an increased red blood cell (RBC) count; the latter is not present in iron deficiency or in any of the other microcytic anemias. The mechanism for the increase in the RBC count may be due to the increase in hemoglobin F, which shifts the oxygen dissociation curve to the left and stimulates the release of erythropoietin. β-Thalassemia minor has an increase in hemoglobin A_2 and hemoglobin F on hemoglobin electrophoresis. Because there is a slight decrease in the synthesis of beta chains, the uninvolved alpha (α), delta (δ), and gamma (γ) chains combine to form hemoglobin A_2 (2α, 2δ chains) and hemoglobin F (2α, 2γ chains). These hemoglobin molecules are normal in iron deficiency. β-Thalassemia minor is a defect in globin chain synthesis, whereas iron deficiency is a defect in heme (iron and protoporphyrin) synthesis.

28. The answer is A. *(Pathology; Hematopoietic/lymphoreticular)*
Approximately 60% of all malignant lymphomas are non-Hodgkin lymphomas (NHL), and the other 40% represent Hodgkin disease (HD). Malignant lymphomas are primary malignancies of lymphoid tissue that often metastasize to the bone marrow and peripheral blood. Approximately 60% of patients are men older than 50 years, with the exception of Burkitt lymphoma and T-cell lymphoblastic lymphomas, which occur in a younger age group. Predisposing factors for NHL include autoimmune disease (e.g., Sjögren syndrome), any immunodeficiency disease (e.g., Wiskott-Aldrich syndrome, acquired immunodeficiency syndrome, Bruton agammaglobulinemia), treatment with alkylating agents, irradiation, and viruses (e.g., Epstein Barr, human T-cell lymphoma virus-I). The classification of NHLs is often confusing. The Rappaport classification is based on the pattern (nodular or diffuse) and

the size of the lymphocytes (i.e., small size, lymphocytic; large size, histiocytic). Although this classification is reproducible among pathologists, it is functionally incorrect because most of the histiocytic lymphomas are really B cell in origin. The Lukes-Collins classification is a more accurate classification because it is based on marker studies. The Working Formulation is primarily a morphologic and a clinical classification that divides the NHLs into low, intermediate, and high grade categories. In this regard, nodular (follicular) lymphomas have a better prognosis than diffuse lymphomas. Lymphocytic lymphomas have a better prognosis than histiocytic lymphomas. Lymphocytic lymphomas are more likely to involve the bone marrow than histiocytic lymphomas. Therefore, a nodular (follicular) lymphocytic lymphoma has the best prognosis (low grade in the Working Formulation). Nodular and diffuse histiocytic lymphomas have an intermediate prognosis (intermediate grade), and Burkitt and immunoblastic lymphomas have a poor prognosis (high grade).

29. The answer is E. *(Pathology; Hematopoietic/lymphoreticular)*
The patient has Burkitt lymphoma (Lukes-Collins; small non-cleaved), which is the most common malignant lymphoma in children. Approximately one third of all childhood lymphomas in the United States are of the Burkitt type. There is an African and a North American variant. The relationship of Burkitt to the Epstein-Barr virus is more secure with the African than the North American variant. The African variant has a predilection for the jaw, while the North American variant is more often an abdominal mass, located in the ileocecal region, ovary, or retroperitoneum. The lymph nodes are diffusely involved with monotonous, round, non-cleaved cells with numerous mitoses (very aggressive cancer) punctuated by clear spaces with macrophages containing debris ("tingible" macrophages) giving the impression of a "starry sky" appearance, the stars representing the macrophages and the black of night the lymphocytes. The neoplastic B cells have a characteristic t(8;14) translocation where the c-*myc* oncogene is translocated from chromosome 8 to chromosome 14. Burkitt cells contain vacuoles that contain lipid, which stains positively with oil red O. It is a high grade lymphoma with a predilection for metastasis to the bone marrow and other organs.

Regarding the other choices in the question:

- A positive non-specific esterase stain is used in identifying neoplastic monocytes in acute or chronic monocytic leukemia.
- A positive specific esterase stain identifies neoplastic granulocytes in acute or chronic myelogenous leukemia.
- A low leukocyte alkaline phosphatase score and a t(9;22) translocation is characteristic of chronic myelogenous leukemia. The c-*abl* oncogene is translocated from chromosome 9 to 22 where it fuses with the break cluster region (*bcr*) on chromosome 22. This latter chromosome is the Philadelphia chromosome.

30. The answer is C. *(Pathology; Hematopoietic/lymphoreticular)*
The patient has Sézary syndrome, which is a malignant helper T-cell lymphoma. The disease starts out with a rash secondary to a dermal infiltrate of neoplastic T cells, which also infiltrate the overlying epidermis to form Pautrier abscesses. However, the "abscess" is not composed of neutrophils but the malignant T cells. From this location, it may spread to other sites, most commonly lymph nodes. The skin involvement progresses through three stages: a rash, raised plaques, and nodular masses that often fungate (thus the term mycosis fungoides). The Sézary syndrome refers to skin involvement plus the presence of abnormal circulating Sézary cells, which have a prominent fold in the nucleus and PAS-positive areas of material in the cytoplasm. The Sézary syndrome is a malignant T-cell disorder and does not originate from B cells, plasma cells, or histiocytes.

31. The answer is E. *(Pathology; Hematopoietic/lymphoreticular)*
All of the diseases involving the spleen that are listed are associated with massive splenomegaly except idiopathic thrombocytopenic purpura. Diseases in the spleen that often result in splenomegaly involve the red pulp, which contains the sinusoids and the cords of Billroth, or the white pulp. In the white pulp, T cells surround the penicillinate artery, and B cells form a mantle around the T cells. Antigenic stimulation of the white pulp results in the formation of germinal

follicles in the B-cell areas. T-cell disorders (e.g., infectious mononucleosis) primarily involve the periarteriolar lymphocyte sheath area of the white pulp. As a rule, most acute and chronic leukemias infiltrate the spleen to produce massive splenomegaly. Primarily, the acute and chronic leukemias localize to the white pulp. The exception to this is hairy cell leukemia, which infiltrates the red pulp. Infiltrative diseases like amyloidosis, Gaucher disease, and Niemann-Pick disease also produce splenomegaly.

32. The answer is B. *(Pathology; Hematopoietic/lymphoreticular)*
Approximately 40% of patients with a malignant lymphoma have Hodgkin disease (HD), which is more common in men than women, adults than children, and whites than blacks. Approximately 40% of the patients initially present with unilateral supraclavicular lymph node enlargement. HD rarely involves Waldeyer's ring (oropharyngeal lymphoid tissue), skin, or the gastrointestinal tract. Patients frequently have problems with cellular immunity (type IV hypersensitivity) when their skin is tested with common antigens. However, this predisposes the patient to bacterial (tuberculosis), fungal, viral, and parasitic infections.

There are four subtypes of HD: lymphocyte predominant (LP) 5%, nodular sclerosing (NS) 60%, mixed cellularity (MC) 30%, and lymphocyte depletion (LD) 5%. The Reed-Sternberg (RS) cell is the neoplastic cell in HD. However, the RS cell is not unique to HD; there are many look-alikes (e.g., immunoblasts in infectious mononucleosis). Unlike non-Hodgkin lymphoma, the lymphocytes in the lymph nodes are not neoplastic. The number of RS cells parallels the type of HD, the age of the patient, and the prognosis. Therefore, the fewer RS cells, the better the prognosis in a younger patient, whereas a greater number of RS cells indicates a poor prognosis in an older patient. The following schematic depicts the types of HD and the number (in parentheses) they represent in the question.

LP (#3) ⟶ NS (#2) ⟶ MC (#4) ⟶ LD(#1)
Least # RS ⟶ Greatest # RS
Young age ⟶ Older age
Good prognosis ⟶ Bad prognosis

33. The answer is D. *(Pathology; Hematopoietic/lymphoreticular)*
The patient has histiocytosis X, which encompasses three diseases: eosinophilic granuloma, Hand-Schuller-Christian disease (the patient has this type), and Letterer-Siwe disease. The histiocytes (Langerhan cells) have cytoplasmic inclusions called Birbeck granules, which on electron microscopy are rod-like tubular structures that look like tennis rackets. The histiocytes are also positive for CD1 antigen.

An eosinophilic granuloma is the benign variant of histiocytosis X that primarily involves bone. It is usually unifocal in its distribution. Lytic bone lesions occur in the skull, ribs, and femur. Curettings of these lesions exhibit in infiltrate of eosinophilic, bland-appearing histiocytes with an indented nucleus. Bone pain and pathologic fractures are commonly observed. Hand-Schuller-Christian disease is a more disabling disease that primarily involves children. It is characterized by a triad of multifocal cystic defects in the skull, diabetes insipidus from involvement of the hypothalamus, and exophthalmos from infiltration of the orbit by neoplastic histiocytes. The prognosis is intermediate between that of eosinophilic granuloma and Letterer-Siwe disease. Letterer-Siwe disease is a disease of infants and children younger than 2 years. It is characterized by a diffuse eczematous rash, organ involvement, and multifocal cystic defects in the skull, pelvis, and long bones. Unfortunately, it is the most aggressive of all of the histiocytoses, and it has a variable prognosis depending on how early chemotherapy starts.

34. The answer is D. *(Pathology; Cardiovascular)*
Alzheimer disease, restrictive cardiomyopathy, and nephrotic syndrome are features associated with amyloid. Amyloidosis refers to diseases that deposit a variety of fibrillary proteins into interstitial tissues, resulting in compression atrophy, organ dysfunction, and vascular compromise. Amyloid is a twisted β-pleated sheet. It has a hyaline, eosinophilic appearance on hematoxylin and eosin stains of tissue. It appears red with the Congo red stain and has an apple-green birefringence under polarized light. On electron microscopy, it is a 7.5–10 nm, linear, nonbranching structure with a hollow core that is often confused with collagen. Amyloid is derived from many different proteins including

light chains, serum associated amyloid, prealbumin, β-proteins, and peptide hormones (e.g., calcitonin).

Amyloidosis can produce a restrictive cardiomyopathy. A localized form involving the heart is seen in the elderly (senile amyloidosis). The amyloid derives from prealbumin. Macroglossia (enlarged tongue) is often the initial sign. Malabsorption and carpal tunnel syndrome may occur. Renal failure is the most common cause of death in amyloidosis. Glomerular involvement is associated with the nephrotic syndrome. Splenomegaly is the rule in primary and secondary amyloidosis. Amyloid deposition in the white pulp is called a sago spleen, whereas deposition in the red pulp is called a lardaceous spleen. The liver frequently is involved (90%) in amyloidosis. The serum alkaline phosphatase is characteristically elevated. Addison disease is a frequent complication, as is a "stocking glove" peripheral neuropathy.

The diagnosis is made by securing biopsy material from the rectal mucosa, gingiva, the omental fat pad, or the organ involved with the disease. The average survival in generalized amyloidosis is only 1 to 4 years.

Less common types of amyloidosis include Alzheimer disease and medullary carcinoma of the thyroid gland. In Alzheimer disease, amyloid derives from β-protein, which is coded for on chromosome 21 (relationship with Down syndrome). The amyloid is toxic to neurons and may be one of the factors involved in dementia. In medullary carcinoma of the thyroid, amyloid derives from calcitonin, which is the tumor marker for this variant of thyroid cancer.

35. The answer is B. *(Pathology; Immune response)*
The patient has idiopathic thrombocytopenic purpura (ITP), which is an autoimmune disease characterized by the development of immunoglobulin G (IgG) antibodies against platelets, with subsequent removal of the sensitized platelets by macrophages in the spleen. ITP is the most common cause of thrombocytopenia in children. Patients present with an abrupt onset of petechia and epistaxis (nose bleed) usually after an upper respiratory infection. Splenomegaly is not present in the majority of patients. Laboratory findings include thrombocytopenia, the presence of platelet-associated IgG, a prolonged bleeding time, and a bone marrow examination exhibiting normal to increased numbers of megakaryocytes. Spontaneous remission is the rule. Prednisone is usually effective in accelerating recovery.

A platelet-adhesion defect, as in von Willebrand disease, is not usually associated with thrombocytopenia. Hypersplenism is usually associated with splenomegaly, which is not present in this patient. Consumption of the platelets in clots implies either disseminated intravascular coagulation (DIC) or the hemolytic uremic syndrome (HUS). The prothrombin and partial thromboplastin times are normal, which rules out DIC. There is no evidence of renal failure or hemolytic anemia, so HUS is also excluded. A platelet production problem is ruled out by the presence of megakaryocytes in the bone marrow aspirate.

36. The answer is C. *(Hematology; Hematopoietic/lymphoreticular)*
The patient has a qualitative platelet disorder because the platelet count is normal, the prothrombin (PT) and partial thromboplastin (PTT) times are normal, fibrinogen degradation products are not present, and the bleeding time is prolonged. She was probably taking a nonsteroidal medication prior to her surgery. Only a platelet transfusion will stop the bleeding in this patient because her platelets are nonfunctional.

Primary fibrinolysis is not present because fibrinogen degradation products are not present. Disseminated intravascular coagulation (DIC) is unlikely because the platelet count and PT and PTT are normal. In addition, fibrinogen degradation products are not present. These are all usually abnormal in DIC. Von Willebrand disease could produce the above findings in this patient; however, the PTT would be prolonged because of the deficiency of factor VIII coagulant. The absence of von Willebrand factor (i.e., platelet-adhesion factor) may result in menorrhagia. The only coagulopathy associated with birth control pills is hypercoagulability, which is caused by an acquired deficiency of antithrombin III.

37. The answer is D. *(Hematology; Human development and genetics)*
Both hemophilia A and von Willebrand disease have a prolonged partial thromboplastin time (PTT). Both disorders have a deficiency of factor VIII coagulant, which is in the intrinsic coagulation system. The PTT

evaluates factors XII, XI, IX, VIII, X, V, II (prothrombin) and I (fibrinogen) down to the formation of a clot.

Hemophilia A is a sex-linked recessive disease transferred to males from a female carrier. Laboratory findings include a deficiency of VIII:C (coagulant factor in intrinsic system), a normal VIII:AG (non-functional antigenic determinant of VIII:C), a normal VIII:vWF (von Willebrand factor; platelet-adhesion factor), a normal bleeding time (no platelet defect), and a prolonged PTT and a normal prothrombin time (PT). Factor VIII is in the intrinsic not the extrinsic system, which is evaluated with the PT. Female carriers have a ratio of VIII:C/VIII:Ag (antigen) of less than 0.75. Mild hemophilia A may be treated with desmopressin acetate (DDAVP), which has been shown to elevate factor VIII complex levels (all factor VIII components). Cryoprecipitate, which contains VIII:C (also fibrinogen and factor XIII) is only used in mild hemophilia. Factor VIII concentrate (lyophilized and heat treated to exclude any risk for contracting hepatitis or AIDS) is used for moderate-to-severe hemophilia A. A recombinant factor VIII preparation has recently been developed.

Von Willebrand disease (vWD) is the most common hereditary coagulation deficiency. The most common variant is autosomal dominant. Epistaxis is the most common presentation. Laboratory findings in classical vWD include low VIII:C, VIII:antigen, and VIII:vWF levels, a prolonged bleeding time, a prolonged PTT, a normal PT, and an abnormal ristocetin platelet-aggregation study. This latter study detects von Willebrand factor, which is deficient in patients with vWD. Desmopressin and cryoprecipitate are the mainstays of therapy.

38. The answer is E. *(Pathology; Hematopoietic/lymphoreticular)*

A regenerating pool of bone marrow multipotential stem cells contains pluripotential stem cells, which further subdivide into lymphoid stem cells and trilineage myeloid stem cells. The committed lymphoid stem cell divides into a pro-T lymphocyte (which enters the thymus and develops into T cells), and a pro-B lymphocyte (which migrates to the bone marrow, where it develops into B cells). The trilineage myeloid stem cells further differentiate into the following committed stem cells: erythrocyte/megakaryocyte (E/Mega), colony-forming units (CFU) for granulocytes and macrophages (CFU-GM), and CFU for eosinophils (CFU-Eo). Although stem cells are not morphologically recognizable in the bone marrow, they have been cultured in vitro. Interleukin-3 is an important stimulator of stem-cell proliferation, thus assuming an important role in hematopoiesis. Stem-cell factor is a growth factor that stimulates the pluripotential stem cell in the marrow. Multipotential stem cells, including the pluripotential, lymphoid, and trilineage stem cells, have the capacity for self renewal. The more committed the stem cell, the less capacity for self renewal. Stem cells live inside and outside the marrow, the latter sites serving as a reserve in case there is injury to the stem cells in the bone marrow. Stem cell diseases include aplastic anemia (destruction of the trilineage myeloid stem cell), acute and chronic leukemias, myeloproliferative diseases (e.g., polycythemia rubra vera, chronic granulocytic leukemia, agnogenic myeloid metaplasia, essential thrombocythemia), pure red blood cell aplasia (association with parvovirus, thymoma), and paroxysmal nocturnal hemoglobinuria.

39. The answer is D. *(Pathology; Hematopoietic/lymphoreticular)*

The reticulocyte count is the best index of effective erythropoiesis, or how well the marrow is responding to an anemia. Peripheral blood reticulocytes require 24 hours to become a mature red blood cell (RBC). They are larger cells than mature RBCs, and they require splenic macrophages for their maturation process. The reticulocyte count is expressed either as a percentage (0.5%–1.5%) or in absolute numbers (percentage reticulocyte count × RBC count), the latter method representing a more accurate assessment of marrow response to an anemia because it is corrected for the degree of anemia, as noted in the following examples:

Normal Adult
- RBC count = 5 million cells/μl
- Reticulocyte count = 1.5%
- Absolute count = 0.015 × 5 million cells/μl = 75,000 reticulocytes/μl

Anemic Adult
- RBC count = 2.5 million cells/μl
- Reticulocyte count = 3%
- Absolute count = 0.03 × 2.5 million cells/μl = 75,000 reticulocytes/μl

Note that the reticulocyte count of 3% in the anemic patient, though increased, is not elevated when corrected for the degree of anemia. In order to correct the percentage of reticulocytes for the degree of anemia, the following formula is utilized:

Corrected reticulocyte count = patient hematocrit (Hct)/45 × reticulocyte count.

Using our patient with a hemoglobin (Hgb) of 10 gm/dl and an uncorrected reticulocyte count of 12% as an example, the 10 gm/dl Hgb must first be translated into a corresponding hematocrit (Hct). According to the rule of 3, the Hgb × 3 should equal the Hct, so this patient should have a 30% Hct. Therefore, the corrected reticulocyte count is: 30/45 × 12 = 8% (choice D).

Due to the importance of determining the bone marrow's response to anemia within a 24-hour time frame, the addition of these shift cells (48 to 72 hours to mature) falsely elevates the total count. To compensate for this effect, an additional correction is applied after the count is corrected for the degree of anemia by simply dividing the number by 2. This additional correction is called the reticulocyte index. In the case previously illustrated, the reticulocyte index would be 4% (8/2), if shift cells (polychromasia) were present in the smear. Most clinicians use a corrected reticulocyte count of less than 2% as a poor response to anemia, while a count more than 3%, is considered a good response.

40. The answer is C. *(Pathology; Hematopoietic/lymphoreticular)*
The mean corpuscular volume (MCV) correlates with red blood cell (RBC) size, thus the terms microcytic, normocytic, and macrocytic anemias. It is calculated as follows: MCV = hematocrit (Hct)/RBC count × 10 (80–100 fl). Peripheral smears with both microcytic and macrocytic RBCs (dimorphic smear) have a normal MCV because the instrument takes an average of the RBC sizes passing through a predetermined aperture. An increased MCV is only normal in newborns. A low MCV is never normal.

A patient who is 2 days post-acute bleed will have a normal mean corpuscular volume (MCV) because the bone marrow has not had enough time to release young red blood cells (reticulocytes) into the peripheral blood. Reticulocytes are larger than a mature RBC and could potentially increase the MCV. The other choices in the question are conditions associated with either an increase or decrease in the MCV. A pregnant woman not taking vitamin supplements will very likely develop folate deficiency because she has only a 3–4-month supply of the vitamin in her liver. A patient with resection of the terminal ileum will develop vitamin B_{12} deficiency, since the terminal ileum has receptors for the vitamin B_{12}–intrinsic factor complex for reabsorption. The patient with β-thalassemia minor as well as the patient with a chronic bleeding peptic ulcer will both have a microcytic anemia; the former from a decrease in globin-chain synthesis, and the latter secondary to iron deficiency.

41. The answer is A. *(Pathology; Hematopoietic/lymphoreticular)*
The most unlikely explanation for the failure of the hemoglobin (Hgb) to increase the expected 1 gram/dl Hgb per unit of packed red blood cells is the patient with severe iron deficiency who has depleted iron stores. This patient should have increased the Hgb from a pretransfusion value of 4 gm/dl to a posttransfusion value of approximately 8 gm/dl. However, in the presence of hemolysis, active bleeding, or splenomegaly with trapping of the transfused cells, the Hgb would not increase to the expected posttransfusion concentration.

42. The answer is D. *(Pathology; Hematopoietic/lymphoreticular)*
When working-up a patient with anemia, the reticulocyte count is the most important initial test because it indicates whether effective erythropoiesis is occurring in the bone marrow. A reticulocyte is a young red blood cell (RBC) that is prematurely released from the bone marrow if accelerated erythropoiesis is occurring in the bone marrow in response to an anemia. For this reticulocyte response to occur, there must be an adequate amount of raw materials (e.g., iron, vitamin B_{12}, folate) in the bone marrow, no intrinsic bone marrow disease (e.g., aplastic anemia), an adequate release of erythropoietin from the kidneys, and no premature destruction of the RBCs before they leave the bone marrow (e.g., B_{12}/folate deficiency). One or more of these factors may explain an inadequate

reticulocyte response, which is called ineffective erythropoiesis.

A serum folate/B_{12} measurement is indicated in the presence of a macrocytic anemia with hypersegmented neutrophils. A Coombs test is indicated if the reticulocyte count is elevated so that an autoimmune hemolytic anemia can be excluded. A bone marrow aspirate is primarily indicated when a patient has a normocytic anemia with an inadequate reticulocyte response, if early iron deficiency and anemia of inflammation have been ruled out. Serum ferritin is primarily indicated in distinguishing iron deficiency (low levels) from the anemia of inflammation (normal to high).

43. The answer is D. *(Pathology; Hematopoietic/lymphoreticular)*
For the occurrence of an adequate reticulocyte response to an anemia, there must be an adequate amount of raw materials (e.g., iron, vitamin B_{12}, folate) in the bone marrow, no intrinsic bone marrow disease (e.g., aplastic anemia), an adequate release of erythropoietin from the kidneys, and no premature destruction of the RBCs before they leave the bone marrow (e.g., vitamin B_{12}/folate deficiency). One or more of these factors may explain an inadequate reticulocyte response, which is called ineffective erythropoiesis.

Iron deficiency treated with ferrous sulfate after 1 week of therapy would be expected to have a good reticulocyte response because the bone marrow needs at least 5–7 days before erythropoiesis is developed enough to release these cells into the peripheral blood. Autoimmune hemolytic anemias characteristically have an elevated reticulocyte count because the RBCs are removed either intravascularly or by macrophages in the spleen. Congenital spherocytosis is a hemolytic anemia in which the spherocytes are removed by the spleen. One would expect a good reticulocyte response because the bone marrow is normal. A patient with thrombotic thrombocytopenic purpura has a microangiopathic hemolytic anemia caused by intravascular destruction of RBCs as they smash into platelet thrombi overlying areas of injury in the microvasculature. The bone marrow responds to this loss with accelerated erythropoiesis and the premature release of reticulocytes into the peripheral blood.

44. The answer is C. *(Pathology; Multisystem processes)*
Because the liver supplies of folate last for only 3–4 months, a pregnant woman or a woman who is lactating, is predisposed to folate deficiency unless she is on folate supplements. Folate deficiency during pregnancy has been implicated in the development of fetal open neural tube defects; therefore, folate supplements are recommended for all women prior to conception. Antrectomy would not predispose to either folate or vitamin B_{12} deficiency. Crohn disease, a pure vegetarian diet, and bacterial overgrowth in the bowel are causes of vitamin B_{12} deficiency. Crohn disease, which involves the terminal ileum in approximately 80% of cases, predisposes to vitamin B_{12} deficiency by interfering with its absorption. A pure vegetarian diet (e.g., Seventh Day Adventists, Hindus) without vitamin B_{12} supplementation often predisposes to deficiency of B_{12}, which is found in animal products. Bacterial overgrowth in the small bowel destroys the intrinsic factor–vitamin B_{12} complex, but has little effect on folate reabsorption. Chronic pancreatitis interferes with cleavage of the R factor, so vitamin B_{12} deficiency is possible. Normally, vitamin B_{12} binds to the R factor in the stomach, which prevents acid destruction of vitamin B_{12}.

45. The answer is C. *(Pathology; Multisystem processes)*
A megaloblastic bone marrow is found in both pernicious anemia (PA) and folate deficiney. Giant band neutrophils are a characteristic finding. Vitamin B_{12} and folic acid are involved in DNA synthesis. Delayed maturation of DNA in hematopoietic cells and other nucleated cells throughout the body results in large, immature nuclei (megaloblastic anemia). PA, which is the most common cause of vitamin B_{12} deficiency, is an autoimmune disease with destruction of parietal cells in the stomach. Parietal cells produce the intrinsic factor (IF) necessary for absorption of vitamin B_{12} in the terminal ileum, where IF receptors are located.

Features characteristic of PA rather than folate deficiency include a low serum vitamin B_{12} level, an increased urine methylmalonic acid because vitamin B_{12} is involved in propionate (odd-chain fatty acid) metabolism. The latter feature is responsible for the neurologic disease that is present in patients with vitamin B_{12} deficiency. The increased concentration of propionate behind the block produces demyelination in the central nervous system and in the spinal cord, where it involves the posterior columns and the lateral corticospinal tract (subacute combined degeneration). Loss of vibratory sensation in the lower extremities is the first manifestation of this disease. Antiparietal cell antibodies occur in 90% of patients with PA. The autoimmune destruction of the parietal cells in the body and fundus produces an atrophic gastritis, with a significant increase in the incidence of gastric adenocarcinoma. The Schilling test helps localize the cause of vitamin B_{12} deficiency. The test is performed to assess the absorption of orally ingested radioactive vitamin B_{12} without IF (stage I) and with IF (stage II). In patients with PA, due to the absence of IF, there is < 7% of the ingested radioactive vitamin B_{12} dose excreted in the urine. However, this is corrected in stage II by the administration of radioactive vitamin B_{12} along with IF. In vitamin B_{12} deficiency caused by bacterial overgrowth, in stage I there is < 7% of the ingested dose excreted in the urine, but this is corrected after 1 week of antibiotic therapy, which destroys the offending bacteria. In pancreatic insufficiency, in stage I there is < 7% of the ingested dose excreted in the urine, but this is corrected by giving the patient oral pancreatic extract prior to ingesting the radioactive vitamin B_{12}. The pancreatic extract cleaves off the R factor and allows radioactive vitamin B_{12} to bind with IF. In small bowel disease, the malabsorption of vitamin B_{12} cannot be corrected.

46. The answer is D. *(Pathology; Hematopoietic/ lymphoreticular)*
The patient most likely has aplastic anemia because all of the hematopoietic elements are reduced, mainly red blood cells (RBCs), white blood cells (WBCs), and platelets. This is called pancytopenia. A bone marrow examination is always the first step in the work-up of a patient with pancytopenia.

Aplastic anemia is a stem cell disorder caused by absent or defective stem cells, suppression of stem cells by macrophages or T lymphocytes, antibody-mediated inhibition of stem cells, or the destruction of a stem cell by a virus (parvovirus). Aplastic crises due to severe hemolytic anemias are self-limited and are commonly associated with a parvovirus infection. Patients with aplastic anemia present with signs of pancytopenia, mainly neutropenia infection, proceeding to thrombocytopenia, causing petechia and ecchymoses, and anemia resulting in fatigue, malaise, and an inability to concentrate. Splenomegaly is conspicuously absent. The diagnosis rests on the performance of a bone marrow examination and noting hypocellularity (< 25%) with absence of RBCs, granulocytes, and megakaryocytes. Lymphocytes are present.

Corticosteroids plus antithymocyte globulin is the treatment of choice if the patient is not a candidate for a bone marrow transplant. Androgens, blood transfusion, recombinant granulocyte macrophage colony-stimulating factor, and antilymphocyte globulin are often used as well. A direct Coombs test is primarily used in documenting an autoimmune hemolytic anemia. Pancytopenia is not usually present in autoimmune hemolytic anemia, and the corrected reticulocyte count is almost always elevated. The sugar water test is a screening test for paroxysmal nocturnal hemoglobinuria (PNH). Although PNH is a possible cause of pancytopenia, the history would have included the presence of hemoglobinuria in the first void of urine in the morning. A serum vitamin B_{12}/folate level measurement is indicated if hypersegmented neutrophils are present and the anemia is macrocytic. A computed tomography scan of the abdomen has no role in the work-up of a patient with pancytopenia.

47. The answer is E. *(Pathology; Hematopoietic/ lymphoreticular)*
The patient has pure red blood cell (RBC) aplasia, which can be inherited (e.g., Blackfan-Diamond syndrome) or acquired, which is most common [e.g., severe hemolytic anemia (sickle cell disease, congenital spherocytosis), thymoma, parvovirus (cytotoxic to erythroid stem cells)].

In pure RBC aplasia, there are no RBC precursors in the marrow and no reticulocytes in the peripheral blood, but granulocytes and megakaryocytes are present in the bone marrow.

Iron deficiency, lead poisoning, and anemia of chronic disease are microcytic, whereas megaloblastic anemias are macrocytic.

48. The answer is C. *(Pathology; Hematopoietic/ lymphoreticular)*
The most common anemia associated with renal failure is associated with decreased production of erythropoietin. It is a normocytic anemia with a corrected reticulocyte count < 2%, because there is very little

stimulation of erythropoiesis in the bone marrow. Recombinant erythropoietin therapy is very effective in correcting the anemia.

Other less common anemias associated with renal disease include iron deficiency from dialysis and blood loss from acute gastritis, folate deficiency from dialysis, and hemolytic anemia (uncommon). In addition, there is a qualitative platelet defect against platelet factor III, which is the most common coagulation abnormality in uremia and contributes to gastrointestinal bleeding and anemia. A prolonged bleeding time identifies this defect.

Ferrous sulfate and folate are useful in treating iron deficiency and folate deficiency, respectively. Vitamin B_{12} deficiency is not usually associated with renal disease. Corticosteroids have no role in the treatment of anemia in renal disease unless there is an autoimmune hemolytic anemia.

49. The answer is A. *(Pathology; Hematopoietic/lymphoreticular)*

The patient has a myelodysplastic syndrome (MDS; preleukemia) with ringed sideroblasts (Prussian blue positive) in the bone marrow. MDS is primarily a disease of elderly men between the ages of 60 and 75 years. It is a clonal stem-cell disorder characterized by maturation defects resulting in ineffective erythropoiesis, granulopoiesis, and thrombopoiesis, as well as an increased risk for progressing to acute leukemia. It is sometimes acquired in patients who were previously treated with chemotherapy or radiation. The French-American-British classification is as follows: refractory anemia (24%), refractory anemia with ringed sideroblasts (24%), refractory anemia with excess blasts (23%), refractory anemia with excess blasts in transformation (9%), and chronic myelomonocytic leukemia (16%). Key features that distinguish these subgroups are the percentage of myeloblasts in the peripheral blood or bone marrow, the percentage of ringed sideroblasts in the bone marrow, the frequency of cytogenetic abnormalities, and the frequency of transformation to an acute myelogenous leukemia (refractory anemia with excess blasts in transformation has the highest rate; refractory anemia with ringed sideroblasts has the lowest).

Laboratory findings include a megaloblastic marrow similar to that seen in patients with vitamin B_{12}/folate deficiency. Ringed sideroblasts are commonly present, indicating a defect in heme synthesis. Myeloblasts are increased but are usually < 30%, unless the patient is progressing to an acute leukemia. Severe pancytopenia is the rule. Chromosome abnormalities are common (50%), particularly 5q⁻ and trisomy 8. A dimorphic RBC population is common (small and large cells). Median survival is 9 to 29 months.

Chronic lymphocytic, chronic myelogenous, and hairy cell leukemias are not associated with ringed sideroblasts. Lead poisoning does have ringed sideroblasts, but the clinical findings are not compatible with those of this patient.

50. The answer is D. *(Pathology; Hematopoietic/lymphoreticular)*

The patient has acute lymphoblastic leukemia (ALL), which is the most common leukemia and overall type of cancer in children. Most children are younger than 15 years, with a mean age of onset of 4 years. ALL is a clonal stem-cell disorder involving neoplastic transformation of the lymphoid stem cell. Approximately 90% of patients have cytogenetic abnormalities. The patients with translocations generally have a poor prognosis, whereas those with hyperdiploidy have a better prognosis. The French-American-British classification is as follows: the L1 subgroup (85% of childhood ALL), the L2 subgroup (65% of adult ALL), the L3 subgroup [frequently have t(8;14) translocation (Burkitt)].

The immunophenotype classification is based on lymphocyte marker studies into the following subtypes: early pre-B [CALLA⁺(most common overall; 55%–60%) or CALLA⁻], pre-B (20%), mature B (1%–2%), and immature T (15%).

Terminal deoxynucleotidyl transferase (TdT) is a marker that is positive in both immature B and T cells, but negative in mature B cells. The early pre-B, CALLA-positive ALL has the best prognosis. The mature B type has the worst prognosis. Immature T/ALL has an intermediate prognosis and involves the mediastinum.

Most patients present with fever and bone pain. Generalized, nonpainful lymphadenopathy (metasta-

ESV – independent of both preload & Afterload

Test 1

sis), epistaxis, ecchymoses, and hepatosplenomegaly are common. Central nervous system spread is common as is testicular involvement in males. The peripheral blood studies reveal a normocytic anemia, thrombocytopenia, and a white blood cell (WBC) count between 10,000 and 100,000 cells/μl, with lymphoblasts present. The bone marrow is usually packed with lymphoblasts. There is a 60% 5-year survival, with the key to survival being the type of ALL the patient has on marker studies of the lymphocytes.

51. The answer is C. *(Physiology; Pharmacodynamic and pharmacokinetic processes)*
A positive inotropic agent increases the force of contraction, and therefore the stroke volume (SV), at any given end-diastolic volume (EDV; preload). The extra blood ejected comes from the end-systolic reserve volume (ESRV). An explanation of ventricular volume terms and a schematic drawing are given below.

ESV = end-systolic volume; RV = reserve volume.

52-53. The answers are: 52-C, 53-B. *(Physiology; Cardiovascular)*
Adding another resistance in parallel simply gives the blood another channel from which to exit the large arteries. Therefore, the total peripheral resistance (TPR) decreases and flow increases. For example, if another vessel with a high resistance (5 mm Hg/ml/min) is added to the circuit, the circulation would be:

$$1/TPR = 1/2 + 1/2 + 1/2 + 1/2 + 1/5$$

$$1/TPR = 2.2$$

$$TRP = 0.45$$

Once TPR is reduced, flow (Q) through the circuit is

$$Q = P1 - P2/TPR$$

$$Q = 100 - 20/.45$$

$$Q = 177 \text{ ml/min}$$

Thus, as TRP falls, flow is greater for the same pressure gradient.

Flow (Q) is proportional to the pressure gradient ($P_1 - P_2$) and is inversely related to the TPR.

$$Q = (P_1 - P_2)/TPR$$

When resistances are parallel (as they are in the systemic circulation), TPR is calculated as follows:

$$1/TPR = 1/R_1 + 1/R_2 + 1/R_3 + 1/R_4$$

$$1/TPR = 1/2 + 1/2 + 1/2 + 1/2$$

$$1/TRP = 2.0$$

56 Body Systems Review I: Hematopoietic/Lymphoreticular, Respiratory, Cardiovascular

$$TPR = 0.5 \text{ mm Hg/ml/min}$$

Thus, in a system of parallel resistances, the TPR is less than any of the individual resistances. Flow through the system is

$$Q = (100 - 20)/0.5$$

$$Q = 160 \text{ ml/min}$$

54. The answer is A. *(Physiology; Cardiovascular)*
Flow through a blood vessel may be laminar (streamlined) or turbulent (disruption of the stream into local eddy currents). The presence of turbulence in a vessel increases the resistance to flow. Turbulence occurs when Reynold's number exceeds 2000.

Reynold's number = (Density × Velocity × Diameter)/Viscosity

Thus, turbulence occurs in large blood vessels in which the velocity is highest (i.e., ascending aorta).

55. The answer is C. *(Physiology; Cardiovascular)*
The pressure gradient ($P_1 - P_2$) responsible for flow across the pulmonary system is the difference between mean pulmonary arterial pressure (P_{PA}) and left atrial pressure (LAP), which equals pulmonary capillary wedge pressure. Flow (Q) is the cardiac output. Thus,

$$Q = (P_1 - P_2)/R$$

$$\text{Cardiac output} = (P_{PA} - LAP)/R$$

$$R = (P_{PA} - LAP)/\text{Cardiac output}$$

$$R = (15 - 5)/5$$

$$R = 2 \text{ mm Hg/L/min}$$

56. The answer is E. *(Physiology; Cardiovascular)*
The Fick equation is a useful method for measuring cardiac output or flow through any organ when the oxygen (O_2) consumption and the difference between arterial and venous O_2 content (vol% = ml/100ml) is known.

Flow = O_2 consumption/arterial (O_2 content − venous O_2 content)

Cardiac output = O_2 consumption/(aorta O_2 − pulmonary artery O_2)

$$= 400/(20 - 15)$$

$$= 8 \text{ L/min}$$

57-59. The answers are: 57-B, 58-B, 59-C. *(Physiology; Cardiovascular)*
Mean arterial blood pressure (MAP) is the pressure in the arterial circulation averaged over time. Because more time is spent in diastole than systole, MAP is calculated using this formula:

MAP = 2/3 (Diastolic pressure) + 1/3 (Systolic pressure)

Before exercise

- MAP = 2/3 (75) + 1/3 (120)
- MAP = 90 mm Hg

During exercise

- MAP = 2/3 (72) + 1/3 (144)
- MAP = 96 mm Hg

MAP can also be determined by these equations:
MAP = Cardiac output × total peripheral resistance

MAP = Stroke volume × heart rate × total peripheral resistance

Stroke volume is a function of venous return and sympathetic activity. In the case above, venous return increases during exercise (primarily because of the muscle pump) and increases preload, as indicated by the increase in right atrial pressure. The increased atrial pressure leads to an increase in the filling of the ventricle (increased end-diastolic volume; increased preload), and the ventricle responds by ejecting a larger stroke volume. However, levels of exercise that increase the heart rate above 150/min decrease preload because ventricular filling time is reduced.

Sympathetic activity to the heart increases contractility and heart rate via β-1 receptors. The increase in contractility causes the heart to eject a greater stroke volume. An increase in contractility is defined as an increase in the force of contraction (and therefore stroke volume) without a change in preload (i.e., sympathetic stimulation of the heart does not increase preload). In the data given in the questions, both the increase in preload and contractility augment the stroke volume, which results in an increase in systolic blood pressure.

The diastolic blood pressure depends on the volume of blood in the arteries at the end of diastole. This volume is determined by the amount of blood that has entered the arteries (a function of cardiac output) compared with the rate at which blood has left the circulation (a function of total peripheral resistance). The diastolic blood pressure changes little during exercise because the rise in cardiac output is offset by the decrease in total peripheral resistance. The increase in systolic pressure is larger than the decrease in diastolic pressure so that a small rise in MAP (an index of afterload) results. For arterial PO_2 to remain constant during exercise (in the face of an increase in the volume of venous blood low in oxygen returning to the heart), alveolar ventilation must increase. Vagal drive to the heart decreases during exercise, and is zero in humans at heart rates above 110/min.

The pulse pressure is determined by the stroke volume and the elasticity of the arteries. Assuming that the elastic properties of the vessel wall do not change during exercise, the increase in pulse pressure is the result of the increased stroke volume.

Pulse pressure = Systolic blood pressure − Diastolic blood pressure
Before exercise: Pulse pressure   = 120 − 75
                                    = 45 mm Hg
During exercise: Pulse pressure   = 144 − 72
                                    = 72 mm Hg

The increased need of oxygen in exercising muscle leads to a metabolically induced arteriolar vasodilation. The resistance of the venules is increased by the compression of the muscle pump and a failure to respond to vasodilator metabolites. Sympathetic vasodilator discharge occurs before the onset of exercise (anticipatory response); it is transient and nonnutritional. Peripheral chemoreceptor activation of the sympathetic nervous system increases arteriolar resistance and is not involved in the vasodilation of skeletal muscle during exercise.

60. The answer is A. *(Physiology; Cardiovascular)*
Resting cardiac output, mean arterial blood pressure, oxygen extraction, and systolic and diastolic pressures are similar in trained athletes and sedentary individuals. Maximum heart rates are the same in sedentary and trained athletes. The trained athlete has a resting bradycardia (enhanced vagal drive) and a high stroke volume. The resting bradycardia allows for the greater increase in cardiac output with exercise.

61. The answer is D. *(Physiology; Cardiovascular)*
The compliance of the large arteries is a major determinant of the pulse pressure. The stiffer the arteries, the wider the pulse pressure. Systolic pressure increases because the stroke volume is ejected into a system of blood vessels in which the capacity to expand and accommodate the increased blood volume has been reduced. Diastolic pressure decreases because the rapid transit of blood out of the large arteries results in a lowered blood volume at the end of diastole. In the elderly, the decrease in metabolic rate leads to arteriolar vasoconstriction and an increase in total peripheral resistance. The increase in total peripheral resistance raises mean arterial blood pressure but does not affect the pulse pressure. The increase in total peripheral resistance reduces venous return and stroke volume. Heart rate does not affect the pulse pressure until it becomes so rapid that ventricular filling time and stroke volume (and therefore pulse pressure) are reduced.

62. The answer is D. *(Physiology; Cardiovascular)*
The intravenous infusion of a large volume of plasma increases ventricular end-diastolic volume, which is an index of preload (i.e., the degree of stretch to which myocardial fibers are subjected prior to contraction). According to the Frank-Starling mechanism, the heart responds by increasing the force of contraction and ejecting a larger stroke volume. The larger stroke volume results in a rise in the systolic blood pressure and, therefore, the pulse pressure (Pulse pressure = Systolic pressure − Diastolic pressure).

The end-diastolic reserve volume is the difference between the end-diastolic volume and the maximal volume the ventricle can hold. An increase in the end-diastolic volume leads to a decrease in the end-diastolic reserve volume. Therefore, the increased stroke volume is the extra volume that was brought to the heart, so that ventricular end-systolic volume remains unchanged.

An increase in venous return leads to a reflex increase in heart rate. Stretch receptors in the cardiopulmonary circulation respond to the increased venous return by increasing the number of impulses ascending to the brain in vagal afferents. This afferent information is inhibitory to efferent vagal discharge, so that vagal suppression of sinoatrial pacemaker activity is withdrawn, and heart rate increases.

63. The answer is A. *(Physiology; Pharmacodynamic and pharmacokinetic processes)*
Venoconstrictor agents act primarily on small veins to increase peripheral venous pressure. This increases the pressure gradient for flow back to the heart so that ventricular end-diastolic volume is increased. An increase in end-diastolic volume results in an increased stroke volume, because the ventricle responds to stretch by developing a greater force of contraction ejecting a greater volume. An increase in end-diastolic volume leads to a decrease in the end-diastolic reserve volume. Conversely, venodilator agents decrease peripheral venous pressure and the gradient for venous return. Therefore, venodilators are used in patients with angina pectoris because they decrease preload and the work of the heart.

64. The answer is B. *(Physiology; Cardiovascular)*
During inspiration, intrapleural pressure becomes more negative, which distends the veins that lead back to the right heart. Central venous pressure decreases, and the pressure gradient for venous return to the right heart increases. End-diastolic volume and stroke volume of the right heart increase. The increased output of the right heart enters a pulmonary circulation in which capacitance has been increased by the decrease in intrapleural pressure. Therefore, venous return to the left atrium does not increase until expiration, at which time intrapleural pressure becomes less negative. Cardiac output from the left stroke volume is maintained because the left ventricle now develops a greater force of contraction (ventricular systolic pressure increases), and cardiac output returns to normal.

65. The answer is C. *(Physiology; Pulmonary/respiratory)*
The volume of ambient fresh air that enters the upper airways with each breath is called the tidal volume (V_T). The volume of fresh air that enters the upper airways per minute is known as the minute volume ($\dot{V}$) and is the product of V_T and respiratory rate (f):

$$\dot{V} = V_T \times f$$

This formula does not describe the volume of fresh air that enters the alveoli per minute, because gas in the conducting airways is interposed between ambient air and alveoli. Gas in the conducting airways occupies the anatomic dead space (V_D). At the end of expiration, the anatomic V_D contains gas that has already engaged in gas exchange; that is, it is low in O_2 and high in CO_2. At the onset of inspiration, the first gas to enter the alveoli comes from the anatomic V_D. The volume of fresh air entering the alveoli per minute [alveolar ventilation per minute ($\dot{V}_A$)] is calculated by subtracting V_D from V_T:

$$\dot{V}_A = (V_T - V_D) \times f$$

A decrease in $\dot{V}_A$ will lead to a fall in arterial P_{O_2} (hypoxemia) and a rise in arterial P_{CO_2} (respiratory acidosis). When the man in the case study adopted a rapid, shallow, breathing pattern, alveolar ventilation decreased even though minute volume was maintained. Hypoventilation of the lung resulted in hypoxemia and CO_2 retention. In the following example of a normal breathing pattern, values typical for a 70 kg man are used: $V_T = 500$ ml; f = 12/min; $V_D = 150$ ml.

$$\dot{V} = V_T \times f$$
$$= 500 \times 12$$
$$= 6000 \text{ ml/min}$$
$$\dot{V}_A = (V_T - V_D) \times f$$
$$= (500 - 150)\,12$$
$$= 4200 \text{ ml/min}$$

When breathing is rapid and shallow, values change as follows: $V_T = 250$ ml; f = 24/min.

$$\dot{V} = V_T \times f$$
$$= 250 \times 24$$
$$= 6000 \text{ ml/min}$$

$$\dot{V}_A = (V_T - V_D) \times f$$

$$= (250 - 150) \, 24$$

$$= 2400 \text{ ml/min}$$

66. The answer is B. *(Physiology; Pulmonary/respiratory)*
At the end of expiration, just prior to inspiration, the pressure of all gases in the intra-alveolar space is equal to the total barometric pressure. By convention, the total barometric pressure is set equivalent to 0 cm H_2O. When the respiratory muscles contract and enlarge the thorax, the lung expands. Because the gases in the intra-alveolar space occupy a larger volume, the pressure they exert falls below the total barometric pressure (i.e., becomes negative). Air rushes into the lungs in response to this newly created pressure gradient and continues to do so until the pressure gradient is eliminated, at the end of inspiration. During a normal respiratory cycle, intra-alveolar pressure starts at 0 cm H_2O and always returns to 0 cm H_2O by the end of inspiration, no matter how large the breath. During a normal respiratory cycle, expiration is passive: the lungs return to their original shape because of elastic recoil.

The curves do not represent intrapleural pressure during a respiratory cycle because intrapleural pressure is negative before the onset of inspiration and becomes more negative as the thorax enlarges.

67. The answer is C. *(Physiology; Pulmonary/respiratory)*
Points A, C, and E represent intrapleural pressure at the beginning of inspiration, the end of inspiration, and the end of expiration. At these times, intra-alveolar pressure is 0, equal to atmospheric pressure, and no air is flowing.

68-69. The answers are: 68-A, 69-C. *(Physiology; Pulmonary/respiratory)*
The partial pressure of a gas (P) is defined as that part of the total barometric pressure (TBP) exerted by that gas, and is calculated as the product of the dry concentration (F) of that gas and the TBP:

$$P_{GAS} = F \, (TBP)$$

At sea level, where the TBP is 760 mm Hg and the concentration of O_2 is 21%, the alveolar P_{O_2} (P_{AO_2}) of ambient air is

$$P_{AO_2} = 0.21(760)$$

$$= 159.6 \text{ mm Hg}$$

As air is inspired, it quickly becomes warmed to 37°C and saturated with H_2O vapor. The partial pressure of H_2O (P_{H_2O}) is dependent only on temperature, and at 37°C is equivalent to 47 mm Hg. Because the partial pressure of a gas is expressed in terms of dry gas, the formula for describing the partial pressure of inspired gas (P_{iGAS}) is

$$P_{iGAS} = F \, (TBP - P_{H_2O})$$

Or, in the example for oxygen used previously,

$$P_{iO_2} = 0.21 \, (760 - 47)$$

$$= 149 \text{ mm Hg}$$

Inspired gas is gas in the respiratory tree that is in the conducting airways but has not reached the alveoli (i.e., in the anatomic dead space). The partial pressure of a 10% helium mixture in the anatomic dead space at the end of inspiration is

$$P_{iHE} = 0.10 \, (760 - 47)$$

$$= 71.3 \text{ mm Hg}$$

A useful equation for determining the ideal or predicted alveolar P_{O_2} (P_{AO_2}) is the alveolar gas equation:

$$P_{AO_2} = P_{iO_2} - P_{ACO_2}/R$$

where P_{ACO_2} = alveolar P_{CO_2}, and R = respiratory exchange ratio. The respiratory exchange ratio is the ratio of CO_2 to O_2 exchanged across the lung (ml/min) in the steady state, and can vary from 0.7 to 1.0. Thus,

$$P_{AO_2} = 149 - 40/0.8$$

$$P_{AO_2} = 99 \text{ mm Hg}$$

The alveolar gas equation allows the prediction of an ideal P_{AO_2} that would be obtained if an enriched O_2 mixture were administered, or if air were breathed at a high altitude. Note also that if R = 1.0, every 1.0 mm Hg rise in P_{ACO_2} would result in a 1.0 mm Hg fall in P_{AO_2}.

70. The answer is B. *(Physiology; Cardiovascular)* Oxygen is carried in two forms in blood. Most is bound to hemoglobin, and a much smaller fraction is physically dissolved. It is the dissolved O_2 that is responsible for the partial pressure of O_2 (P_{O_2}); that is, only gases in solution exert a partial pressure. The P_{O_2} determines what percentage of the four sites on each hemoglobin that are available for O_2 binding are occupied by an O_2 molecule. If one of the four sites is occupied, the hemoglobin is 25% saturated; if two of the four are occupied, the hemoglobin is 50% saturated, and so forth. The following graph depicts the relationship between P_{O_2} and percent saturation of hemoglobin.

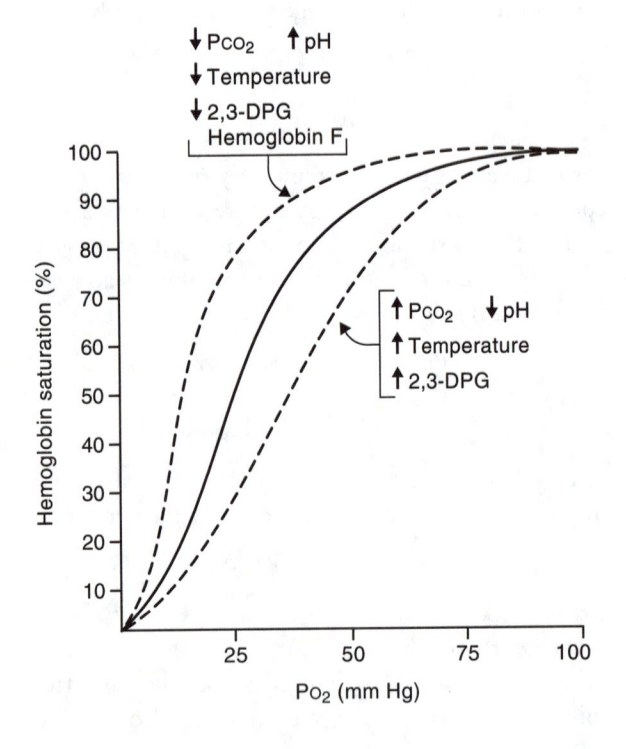

Some important characteristics are worth noting. The shape of the curve is sigmoidal, which implies that as fully saturated hemoglobin enters a tissue, the unloading of the first O_2 molecule facilitates the release of the second and third. Under normal conditions, hemoglobin is 50% saturated when P_{O_2} is 25 mm Hg, 97.5% saturated when P_{O_2} is 100 mm Hg, and does not become 100% saturated until P_{O_2} reaches 150 mm Hg.

The affinity of hemoglobin for O_2 can vary. If P_{CO_2}, temperature, or H^+ concentration of blood is increased, the curve shifts to the right. A rightward shift indicates a decrease in the affinity of hemoglobin for O_2 (i.e., less saturation at the same P_{O_2}). In exercising muscle, production of CO_2, acid, and heat shifts the O_2 dissociation curve to the right. At a P_{O_2} of 25 mm Hg, hemoglobin is only 25% saturated. Thus, a rightward shift increases the amount of O_2 released to a tissue. If P_{CO_2}, H^+ concentration, or temperature of blood falls, a leftward shift results (i.e., decreased affinity).

Another factor that shifts the position of the curve on its coordinates is 2,3-diphosphoglyceric acid (2,3-DPG). An increase in 2,3-DPG shifts the curve to the right. The most important stimulus for increased 2,3-DPG production is chronic hypoxemia. The leftward shift that is associated with fetal hemoglobin results from its failure to bind 2,3-DPG as strongly as does adult hemoglobin.

71. The answer is A. *(Physiology; Cardiovascular)* The alveolar-arterial (A-a) gradient is one of the most important calculations to make when interpreting arterial blood gases, particularly when evaluating the etiology of hypoxemia (low Pa_{O_2}). Partial pressure of alveolar oxygen (P_{AO_2}) rarely matches partial pressure of arterial oxygen (Pa_{O_2}) because of small ventilation/perfusion mismatches throughout the lungs and the thebesian veins that empty unoxygenated blood into the left side of the heart. The A-a gradient is the difference between the two. A medically significant difference is 30 mm Hg or more, which indicates problems in the lungs with ventilation, perfusion, diffusion, or a combination of the three.

The formula for calculating P_{AO_2} is as follows:

$$P_{AO_2} = \% \text{ oxygen } (713) - Pa_{CO_2} \div 0.8$$

For this patient, therefore:

$$P_{AO_2} = 0.21 \, (713 \text{ mm Hg}) - 80 \text{ mm Hg} \div 0.8$$

$$P_{AO_2} = 50 \text{ mm Hg}$$

The A-a gradient is the $P_{AO_2} - Pa_{O_2}$: 50 mm Hg − 40 mm Hg = 10 mm Hg, which is normal.

72. The answer is C. *(Physiology; Cardiovascular)* The law of diffusion for gases states that a gas moves from a point of high to low concentration. The P_{O_2}

of room air is 0.21 × 760 mm Hg, or 160 mm Hg. Assuming that the patient has a normal PaO_2 of 95 mm Hg, oxygen will diffuse into the blood and falsely elevate the PaO_2. Atmospheric CO_2 is 0.0003 × 760 mm Hg, or 0.23 mm Hg. Assuming the patient's PaCO_2 is normal (40 mm Hg), CO_2 will diffuse out of the blood into the atmosphere, thus falsely lowering the concentration.

73. The answer is C. *(Pathology; Pulmonary/respiratory)*
The alveolar-arterial (A-a) gradient is useful for evaluating the source of hypoxemia. A medically significant difference of 30 mm Hg or more between the partial pressure of alveolar oxygen (PAO_2) and arterial oxygen (PaO_2) characterizes diseases of the lungs that interfere with ventilation, perfusion, or diffusion in the lung. Hypoxemia unassociated with problems in the lungs occurs if there is depression of the respiratory center, such as that caused by heroin or barbiturates, or a problem with the chest bellows, such as paralysis of the diaphragms. Because the lungs are not affected in these conditions, the A-a gradient is normal.

74-75. The answers are: 74-D, 75-A. *(Pathology; Pulmonary/respiratory)*
The respiratory unit in the lungs is where gas exchange occurs. In the diagram, A represents a respiratory bronchiole; B, an alveolar duct; and C, alveoli. Both of these patients have types of emphysema, a disease that involves the respiratory unit of the lungs. It is characterized by permanent enlargement of all or part of the unit, with destruction of the elastic tissue support in these structures.

The 28-year-old, nonsmoking man with chronic lung disease limited to the lower lobes, a negative sweat test, and a serum protein electrophoresis abnormality most likely has $α_1$-antitrypsin (AAT) deficiency, an autosomal recessive disease. The entire respiratory unit is susceptible to damage caused by ATT deficiency, thus the term panacinar emphysema. The lower lobes are most commonly involved. The $α_1$-peak on a normal serum protein electrophoresis mainly represents AAT. An abnormal electrophoresis in which this peak is missing would indicate ATT deficiency. The disease involves a defect in the movement of AAT from the endoplasmic reticulum (ER) to the Golgi apparatus, hence ATT accumulates in the ER.

The 49-year-old coal mine worker with severe dyspnea and a productive cough with numerous dust cells in sputum most likely has coal worker's pneumoconiosis (CWP), or black lung disease. The proximal respiratory bronchioles are affected in CWP, but the alveolar ducts and alveoli are spared. Immunologic mechanisms are responsible for the damage associated with this disease. Alveolar macrophage release of interleukin 1 (IL-1) stimulates fibrogenesis, which decreases lung compliance and initiates the release of proteases and free radicals from other macrophages and neutrophils. CWP is a restrictive lung disease that varies in severity from mild to disabling. Mild disease involves the deposition of coal dust (anthracotic pigment) in the parenchyma, hilar lymph nodes, and alveolar macrophages (dust cells) without producing significant ventilatory impairment. Severe disease is characterized by progressive massive fibrosis associated with upper lobe focal "dust" emphysema in association with the fibrotic areas in the lungs.

76. The answer is B. *(Pathology; Pulmonary/respiratory)*
Fever that develops within 24–48 hours after surgery is due to atelectasis. Absorption atelectasis is the collapse of alveoli secondary to obstruction of the airway (e.g., by mucous plugs or tumor) followed by the resorption of air distal to the obstruction through the pores of Kohn. Grossly, the lung surface is depressed because of loss of air, and has a purple-red, meaty appearance. Consequences of atelectasis include focal areas of intrapulmonary shunting, which causes hypoxemia and a predisposition to bronchopneumonia if the lung remains atelectatic for more than 72 hours. Clinically, patients present with fever, tachypnea, sinus tachycardia, elevation of the diaphragms, and scattered rales and decreased breath sounds.

77. The answer is B. *(Pathology; Pulmonary/respiratory)*
The patient has a lobar pneumonia most likely due to *Klebsiella pneumoniae*, which is a gram-negative rod with a capsule. Patients with bacterial pneumonias have an abrupt onset of fever, chills, productive cough, and, in some cases, pleuritic chest pain. The two types

of pneumonia that involve the alveoli are lobar and bronchopneumonia. Lobar pneumonia involves the entire lobe of a lung without bronchial involvement and tends to be associated with more virulent organisms than in bronchopneumonia. Bronchopneumonia first involves the bronchus and then spreads out into the alveoli. Although *Streptococcus pneumoniae* is the most common etiologic agent in community-acquired pneumonias, *K. pneumoniae* is a frequent agent in alcoholics and tends to involve the upper lobes.

78. The answer is C. *(Pathology; Pulmonary/respiratory)*
In respiratory acidosis, carbon dioxide is retained in the alveoli, causing increased partial pressure of alveolar CO_2 (P_{ACO_2}) and arterial CO_2 (P_{aCO_2}), with subsequent alteration in the partial pressure of the alveolar oxygen that is available for gas exchange. According to Dalton's law, the sum of the partial pressures of a gas must equal 760 mm Hg. Because nitrogen is not exchanged in the alveoli, an increase in P_{ACO_2} (normally 40 mm Hg) automatically results in a corresponding decrease in P_{AO_2} (normally 100 mm Hg). For example, if P_{ACO_2} is 80 mm Hg (40 mm Hg above normal), the P_{AO_2} must decrease from 100 mm Hg to 60 mm Hg (100 − 40 = 60).

Oxygen saturation (S_{aO_2}) refers to the percentage of the four heme groups on the hemoglobin molecule that are occupied by oxygen. If only half of the heme groups are occupied by oxygen, S_{aO_2} is 50%. In respiratory acidosis, the decreased P_{AO_2} results in less oxygen diffusing through the alveolar–capillary interface into the pulmonary capillary blood. This not only decreases the amount of oxygen dissolved in the blood (i.e., hypoxemia), reducing P_{aO_2}, but also decreases the amount of oxygen available to bind with the heme groups in hemoglobin, thus reducing S_{aO_2} as well.

79. The answer is B. *(Pathology; Cardiovascular)*
The patient has the superior vena caval syndrome (SVCS). Malignancy accounts for 80% to 90% of cases of SVCS. Most cases are caused by a primary lung cancer extending into the neck and obstructing the superior vena cava. Clinically, patients have a puffy face with a blue to purple discoloration of the face, arms and shoulders. Retinal vein engorgement results from increased venous pressure. Central nervous system findings include dizziness, convulsions, and visual disturbances. Increased venous pressure is noted in the upper extremities. The diagnosis is confirmed by venography, and the prognosis is poor.

80. The answer is B. *(Pathology; Cardiovascular)*
The patient has a thoracic outlet syndrome (TOS), which is an abnormal compression of the arteries, veins, or nerves in the neck. TOS is more common in adults than in children, and may occur in either sex. The compression may be caused by a cervical rib, the scalenus anticus muscle, or positional changes. The diagnosis is confirmed with an arteriogram. Nerve conduction studies do not distinguish TOS from cervical disc disease, since the nerves are also involved in TOS.

81. The answer is A. *(Pathology; Pulmonary/respiratory)*
Cystic fibrosis (CF) is an autosomal recessive disease that affects 1 in 2000 live births. Recurrent pulmonary infections with chronic obstructive pulmonary disease, pancreatic insufficiency, and infertility (in males) are a classic triad for a young adult with CF.

Most patients with CF (80%–90%) die of respiratory complications related to pulmonary infections associated with *Staphylococcus aureus*, *Haemophilus influenzae*, *Pseudomonas aeruginosa* (particularly the mucoid strain), and *P. cepacia*. CF is the most common cause of bronchiectasis in the United States.

82. The answer is C. *(Pathology; Human development and genetics)*
A person with situs inversus, recurrent sinopulmonary infections, and sterility has Kartagener syndrome. In this autosomal recessive disease, a defect in ciliary motility is caused by the absence of the dynein arm, which provides the movement for cilia. This interferes with mucociliary clearance of pathogens and foreign bodies, thus the tendency for recurrent sinopulmonary infections. Disturbances of cell motility during embryogenesis result in situs inversus, in which the organs in the chest and abdomen are located on the side opposite from their normal location (e.g., the liver is on the left side of the abdomen). Dextrocardia, in which the heart is located in the right chest cavity,

may also occur. Males are frequently sterile as well, because sperm tails, which are modified cilia, are nonfunctional.

83. The answer is A. *(Pathology; Pulmonary/respiratory)*

The photograph reveals multiple thin-walled abscess cavities in the lower lobe of the lung. Patchy areas of consolidation are also visible throughout all lung fields. These most likely represent bronchopneumonia. Alcoholism is a predisposing condition of bronchopneumonia.

The organisms isolated in lung abscesses include both aerobic and anaerobic bacteria, although anaerobes are most common. *Staphylococcus aureus* and *Klebsiella pneumoniae* are frequently isolated. Mixed infections are the most common cause of lung abscesses. They arise from aspiration of infected oropharyngeal material—even carious teeth—in patients with poor dental hygiene.

Clinically, patients present with fever and expectoration of foul-smelling sputum, particularly when anaerobes are involved. Gram stains of sputum in anaerobic abscesses show a mixture of gram-negative rods, gram-positive cocci, and fusobacterium (gram-negative rods with tapered ends). Fluid-filled cavities are frequently visible on radiographs of the chest. Regarding the other choices in the question:

- The multiple cavities are in the wrong location for tuberculosis. Reactivation tuberculosis is most commonly located in the upper lobe. Caseous material should also be present, which normally gives cavitary lesions a shaggy appearance.
- Squamous cancers often cavitate, but they are more centrally located. A clear cut relationship with a mainstem bronchus would also be expected, since obstruction of a bronchus is a predisposing cause for lung abscesses.
- Small cell carcinomas are non-cavitary, centrally located cancers. Hyponatremia, from ectopic secretion of antidiuretic hormone, is a frequent finding.
- Metastatic infiltrating ductal carcinoma of the breast does not usually cavitate.

84. The answer is A. *(Pathology; Pulmonary/respiratory)*

The patient has triad asthma, which is a type of intrinsic asthma characterized by aspirin sensitivity, wheezing, and nasal polyps. Aspirin inhibits cyclooxygenase, thus blocking prostaglandin synthesis. This leaves the leukotriene pathway unchecked, resulting in subsequent synthesis of leukotrienes C_4, D_4, and E_4, which are powerful bronchoconstrictors.

85–87. The answers are: 85-D, 86-B, 87-B. *(Pathology; Pulmonary/respiratory)*

The patient has an acute exacerbation of bronchial asthma, most likely associated with type I, immunoglobulin E (IgE)-mediated hypersensitivity. Bronchial asthma is the most common chronic pulmonary disease in the United States. It is characterized by increased reactivity of the small airways to a variety of stimuli, including chemical mediators, physicochemical agents, or allergens. An exaggerated bronchoconstrictor response to the stimulus produces dyspnea and inspiratory/expiratory wheezing caused by increased small-airway resistance to air flow. Although the small-airway damage is reversible in the majority of cases, more recent studies indicate that fixed airway disease occurs in patients with chronic disease.

Because the patient has a respiratory rate of 30/min, her arterial blood gases would be expected to indicate primary respiratory alkalosis. Respiratory alkalosis in asthma develops initially as a result of hyperventilation. When the patient tires, respiratory acidosis develops, portending possible respiratory failure.

The findings of an increased anteroposterior diameter, hyper-resonance to percussion, and flattened diaphragms on chest x-ray are most closely related to an increase in residual volume, because the patient is trapping air on expiration behind the inflamed small airways. Her tidal volume, vital capacity, and FEV_1 would be decreased, not increased.

The pathogenesis of this patient's clinical findings are most closely related to inflammation in the small-caliber airways. Normally, the small-caliber airways offer the least resistance to air flow, because their cross-sectional area increases as branching increases the number of airway conduits. In bronchial asthma, however, the small airways are inflamed, and are the primary site for resistance to normal airflow.

88. The answer is D. *(Pharmacology; Cardiovascular)*

Both aspirin and ticlopidine can be used to prevent vascular events that result from the formation of arterial thrombi in patients at risk. Ticlopidine may cause leukopenia, so the white blood cell count must be monitored periodically. Ticlopidine acts by inhibiting ADP-induced platelet aggregation, whereas aspirin inhibits the synthesis of thromboxane, another potent platelet-derived aggregation factor. Neither aspirin nor ticlopidine is particularly effective in preventing venous thrombosis, and heparin or warfarin should be used for that purpose.

89. The answer is B. *(Pharmacology; Cardiovascular)*
Oral amoxicillin should be administered to cardiac valve patients one hour before and 6 hours after dental surgery in order to prevent endocarditis resulting from surgical bacteremia. Gram-positive organisms sensitive to penicillins are the most likely cause of endocarditis in this setting. Erythromycin should be administered to most penicillin-allergic patients, but high-risk penicillin-allergic patients should receive vancomycin. Ampicillin or clindamycin may be given parenterally to low-risk patients who can not take oral amoxicillin.

90. The answer is C. *(Pharmacology; Cardiovascular)*
Angiotensin-converting enzyme (ACE) inhibitors not only inhibit the formation of a vasoconstrictor, angiotensin II, they also inhibit the degradation of a vasodilator, bradykinin. This latter effect results from inhibition of the kininases that inactivate bradykinin. Other peptides that cause vasodilation include substance P and atrial natriuretic peptide. Endothelins and vasopressin cause vasoconstriction.

91. The answer is D. *(Pharmacology; Pulmonary/respiratory)*
Terfenadine and astemizole are relatively new, nonsedating antihistamines that may prolong the QT interval of an electrocardiogram and cause cardiac arrhythmias. Erythromycin, ketoconazole, and itraconazole may inhibit the metabolism of terfenadine, resulting in higher blood levels and increased risk of arrhythmia. Loratadine, another nonsedating antihistamine, apparently lacks the QT prolongation effect and no drug interactions with either erythromycin or the antifungal agents listed have been established. Diphenhydramine, clemastine, and chlorpheniramine are also free of these drug interactions. However, many antihistamines have some potential to cause tachycardia and palpitations because of their antimuscarinic activity.

92. The answer is E. *(Pharmacology; Cardiovascular)*
Propranolol and prazosin act as competitive receptor antagonists at β- and α-adrenergic receptors, respectively, and both agents reduce arterial blood pressure. β-blockers decrease arterial pressure by reducing cardiac output and renin secretion. α-Blockers decrease vascular resistance. Neither drug increases vagal tone (an effect of digitalis).

93. The answer is A. *(Pharmacology; Cardiovascular)*
Verapamil has approximately equal effects on calcium channels in smooth muscle and cardiac tissue, whereas nifedipine and other dihydropyridines selectively block calcium channels in smooth muscle. Both agents can lower vascular resistance, but only verapamil decreases heart rate. Nifedipine may cause reflex tachycardia.

94-95. The answers are: 94-D, 95-E. *(Pharmacology; Cardiovascular)*
Enalapril inhibits angiotensin-converting enzyme (ACE), which converts angiotensin I to angiotensin II. The reduced angiotensin II leads to decreased aldosterone secretion and increased potassium levels. Secondarily, the fall in arterial pressure causes renin secretion to increase, leading to elevated angiotensin I levels. Enalapril has no effect on the renin substrate angiotensinogen, which is synthesized in the liver.

96. The answer is C. *(Gross anatomy; Hematopoietic/lymphoreticular)*
Lymph from the left bronchomediastinal trunk enters the thoracic duct, whereas lymph from the right bronchomediastinal trunk enters the right lymphatic duct. All lymph from below the diaphragm enters the thoracic duct. Lymph from the right side of the body above the diaphragm enters the right lymphatic duct. Because the thoracic duct is close to the mediastinal

pleura, damage to the thoracic duct can cause a chylothorax.

97. The answer is E. *(Gross anatomy; Cardiovascular)*
The atrioventricular nodal artery and the sinoatrial nodal artery are typically branches of the right coronary artery, making sinus bradycardia more common in right coronary artery thrombosis. The left coronary artery divides into the circumflex artery and the anterior interventricular artery after passing behind the pulmonary trunk. The circumflex artery enters the coronary sulcus, and the anterior interventricular artery enters the anterior interventricular sulcus.

98. The answer is C. *(Gross anatomy; Pulmonary/respiratory)*
Contraction of the rectus abdominis increases intra-abdominal pressure, thus assisting with expiration. Because the pectoralis major, pectoralis minor, anterior scalene, and middle scalene pull the ribs upward, they assist with inspiration. A patient who is having difficulty during inspiration may grasp a stable object above his head (e.g., a bedpost) to fix the upper limbs in flexion, allowing muscles that attach the limbs to the chest wall (pectoralis major) to help expand the chest wall. The scalene muscles connect the cervical vertebrae to the upper two ribs. These muscles contract on inspiration.

99. The answer is E. *(Gross anatomy; Nervous/special senses)*
Preganglionic sympathetic neurons arise only from the thoracic and upper lumbar spinal cord segments. The preganglionic sympathetic neurons associated with innervation of the heart arise only from the upper thoracic spinal cord segments. These fibers synapse in the cervical sympathetic ganglia and the upper four or five thoracic sympathetic ganglia. The postganglionic cardiac nerves exit from these ganglia and contribute to the cardiac plexus to innervate the sinoatrial and atrioventricular nodes, the cardiac muscle fibers, and the coronary vasculature. Parasympathetic innervation to the heart is generated by preganglionic cardiac nerves, which are branches of the vagus nerve and synapse on postganglionic neurons in the cardiac plexus.

100. The answer is C. *(Gross anatomy; Musculoskeletal)*
The intercostal neurovascular bundle lies along the inferior border of the rib, forming the upper boundary of the intercostal space. The neurovascular bundle is arranged, moving downward, as vein, artery, and nerve (VAN). Because the intercostal artery lies against the inferior border of the rib and is used as a collateral channel (e.g., coarctation of the aorta), enlargement of this artery can cause thinning of the inferior border of the rib (i.e., notching).

101. The answer is B. *(Gross anatomy; Hematopoietic/lymphoreticular)*
The sinoatrial node receives a nerve supply; however, this node initiates the heartbeat autonomously. The autonomic innervation modulates the rate of depolarization of the sinoatrial node, thereby modulating the heart rate. The sinoatrial node is composed of specialized cardiac muscle cells that depolarize spontaneously at a greater rate than that of other cardiac muscle cells. Because cardiac muscle cells contact each other electrically through gap junctions at the intercalated disks, the depolarization of the sinoatrial node drives the other cardiac muscle cells of the atrium to depolarize at the same rate as the node.

102. The answer is E. *(Behavioral science; Psychosocial, cultural, and environmental influences)*
A clinician is least likely to see a personality disorder in this patient. Noncompliance with medical advice, insomnia, disorientation, and agitation are all commonly seen in patients in the intensive care unit.

103. The answer is D. *(Pathology; Cardiovascular)*
Diastolic dysfunction is an impairment in the filling of the left ventricle ("stiff ventricle") caused by decreased compliance at a normal left atrial pressure. Systolic heart failure, which is an impairment in contractility, has a better response to positive inotropic agents. Most patients have a combination of both systolic and diastolic dysfunction.

In diastolic heart failure, filling of the left ventricle is slow or incomplete (i.e., decreased compliance) unless left atrial pressures increase to maintain the ejection fraction at the expense of increasing pulmonary and systemic venous congestion. Clinical findings include dyspnea, which is the most common symptom, resulting from interstitial and pulmonary edema. Pathophysiologically, there is left atrial hypertension and pulmonary venous hypertension, leading to pulmonary congestion. Therapy is aimed at slowing the heart rate, which promotes filling of the left ventricle at low pressures. β-Adrenergic blockers and calcium channel blockers are effective. Patients with diastolic dysfunction have a better survival rate than those with systolic dysfunction, because the ejection fraction is normal to slightly decreased, albeit at the expense of increased left atrial pressures.

Systolic dysfunction is an impairment in contractility of the left ventricle, or a defect in the ability of myofibrils to shorten against a load. The left ventricle loses its ability to eject blood into the aorta, so the ejection fraction (i.e., stroke volume/left ventricular end-diastolic volume) is decreased (usually < 40%); the normal ejection fraction is 80/120, or 66%, with a range of 55% to 75%. The ejection fraction is measured by echocardiography or radionuclide ventriculography. Signs of isolated systolic dysfunction include fatigue, prerenal azotemia, cool skin, and mental obtundation. The left ventricular chamber eventually dilates and patients develop fatigue, dyspnea, and peripheral edema, which are signs of right-sided heart failure. Examples of systolic dysfunction include post-myocardial infarction, ischemic injury (e.g., acute myocardial infarction), and congestive cardiomyopathy. Therapy is aimed at improving the performance of the left ventricle with administration of positive inotropic agents and peripheral vasodilators [e.g., angiotensin-converting enzyme (ACE) inhibitors] to decrease peripheral resistance.

104. The answer is B. *(Pathology; Cardiovascular)*
Approximately 70% of patients with acute myocardial infarction have classic electrocardiogram (ECG) findings; sensitivity improves with repeat ECGs. The following is the classic evolution of an acute myocardial infarction in sequence:

1. Peaked (i.e., hyperacute) T waves, which represent the area of ischemia
2. ST segment elevation, which represents the area of injury, not cell death
3. Q waves, which represent the area of infarct and cell death
4. T-wave inversion, which represents the area of ischemia

The classic ECG patterns are as follows: In anterior acute myocardial infarction, Q waves are seen in leads V_1 through V_4. In lateral wall acute myocardial infarction, Q waves are seen in leads I and aVL. In inferior acute myocardial infarction, Q waves are seen in leads II, III, and aVF. There is a better prognosis in this type than in anterior acute myocardial infarction because of overlap from the right coronary artery blood supply. In posterior wall acute myocardial infarction, large R wave and S-T segment depression are seen in leads V_1, V_2, and/or V_3.

105. The answer is B. *(Pathology; Cardiovascular)*
A hepatojugular reflux indicates liver congestion caused by the backup of blood into the venous circulation, secondary to problems in the right heart (e.g., right-sided heart failure, tricuspid stenosis, tricuspic regurgitation).

An S_3 heart sound occurs during the rapid-filling phase of the ventricle in early diastole. In mitral valve incompetence and increased preload in the left ventricle, an S_3 heart sound is an excellent indicator of ventricular failure. The increased left ventricular end-diastolic pressure is reflected back into the left atrium, with a subsequent increase in the hydrostatic pressure in the pulmonary veins. When pulmonary venous pressure is greater than the pulmonary capillary oncotic pressure, a transudate leaks first into the interstitium (Kerley lines on a chest radiograph) and eventually into the alveoli, resulting in pulmonary edema and congestion.

Dyspnea, or the sensation of difficult or uncomfortable breathing, is the most common symptom of left-sided heart failure. When a patient with left-sided heart failure lies down, the reduced gravitational force increases venous return to the heart. Because the left ventricle cannot handle the increased load, the blood backs up into the lungs, producing dyspnea. Paroxysmal nocturnal dyspnea specifically refers to a choking sensation that wakens the patient at night. It usually subsides 5 to 20 minutes after the patient stands up, opens the window, and takes deep breaths.

Another finding associated with left-sided heart failure is reduced renal perfusion, which occurs during the day due to the decreased cardiac output. Reduced cardiac output decreases the effective arterial blood volume and the glomerular filtration rate, allowing a greater reabsorption of urea in the proximal tubule. Because creatinine is neither reabsorbed nor secreted in the tubules, a reduction in the glomerular filtration rate causes a minor increase in serum creatinine. The disproportionate increase in the blood urea nitrogen level over that of the serum creatinine is called prerenal azotemia.

106. The answer is C. *(Pathology; Pulmonary/respiratory)*
Wegener granulomatosis is not associated with Raynaud phenomenon. Wegener granulomatosis is a necrotizing granulomatous vasculitis that primarily involves small arteries and veins in the upper airways (e.g., sinusitis, upper airway collapse), lungs (e.g., nodular densities, recurrent pneumonia), and kidneys (e.g., crescentic glomerulonephritis).

Kawasaki disease is a febrile illness in children associated with a systemic vasculitis, involving skin, mucous membranes, and the heart (e.g., coronary arteritis).

Polyarteritis nodosa is a male-dominant, necrotizing immune vasculitis of small-to-medium size arteries. Hepatitis B antigenemia is present in 30% to 40% of cases.

Takayasu disease is a granulomatous vasculitis that involves the aortic arch vessels in young, typically Asian, women.

Thromboangiitis obliterans is an inflammatory vasculitis involving the neurovascular compartment, including medium-size vessels, principally the tibial and radial arteries. It is usually seen in young to middle age, cigarette-smoking males.

107. The answer is C. *(Pathology; Cardiovascular)*
Immune complexes are least likely involved in the pathogenesis of dissecting aortic aneurysms, which are often associated with elastic tissue fragmentation and cystic medial necrosis. The addition of hypertension, which adds stress to the weakened wall of the elastic artery, results in an intimal tear and dissection of blood proximally, distally, or both through areas of weakness.

Bacterial antigens combine with antibodies to form immune complexes in infective endocarditis. Immune complexes are also involved in the pathogenesis of the hypersensitivity vasculitis syndromes, which are a group of syndromes involving small vessels (principally postcapillary venules) that present with skin manifestations of palpable purpura. Polyarteritis nodosa is a necrotizing immune vasculitis of small-to-medium size arteries. Key mechanisms involve immune complex deposition (type III hypersensitivity) and activation of neutrophils and monocytes by antineutrophil cytoplasmic antibodies.

108. The answer is E. *(Pathology; Cardiovascular)*
Neutrophils do not have a significant role in the development of a fibrofatty plaque in a vessel.

The response to injury theory of atherosclerosis states that an injurious stimulus [e.g., carbon monoxide in cigarette smoke, increased native low-density lipoproteins (LDL), oxidized LDL, infectious agents] causes endothelial damage. Platelet and monocyte adhesion to the injured surface occurs. Platelets and monocytes release growth factors, which stimulate medial hyperplasia of smooth muscle cells and subsequent migration of smooth muscle cells to a subintimal location beneath the endothelium. Monocytes/macrophages and injured endothelial cells release free radicals, which form oxidized LDL. Macrophages interact with T lymphocytes, which release cytokines that also promote smooth muscle cell proliferation, the release of growth factors, and cell adhesion. Oxidized LDL is taken up by "scavenger receptors" on macrophages to form foam cells. There is insudation of LDL and triacylglycerol into the damaged vessel wall. Smooth muscle LDL receptors imbibe LDL to form foam cells, which collect beneath the endothelium to form the fatty streak, the earliest morphologic evidence of atherosclerosis.

109. The answer is E. *(Biochemistry; Biochemistry and molecular biology)*
Although an increase in low-density lipoproteins (LDLs) is a major risk factor for coronary artery disease, it is not the most important risk factor. According to a study on lipids by the National Cholesterol Education Panel in 1993, a person's age is considered the single most important risk factor for coronary artery

disease. This study defined age as a risk factor for men 45 years of age or older and for women 55 years of age or older.

LDL is calculated rather than measured in most lipid profiles. The calculation is as follows:

LDL = Cholesterol − high-density lipoprotein (HDL) − triacylglycerol/5

For example,

Cholesterol = 350 mg/dl, HDL = 20 mg/dl, triacylglycerol = 450 mg/dl

LDL = 350 − 20 − 450/5 = 240 mg/dl

An LDL concentration less than 130 mg/dl is desirable; 130 to 160 mg/dl is borderline high risk; and 160 mg/dl or greater is high risk.

110. The answer is D. *(Pathology; Hematopoietic/lymphoreticular)*
The patient has disseminated intravascular coagulation (DIC), which was most likely precipitated by endotoxemia from gram-negative sepsis. A bone marrow examination is not indicated in the work-up of a patient with DIC. DIC refers to intravascular consumption of clotting factors, which include platelets, fibrinogen, factor V, and factor VIII. Causes include sepsis (most common), malignancy, amniotic fluid embolism, and snake bite. The clinical presentation is usually oozing from all puncture sites, as in this patient. Laboratory findings include thrombocytopenia (commonly present), decreased fibrinogen (serial assays may be needed to demonstrate a decline), and a prolonged prothrombin time (PT) and partial thromboplastin time (PTT) caused by consumption of factors V, VIII, and fibrinogen. The prolonged thrombin time test evaluates the conversion of fibrinogen into fibrin after adding thrombin to the test tube. Low fibrinogen levels prolong the test. Increased fibrin-degradation products (from the effect of plasmin on breaking up clots) is the best indicator of DIC. A prolonged bleeding time is caused by thrombocytopenia and interference of the fibrin-degradation products with platelet aggregation. Schistocytes appear in the peripheral blood from damage to the red blood cells by intravascular clots.

The most important treatment of DIC is to treat the underlying disease. Component therapy is used to maintain adequate hemostasis. Cryoprecipitate (which replaces fibrinogen and factor VIII), fresh frozen plasma (which contains all the factors), platelet concentrates, and packed red blood cells are commonly infused.

111. The answer is E. *(Physiology; Cardiovascular)*
The viscosity of a liquid is a measure of its internal resistance to flow. The greater the viscosity, the greater the resistance. The viscosity of blood is determined by the hematocrit: increasing the hematocrit increases the viscosity, the size of the blood vessel, and the velocity. In vessels smaller than 200 μ, red blood cells congregate in the center of the vessel (axial streaming), which decreases the viscosity. At low velocities, red blood cells clump together forming rouleaux, which increases the viscosity. Plasma proteins do not contribute significantly to the viscosity of blood, unless they are markedly increased (as in Waldenstrom macroglobinemia with a monoclonal increase in IgM).

112. The answer is E. *(Physiology; Cardiovascular)*
An increase in total peripheral resistance leads to an increase in mean arterial blood pressure (Mean arterial blood pressure = Cardiac output × Total peripheral resistance). Mean arterial blood pressure is a good measure of afterload, which is a measure of the force the heart must develop during ejection. An increase in total peripheral resistance therefore increases afterload. A sudden increase in total peripheral resistance momentarily reduces stroke volume and, therefore, cardiac output. The reduced ejection leads to an increase in end-systolic volume and a subsequent increase in end-diastolic volume. In the long term, stroke volume is maintained because the left ventricle now develops a greater force of contraction (ventricular systolic pressure increases) and cardiac output returns to normal.

113. The answer is A. *(Pharmacology; Cardiovascular)*
Warfarin easily crosses the placenta and may cause birth defects because of its antagonism of vitamin K-dependent bone development. It may also cause fetal hemorrhage. Heparin is a large polymer that does not easily cross the placenta. The action of warfarin is delayed because of the time required to deplete circulating coagulation factors, whereas heparin acts immediately to inactivate circulating coagulation factors (via

antithrombin III activation). For this reason, heparin is active in vitro as well as in vivo. Warfarin is administered orally and is suitable for chronic use, whereas heparin is given intravenously or subcutaneously and treatment is usually limited to a duration of a few days.

114. The answer is D. *(Pharmacology; Cardiovascular)*
Nitric oxide (NO) is a gas that mediates the vasodilating effect of several endogenous substances and drugs. This includes the vasodilation responsible for penile erection that is activated by the parasympathetic nerves. It also inhibits platelet aggregation and is involved in nervous control of gastric and colonic function (i.e., the adaptive relaxation of the stomach and relaxation of the internal anal sphincter). NO activates guanylate cyclase and increases cyclic guanosine monophosphate production, accounting for its smooth-muscle relaxing effects. NO is formed from arginine, which is converted to NO and citrulline by NO synthetase, an enzyme that is activated by the calcium–calmodulin complex following activation of receptors for various neurotransmitters and neuromodulators.

115. The answer is D. *(Pharmacology; Pulmonary/respiratory)*
Terbutaline and albuterol are β_2-adrenergic agonists that cause bronchodilation by increasing cyclic AMP levels in smooth muscle. Epinephrine is a nonspecific adrenergic agonist that acts by the same mechanism. Ipratropium is an atropine analog that blocks muscarinic receptors that mediate bronchial muscle contraction via inositol 1,4,5-triphosphate (IP_3) and calcium release. Cromolyn sodium does not cause bronchodilation, but acts prophylactically to prevent allergic and asthmatic reactions by stabilizing mast cells. It is of no value in treating an acute asthmatic attack.

116. The answer is B. *(Pharmacology; Pulmonary/respiratory)*
Terfenadine is a long-acting, non-sedating antihistamine that is often effective in preventing allergic reactions. Pseudoephedrine is an adrenergic agonist that causes vasoconstriction and decongestion of nasal passages in patients with allergic rhinitis. Cromolyn sodium stabilizes mast cells and is an effective prophylactic of allergic reactions. Beclomethasone, a potent glucocorticoid, can be administered by nasal inhalation and effectively prevents inflammation of the nasal passages associated with respiratory allergies. Albuterol, a β_2-adrenergic agonist and bronchodilator, has no use in treating allergic rhinitis.

117. The answer is C. *(Pharmacology; Pulmonary/respiratory)*
Tachycardia and palpitations may occur due to activation of β_1- and β_2-receptors in the heart, because these drugs are not absolutely selective for β_2-receptors. Also, approximately 20% of cardiac β-receptors are β_2-receptors, and these receptors also mediate increased heart rate and force. In addition, these receptors mediate central nervous system stimulation, causing anxiety, restlessness, and even convulsions in a few patients. Skeletal muscle tremor may occur due to the presence of β_2-receptors in skeletal muscle. Oral candidiasis is not usually associated with β_2-agonists, but often occurs during chronic use of corticosteroid inhalers.

118. The answer is A. *(Pharmacology; Pulmonary/respiratory)*
Guaifenesin is an expectorant that reduces the viscosity of sputum and facilitates expectoration, but it does not suppress coughing. Codeine and hydrocodone are opiates with excellent cough-suppressing activity, and are available in liquid cough preparations. The dextro isomer of methorphan is an effective antitussive but has no other significant opiate activity in the central nervous system. Benzonatate is a nonopiate antitussive related to local anesthetics, and may be preferred in patients who can not receive opiate drugs.

119-120. The answers are: 119-C, 120-C. *(Pathology; Cardiovascular)*
Atherosclerotic coronary artery disease (thrombosis) most commonly involves, in descending order of frequency, the left anterior descending coronary artery, right coronary artery, and the left circumflex coronary artery. The epicardial coronary arteries are primarily involved. The left anterior descending coronary artery has diagonal and septal branches, which supply the anterior wall of the left ventricle and the anterior two thirds of the interventricular septum, respectively. The

right coronary artery, via its marginal branches, supplies the right ventricle. In 80% of cases, the posterior descending artery arises from the right coronary artery and supplies the posterior–inferior wall of the left ventricle, including the posteromedial papillary muscle and the posterior one third of the interventricular septum. The right coronary artery also supplies the atrioventricular node in 95% of cases. The sinoatrial node is supplied from both the left (45% of cases) and right coronary arteries (55% of cases), so it is less affected in coronary artery disease than the atrioventricular node. The left circumflex artery has marginal branches, which supply the lateral free wall of the left ventricle. Coronary arteries are normally end-arteries, but atherosclerosis opens the collateral circulation; thus occlusion of the right coronary artery serving a collateral role can cause infarction of the left ventricle.

Most papillary muscle ruptures involve the posteromedial muscle, which is supplied by the right coronary artery. In an acute myocardial infarction, the myocardium is softest from day 3 to 7, thus most ruptures occur during this time. Acute mitral insufficiency and heart failure ensue.

Sinus bradycardia is most commonly due to disruption of the atrioventricular node in thrombosis of the right coronary artery with infarction of the inferior wall of the left ventricle.

121-123. The answers are: 121-A, 122-C, 123-B. *(Pathology; Hematopoietic/lymphoreticular)*
Area A in the schematic drawing represents a germinal follicle, which is where B cells are located. Area B depicts the paracortex, which is where T cells are located, and area C identifies the sinuses of the lymph node, which is where histiocytes are located. The morphologic changes caused by antigen stimulation of the B cells give the primary follicle a different appearance under the microscope, thus the term germinal follicle. The presence of germinal follicles indicates an antigenically stimulated lymph node, which is also called reactive hyperplasia. Most malignant lymphomas derive from B cells (80%–85%) located in the follicular area of the lymph node (option A), thus the term follicular lymphoma. Follicular lymphomas may progress into a diffuse pattern as the follicle becomes obliterated by neoplastic cells streaming into the paracortical area of the node.

Proliferation of benign or neoplastic T cells begin in the paracortex (option B) located around the follicles. Therefore, in patients with DiGeorge syndrome, in whom failure of development of the third and fourth pharyngeal pouches results in absence of the parathyroid glands and thymus, there is an absence of paracortical T lymphocytes, although the primary follicles are normal. Antigenic stimulation of T cells (e.g., infectious mononucleosis) or neoplastic transformation of T cells also involve the paracortical area of the node.

Histiocytes are normally present in the sinuses of lymph nodes. Therefore, benign (sinus histiocytosis) or neoplastic proliferation of histiocytes (malignant histiocytosis) are primarily located in the sinuses of the lymph nodes. Since carcinomas first spread to lymph nodes, the first site of metastasis is in the peripheral sinuses (option C). From this location, they often invade and destroy the remainder of the node.

124-127. The answers are: 124-C, 125-B, 126-D, 127-A. *(Pathology; Hematopoietic/lymphoreticular)*
Low serum haptoglobin levels are a useful monitor of acute intravascular hemolysis. Chronic intravascular hemolysis, in addition to having low serum haptoglobin, also has an increase in urine hemosiderin. Extravascular hemolysis by macrophages leads to an increase in indirect bilirubin.

Hemolytic anemias are pathophysiologically subdivided into intrinsic, where something is wrong with the RBC, or extrinsic, where something outside the RBC is responsible for hemolysis. As previously discussed, the mechanism of hemolysis may be intravascular (within the circulation) or extravascular (RBCs removed by macrophages).

Paroxysmal nocturnal hemoglobinuria (PNH) is an acquired stem-cell disorder characterized by increased sensitivity of hematopoietic cells, including RBCs, platelets, and white blood cells, to intravascular destruction by activation of the classical complement system. The hematopoietic cells have a reduction of decay-accelerating factor on their membrane, which renders them susceptible to complement destruction. Therefore, PNH is an intrinsic hemolytic anemia with intravascular hemolysis (choice C).

The majority of cases of warm autoimmune hemolytic anemia involve the deposition of immunoglobulin G (IgG) or C3b on the RBC membrane with extravascular removal by macrophages, which have Fc receptors for IgG and receptors for C3b. Therefore, warm autoimmune hemolytic anemia is an extrinsic hemolytic anemia with extravascular hemolysis (choice B).

Sickle cell anemia is a hemoglobinopathy with substitution of valine for glutamic acid in the sixth position of the beta chain. The sickle cells are unable to exit the Billroth cords in the spleen and are removed extravascularly by the splenic macrophages. Therefore, sickle cell anemia is an intrinsic hemolytic anemia with extravascular hemolysis (choice D).

Anemia associated with a prosthetic heart valve is the result of intravascular damage of the RBCs after they pass through the prosthetic valve. This produces a microangiopathic hemolytic anemia. Schistocytes (fragmented RBCs) are present in the peripheral blood. This is an example of an extrinsic hemolytic anemia with intravascular hemolysis (choice A).

128-131. The answers are: 128-B, 129-D, 130-C, 131-A. *(Pathology; Hematopoietic/lymphoreticular)* Polycythemia refers to an increase in the red blood cell (RBC) count, hemoglobin (Hbg) concentration, or hematocrit (Hct). In an adult, it is defined as Hct > 55% (male) or > 48% (female). An increased RBC mass, which is a radioactive test reported in ml/kg, is not the same as an increase in the RBC count, which is reported as the number of cells/μl. The RBC mass and RBC count do not always parallel each other. For example, a decrease in plasma volume increases the RBC count and Hct due to hemoconcentration, but this does not affect the RBC mass.

The differential diagnosis of polycythemia is based on whether there is an absolute or relative increase in the RBC mass, the latter referring to a decrease in plasma volume contracting around a normal RBC mass. Next, a determination must be made of whether the polycythemia is an appropriate erythropoietin response, caused by the stimulus of hypoxemia/tissue hypoxia (e.g. chronic obstructive pulmonary disease), an inappropriate response secondary to ectopic secretion of erythropoietin (e.g., renal adenocarcinoma), or an inappropriate autonomous production of RBCs in the bone marrow [e.g., polycythemia rubra vera (PRV)]. An arterial blood gas is the first step in the work-up of polycythemia, because it characterizes the polycythemia as appropriate or inappropriate. RBC mass and plasma volume are also useful because they define whether the polycythemia is absolute, where the RBC mass is increased or relative, and whether the RBC mass is normal. Erythropoietin levels are increased if the polycythemia is appropriate or ectopically secreted. It is normal if the problem is related to hemoconcentration, and it is low if the increased RBC mass is caused by a myeloproliferative disease, such as PRV.

The 62-year-old man with a plethoric face, neutrophilic leukocytosis, thrombocytosis, splenomegaly, and a zero erythrocyte sedimentation rate most likely has PRV. The expected laboratory findings are an increase in RBC mass and plasma volume, a normal oxygen saturation, and a low erythropoietin level (choice B). Splenomegaly is a feature of all the myeloproliferative diseases, which include PRV, chronic myelogenous leukemia, agnogenic myeloid metaplasia, and essential thrombocythemia. The clinical findings in PRV correlate with the 4 H's:

- Hypervolemia is caused by an increase in plasma volume. PRV is the only polycythemia with increased plasma volume.
- Hyperviscosity is caused by the increased RBC mass, which predisposes the patient to thrombotic episodes (the most common cause of death) such as acute myocardial infarction, strokes, and hepatic vein thrombosis.
- Hyperuricemia occurs from increased breakdown of cells and subsequent increase in purine metabolism. Symptomatic gout occurs in 5%–10% cases.
- Histaminemia is caused by increased release of histamine from basophils and mast cells, which are increased in all of the myeloproliferative diseases. Histamine is responsible for vasodilation (plethoric face) and pruritus after taking warm showers. Histamine also stimulates gastric acid secretion, which predisposes the patient to peptic ulcer disease.

The erythropoietin level is low because the total oxygen content is actually increased (oxygen content = 1.34 × Hgb × oxygen saturation × PaO_2), so there is no stimulus for erythropoietin. The erythrocyte sedimentation rate is zero because the crowding of the RBCs prevents the settling of the RBCs.

The 72-year-old woman with a 4-cm cystic renal mass, cannon-ball metastases in the lung and polycythemia most likely has a renal adenocarcinoma with inappropriate production of erythropoietin. Renal adenocarcinoma and other disorders characterized by inappropriate production of erythropoietin (including hepatocellular carcinoma, cerebellar hemangioblastoma, uterine leiomyomas, and renal cysts) have an increased RBC mass, normal plasma volume, normal oxygen saturation, and increased erythropoietin levels (choice D).

A 4-year-old child with uncorrected tetralogy of Fallot would likely have hypoxemia (low PaO_2) caused by a right-to-left shunt through a ventricular septal defect. Hypoxemia stimulates the release of erythropoietin resulting in an appropriate polycythemia with an increase in RBC mass, normal plasma volume, low oxygen saturation, and an increased erythropoietin concentration (choice C).

A 22-year-old female marathon runner who is dehydrated from lack of water supplementation during the race is hemoconcentrating her RBCs. This is a relative polycythemia, which is characterized by a normal RBC mass, low plasma volume, normal oxygen saturation, and normal erythropoietin concentration (choice A). This constellation of findings is also characteristic of stress polycythemia, which is typically seen in obese males with hypertension.

132-135. The answers are: 132-A, 133-C, 134-A, 135-B. *(Physiology; Cell biology)*
An increase in cytosolic calcium ion (Ca^{2+}) is required for excitation–contraction coupling in both skeletal and cardiac muscle. In skeletal muscle, the source of Ca^{2+} is the sarcoplasmic reticulum. In cardiac muscle, the source of Ca^{2+} is both the extracellular fluid and the sarcoplasmic reticulum. Thus, the strength of contraction of cardiac muscle depends on the extracellular fluid Ca^{2+} concentration. The influx of Ca^{2+} accounts for approximately 1% of the total increase in cytosolic Ca^{2+}, and is called the "trigger Ca^{2+}." Blockade of the fast sodium ion channels prevents an action potential from occurring in skeletal muscle, but does not prevent the Ca^{2+} current from developing in cardiac muscle.

136-137. The answers are: 136-C, 137-A. *(Pathology; Hematopoietic/lymphoreticular)*
Anemia in a patient taking methyldopa (Aldomet) is most likely a drug-induced autoimmune hemolytic anemia (AIHA). The direct Coombs test is the drug of choice to document the autoimmune nature of the anemia. An acute onset of anemia precipitated by eating fava beans is caused by the Mediterranean variant of glucose-6-phosphate dehydrogenase (G6PD) deficiency, which can be verified by a Heinz body preparation.

Drug-induced AIHA accounts for 16%–18% of all cases of AIHA and includes penicillin type, innocent-bystander mechanism, and a true type (see Table). An example of a true autoimmune hemolytic anemia is produced by methyldopa (Aldomet). The drug alters Rh antigens on the red blood cell (RBC) membrane, resulting in the formation of an immunoglobulin G (IgG) autoantibody against the Rh antigen. The RBC is removed extravascularly, which is a type II hypersensitivity reaction. The direct Coombs test is positive in 10% of patients on methyldopa, but only 1%–2% of patients actually develop a hemolytic anemia.

	Direct Coombs Test	Antibody	Hemolysis	Drug(s)
Hapten model	+	IgG	Extravascular	Penicillin
Immune complex (innocent bystander)	+	IgM	Intravascular	Quinidine, isoniazid, phenacetin
Autoantibody	+	IgG	Extravascular	Methyldopa

The sine qua non for diagnosing AIHA is the direct and indirect Coombs test (see diagram). The direct Coombs test detects antibody or complement on the surface of RBCs, whereas the indirect Coombs test identifies specific IgG antibodies in the serum. Coombs reagent is formed by injecting human IgG and C3b into rabbits. The rabbits develop antibodies against these components, which are subsequently removed and purified. In a direct Coombs test, Coombs reagent containing a mixture of anti-IgG or C3b antibodies is added to a test tube containing the patient's RBCs. If IgG or C3b are on the surface of the RBCs, the IgG/C3b antibodies cause the RBCs to clump as they form a bridge between subjacent cells. The indirect Coombs test is a two-step procedure. IgG antibodies in the patient's serum are first reacted with type O blood cells, which have all the known antigens on their surface. The Coombs reagent is added to determine if any antibodies are attached to antigens on the test cell (e.g., anti D against D antigen on type O test cells). In most cases, the serum antibodies have a specificity against a known antigen (e.g., anti-D, anti-Kell) that may be identified in the blood bank by using a panel of test RBCs of known antigenic composition and reacting the patient serum against them. Through a process of elimination, the identity of the antibody is usually discovered.

Coombs test

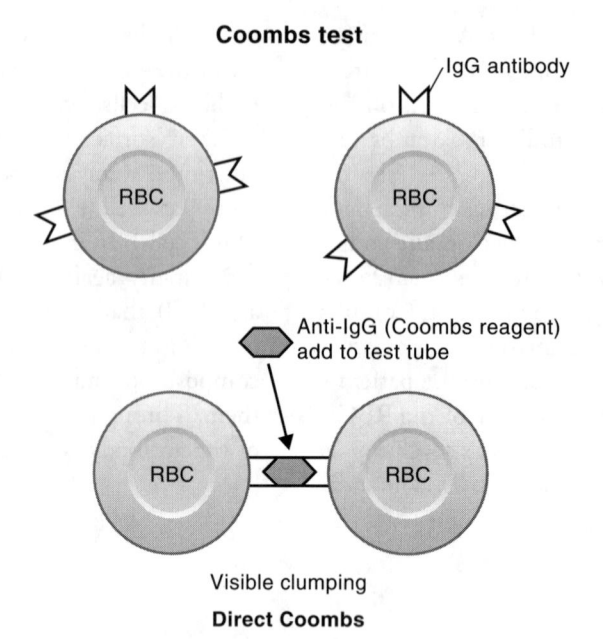

Direct Coombs

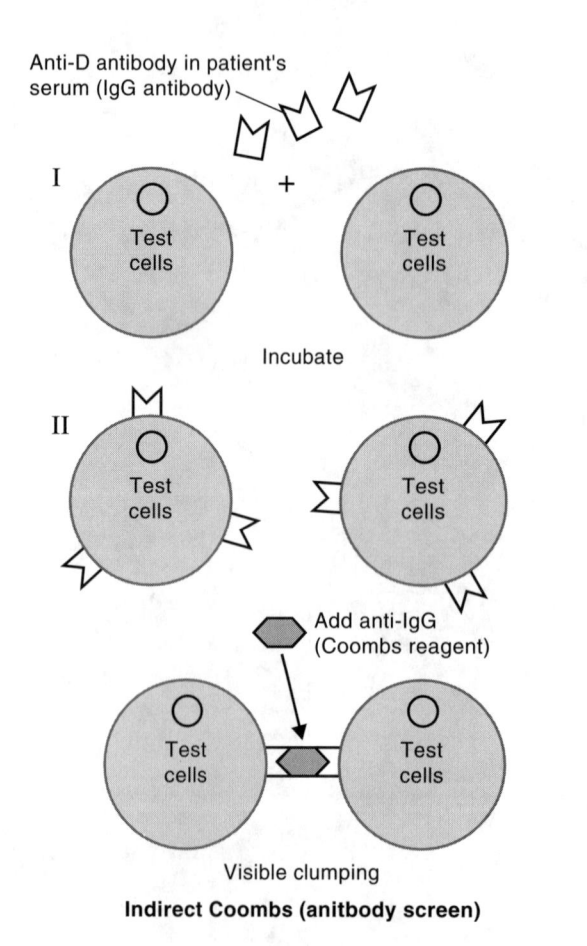

Indirect Coombs (anitbody screen)

G6PD deficiency is the most common hereditary enzyme deficiency associated with a hemolytic anemia. It is a sex-linked recessive disease, which is transmitted to the male by a female carrier, who received the abnormal gene from her affected father. G6PD is the key enzyme of the pentose–phosphate shunt (see diagram).

$$\text{Denatures Hgb} \rightarrow \text{Heinz bodies}$$
$$\uparrow$$
$$\text{Oxidant stress or infection} \rightarrow H_2O_2 \text{ ---//---> } (2) H_2O$$
$$\text{(glutathione)} \downarrow \mathbf{GSH} \text{ <--//---- } GSSG$$
$$NADP \text{ ---//----> } NADPH$$
$$\text{Glucose 6-phosphate ---------//--------> 6-phosphogluconate}$$
$$\nearrow$$
$$\textit{Glucose-6-phosphate dehydrogenase}$$

An absent or defective enzyme with a short half-life ultimately results in a deficiency of glutathione (GSH), which is necessary to detoxify the hydrogen peroxide (H_2O_2) that accumulates in RBCs. Peroxide is a product of oxidant injury and infections. If left unneutralized, it denatures hemoglobin (Hgb), which forms Heinz bodies. Like reticulocytes, they require a supravital stain for identification in the peripheral blood. Macrophages in the spleen destroy the RBCs while attempting to remove the Heinz bodies. The Heinz bodies closely applied to the RBC membrane are removed by the macrophages producing a bite cell. G6PD deficiency is an example of an intrinsic hemolytic anemia, since something is wrong with the RBC, and the mechanism of hemolysis is primarily extravascular.

The two main types of G6PD deficiency are the A- variant in blacks and the Mediterranean variant, in Greeks and Italians. The A- variant occurs in 10% of blacks and is characterized by a defective enzyme with a decreased half-life. The enzyme is only present in older RBCs, so only the older cells are destroyed in acute hemolytic episodes. The Mediterranean variant has reduced synthesis of the enzyme, as well as production of a defective enzyme with a short half-life in both young and old RBCs, thus making the anemia more severe than the A- variant. Oxidant drugs and infection are the most common precipitants of acute hemolysis, which usually has a lag phase of 2–3 days. Oxidant drugs include antibiotics (sulfur-containing drugs), antimalarials (primaquine, not chloroquine), and antipyretics (acetanilid, not aspirin or acetaminophen). Fava beans may also precipitate hemolysis in the Mediterranean variant, thus the term favism. G6PD deficiency offers some protection against falciparum malaria. The Heinz body preparation is positive during active hemolysis. An enzyme assay is only abnormal in the quiescent period because active hemolysis removes RBCs with defective or deficient enzyme, thus leaving normal RBCs behind with normal G6PD. Treatment is symptomatic. Patients should avoid drugs that are known to precipitate hemolysis. Splenectomy is of no benefit.

138-142. The answers are: 138-A, 139-B, 140-E, 141-C, 142-D. *(Physiology; Cardiovascular)*
Cardiac muscle responds to stretch by developing a greater force of contraction and ejecting a larger stroke volume (A to B). End-diastolic volume is an index of the degree of stretch prior to contraction (preload). A decrease in cardiac filling (D to B), which might occur during rapid pacing of the heart or with venodilating agents, decreases preload and, therefore, stroke volume. An increase in the force of contraction with no change in preload (A to C) indicates an increase in contractility (positive inotropic agents), whereas negative inotropic agents (β-1 blockers and calcium-channel blockers) reduce the stroke volume for the same degree of ventricular filling. During moderate exercise, preload increases if the rate of filling (venous return) is greater than the rate of ventricular emptying (heart rate). Contractility also increases (sympathetic activation), so that during exercise the increase in stroke is the result of moving from A to D.

143-145. The answers are: 143-C, 144-B, 145-A. *(Physiology; Pulmonary/respiratory)*

During a normal inspiration, air enters the lungs because the pressure exerted by all of the gases in the intra-alveolar space falls below atmospheric pressure, which is set to zero. Pressure returns to 0 at the end of inspiration. If inspiration is accomplished by using a mechanical ventilator (i.e., air is forced into the lungs) intra-alveolar pressure during and at the end of inspiration will be positive. If expiration is passive, intra-alveolar pressure will return to atmospheric pressure.

In positive end-expiratory pressure (PEEP) breathing, the lungs are mechanically ventilated in a manner that prevents intra-alveolar pressure from returning to 0 at the end of expiration. This ensures that the lungs remain relatively more inflated at the end of expiration than is usual.

At functional residual capacity (prior to the onset of inspiration), intrapleural pressure is negative (i.e., subatmospheric). The natural tendency of the lungs is to rest at a smaller volume while the chest wall rests at a larger volume. Thus, a distending force is applied to the intrapleural space, causing its pressure to become subatmospheric. When the respiratory muscles contract, an even greater distending force is exerted and intrapleural pressure becomes more negative.

146-155. The answers are: 146-B, 147-G, 148-C, 149-E, 150-A, 151-J, 152-F, 153-D, 154-H, 155-I. *(Pathology; Renal/urinary)*

A simplified version of the Henderson-Hasselbalch equation is useful in answering these questions:

$$pH = HCO_3^-/P_{CO_2}$$

Increased pH (> 7.45) indicates alkalosis, and decreased pH (< 7.35) indicates acidosis. Normally, the ratio $HCO_3^-:P_{CO_2}$ is 20:1, which equals a pH of 7.40. A nonlogarithmic way of expressing the same relationship is as follows:

$$H^+ \text{ concentration} = 24 (P_{aCO_2}) / HCO_3^-$$

Increased H^+ ions indicates acidosis, and a decrease indicates alkalosis.

Altered HCO_3^- indicate a metabolic disorder: increased HCO_3^- (> 28 mEq/L) is metabolic alkalosis, and decreased HCO_3^- (< 22 mEq/L) is metabolic acidosis. Altered arterial P_{CO_2} indicates a respiratory disorder: increased P_{CO_2} (> 44 mm Hg) is respiratory acidosis, and decreased P_{CO_2} (< 33 mm Hg) is respiratory alkalosis. Note that if either HCO_3^- increases or P_{CO_2} decreases, pH becomes alkalotic (> 7.45). If HCO_3^- decreases or P_{aCO_2} increases, pH becomes acidotic.

Compensation is the body's attempt to bring the ratio of $HCO_3^-:P_{CO_2}$ back to 20:1 (pH of 7.40). Therefore, if the numerator (HCO_3^-) increases as a primary disorder (i.e., metabolic alkalosis), then the denominator (P_{CO_2}) must increase (i.e., respiratory acidosis) as compensation. If P_{CO_2} decreases as a primary disorder (i.e., respiratory alkalosis), then HCO_3^- must decrease (i.e., metabolic acidosis). Note that compensation always moves in the same direction as the primary disorder. Furthermore, compensation for primary metabolic problems must be respiratory, and vice versa. The terms used for compensation include:

- Uncompensated, in which compensation does not go outside of its reference interval
- Partially compensated, in which compensation is outside the normal reference interval
- Fully compensated, in which compensation brings pH back into the normal range. This does not occur in the United States. Full compensation occurs in people living in the Andes Mountains, where, at high elevations, chronic respiratory alkalosis is associated with normal pH.

Blood gas disorders may be simple or mixed. In simple blood gas disorders only one disorder is present, with or without compensation. In mixed acid–base disorders, there is a combination of two or more acid–base disorders at the same time.

The key to interpreting acid–base disorders is to know what to expect with various clinical situations. The history should be considered first, the conditions that should be present should be defined, and then laboratory results should be evaluated. A method for interpreting laboratory results is as follows:

1. Evaluate arterial P_{CO_2} and HCO_3^- levels, looking for indications of respiratory or metabolic acidosis or alkalosis.
2. Evaluate pH level. pH always reflects the primary disorder, and can be used as a guide to determine the primary disorder versus compensation. For example, if the pH is acid, respiratory acidosis is the primary disease, and metabolic alkalosis is the compensation.

3. Determine whether the disorder is uncompensated, partially compensated, or fully compensated (the latter does not occur in a simple disorder); and whether it is simple or mixed. Clues for mixed disorders include an acid–base disorder with a normal pH, because pH never corrects into the normal range with a single acid–base disorder, and an exaggerated pH (i.e., extreme acidosis or alkalosis).

The medical student with an acute anxiety attack and a respiratory rate of 26/min would most likely have a respiratory alkalosis resulting from hyperventilation. The values in choice B reflect an uncompensated acute respiratory alkalosis.

The pregnant woman with excessive vomiting would most likely have a mixed acid–base disorder. Pregnant women normally have respiratory alkalosis due to the stimulatory effect of progesterone on the respiratory center. Hyperemesis gravidarum is associated with metabolic alkalosis. The combination of respiratory and metabolic alkalosis produces an exaggerated alkalemia, and normal to slightly decreased arterial P_{CO_2} and HCO_3^-. Choice G correlates best with the patient's condition, because pH is highly alkaline, arterial P_{CO_2} is consistent with respiratory alkalosis, and HCO_3^- is compatible with metabolic alkalosis.

The 5-year-old child with polycystic kidney disease and chronic renal failure most likely has an increased anion gap metabolic acidosis resulting from retention of organic acids, and respiratory alkalosis as compensation. Choice C best represents this clinical picture. pH is acid, arterial P_{CO_2} exhibits respiratory alkalosis, and HCO_3^- reveals metabolic acidosis.

The 45-year-old woman with chronic renal failure who is on a loop diuretic for heart failure would most likely have a mixed acid–base disorder. Chronic renal failure results in an increased anion gap metabolic acidosis, and a loop diuretic produces a metabolic alkalosis. The combination of metabolic acidosis and metabolic alkalosis should normalize pH, arterial P_{CO_2}, and HCO_3^-, so choice E is the best answer.

The semi-comatose man with heroin overdose would most likely have a respiratory acidosis resulting from depression of the respiratory center in the medulla. This leads to hypoventilation and retention of CO_2, and either a normal or slightly elevated HCO_3^-, representing compensatory metabolic alkalosis. The best choice is A, because pH is acidemic, arterial P_{CO_2} exhibits respiratory acidosis, and HCO_3^- is in the normal range, indicating an uncompensated state.

The 65-year-old man with cardiorespiratory arrest would most likely have a mixed acid–base disorder. Absence of breathing results in respiratory acidosis, and cardiac standstill produces tissue hypoxia with lactic acidosis from anaerobic glycolysis. Choice J best fits the expected findings: pH is extremely acidemic, arterial P_{CO_2} indicates respiratory acidosis, and HCO_3^- is compatible with metabolic acidosis.

The woman with disabling rheumatoid arthritis who is on toxic doses of salicylates most likely has a mixed disorder resulting from salicylate intoxication. Salicylates overstimulate the respiratory center, causing respiratory alkalosis, and salicylic acid itself produces an increased anion gap metabolic acidosis. This should normalize the pH, and lower both arterial P_{CO_2} and HCO_3^-. Choice F best fits these findings. The pH is normal, which should never occur in a simple disorder.

The man with pyloric obstruction secondary to chronic duodenal ulcer disease would most likely have a metabolic alkalosis with compensatory respiratory acidosis. This is best represented by choice D.

The female smoker with chronic bronchitis would most likely have chronic respiratory acidosis caused by the retention of CO_2. Because the condition is chronic, there is an opportunity for renal compensation via the reabsorption of HCO_3^-, producing a compensatory metabolic alkalosis. This should bring the pH near—but not into—normal range. (Note that in the previous example of heroin overdose, respiratory acidosis is more acute because the kidney would not have enough time to reabsorb HCO_3^-.) Choice H best correlates with chronic respiratory acidosis. pH is slightly acidic, arterial P_{CO_2} is consistent with respiratory acidosis, and HCO_3^- represents compensatory metabolic alkalosis.

The woman with long-term chronic obstructive pulmonary disease and vomiting would be expected to have a mixed disorder. Patients with chronic obstructive pulmonary disease retain CO_2, resulting in chronic respiratory acidosis. Vomiting results in metabolic alkalosis. These combine to produce a normal pH and markedly elevated levels of arterial P_{CO_2} and HCO_3^- (Choice I).

156-159. The answers are: 156-F, 157-D, 158-A, 159-G. *(Pharmacology; Cardiovascular)*
Digitalis glycosides have no direct effect on heart rate but increase contractility moderately. Their small vasoconstrictive effect is usually masked by a larger reduction in reflex sympathetic and angiotensin-mediated vasoconstriction.

Isoproterenol is a potent β-adrenoceptor agonist and markedly increases heart rate and contractility while causing vasodilation due to $β_2$-activation.

Dopamine and dobutamine have similar effects on hemodynamics. Dopamine may increase renal blood flow more than dobutamine because of its specific activation of renal dopaminergic receptors. Dopamine also produces dose-dependent vasoconstriction and activation of α-adrenergic receptors, whereas dobutamine has little vasoconstrictive effect.

Phenylephrine has little effect on the heart, though a direct stimulating effect can be demonstrated in isolated tissues. It produces vasoconstriction. Reflex bradycardia may occur after large doses.

Epinephrine usually has both vasoconstrictive and vasodilating effects and usually increases systolic pressure and decreases diastolic pressure. It also increases heart rate and contractility. As a drug, norepinephrine produces vasoconstriction and reflex bradycardia, particularly at larger doses. It also increases cardiac contractility.

160-163. The answers are: 160-D, 161-A, 162-B, 163-C. *(Pharmacology; Pharmacodynamic and pharmacokinetic processes)*
Bromocriptine is a potent dopamine-receptor agonist and is used to activate pituitary dopamine receptors to mediate inhibition of prolactin secretion. It is also used in Parkinson disease.

Ergonovine, ergotamine, and methysergide are all partial agonists at serotonin $5-HT_2$ receptors, and act to constrict most blood vessels. Ergonovine has the most selective effect on uterine smooth muscle and is usually preferred for obstetric uses, including control of postpartum hemorrhage, partly because of its lower toxicity in this setting. Ergotamine is usually preferred for terminating acute migraine attacks. Methysergide has been used for prophylaxis of migraine, despite its extensive chronic toxicity.

Test 2

QUESTIONS

DIRECTIONS:
Each of the numbered items or incomplete statements in this section is followed by answers or by completions of the statement. Select the ONE lettered answer or completion that is BEST in each case.

1. In the heart, the septomarginal trabecula contains which of the following structures?

   (A) The atrioventricular node
   (B) The sinoatrial node
   (C) The atrioventricular bundle
   (D) Conducting fibers from the left bundle branch
   (E) Fibers from the right bundle branch

2. A chest wound results in bleeding into the pericardial space. Blood accumulates between which of the following structures?

   (A) Fibrous pericardium and parietal pericardium
   (B) Parietal pericardium and visceral pericardium
   (C) Visceral pericardium and epicardium
   (D) Visceral pericardium and myocardium
   (E) Epicardium and myocardium

3. If a knife enters the fifth intercostal space, immediately to the left of the sternum, the structure it is most likely to penetrate is the

   (A) right atrium
   (B) right ventricle
   (C) left atrium
   (D) left ventricle
   (E) left lower lobe of the lung

4. Shortly after birth, it is noted that an infant becomes cyanotic while crying. The developmental defect that this infant most likely has is

   (A) primum type atrial septal defect
   (B) secundum type atrial septal defect
   (C) ventricular septal defect
   (D) patent ductus arteriosus
   (E) tetralogy of Fallot

5. Which of the following scenarios for the development of the interatrial septum in the heart is true?

   (A) The ostium primum develops as an opening within the septum primum
   (B) The ostium primum closes before the ostium secundum develops
   (C) The septum secundum develops on the right side of the septum primum and covers the ostium secundum
   (D) The ostium primum gets larger as the septum primum grows
   (E) When the ostium primum closes, blood can no longer pass from the right atrium to the left atrium

6. Which of the following statements concerning respiration is true?

   (A) Contraction of the diaphragm is controlled by the vagus nerve
   (B) Elevation of the ribs increases the anteroposterior and transverse dimensions of the thoracic cavity
   (C) Air enters the pleural cavities through the bronchi
   (D) Contraction of the diaphragm reduces pressure in the abdominal cavity during inspiration
   (E) Forced inspiration can be assisted by contraction of the abdominal wall muscles

7. While passing an endoscope through a patient's esophagus, a physician observes the pulsation of an artery through the wall of the esophagus. The artery that is in direct contact with the esophagus is the

   (A) ascending aorta
   (B) aortic arch
   (C) pulmonary trunk
   (D) left pulmonary artery
   (E) brachiocephalic artery

8. As a result of trauma, a patient has a severely damaged larynx and a compromised airway. The physician decides to perform a tracheostomy in order to create an airway, and chooses to enter the trachea in the midline at the level of the fifth or sixth tracheal ring. At this level, a blood vessel that one would be concerned about encountering would be the

(A) superior thyroid artery
(B) inferior thyroid artery
(C) thyroidea ima artery
(D) superior thyroid vein
(E) middle thyroid vein

9. A 37-year-old man in diabetic ketoacidosis has yellowish papules scattered over his trunk and extremities. A tube of plasma that is collected from the patient and refrigerated overnight at 4°C develops a turbid supranate and infranate. This is consistent with which of the following types of hyperlipoproteinemia?

(A) Type I
(B) Type IIa
(C) Type III
(D) Type IV
(E) Type V

10. Women are more likely than men to

(A) smoke cigarettes
(B) abuse alcohol
(C) experience spinal cord injuries
(D) take illegal drugs
(E) engage in risk-taking behavior

11. The most common reason for ordering a stress electrocardiogram is to

(A) evaluate the cause of chest pain
(B) document the presence of acute myocardial infarction
(C) diagnose acquired valvular diseases
(D) evaluate the extent of a myocardial infarction
(E) rule out a cardiomyopathy

12. Which of the following test procedures is most useful in measuring ejection fraction, detecting a pericardial effusion, and differentiating systolic from diastolic heart failure?

(A) Standard electrocardiogram
(B) Coronary angiography
(C) Stress electrocardiogram
(D) Echocardiography
(E) Chest x-ray

13. Which of the following statements regarding the pathophysiology of a transmural acute myocardial infarction involving the left ventricle is true?

(A) Compliance of damaged myocardial tissue is decreased
(B) Catecholamine release results in increased stroke volume
(C) Contractility of damaged tissue is increased
(D) Adenosine triphosphate (ATP) stores in cardiac muscle are maintained by oxidative phosphorylation
(E) Reperfusion reverses the damage incurred by injured muscle surrounding the necrotic area

14. A Valsalva maneuver would be expected to have which one of the following physiologic effects?

(A) Decrease cerebrospinal fluid pressure
(B) Diminish intrathoracic pressure
(C) Increase cardiac output
(D) Increase central venous pressure
(E) Increase venous return to the heart

15. An 8-year-old child has a history since early childhood of malabsorption, ataxia, acanthocytes in the peripheral blood, and very low cholesterol and triglyceride levels. In addition, the patient has been developing progressive, bilateral, concentric contraction of the visual fields and loss of central vision. The underlying pathogenesis of this patient's disease is

(A) a degenerative disease involving the cerebellum
(B) a defect in the synthesis of apolipoprotein B
(C) degeneration of the posterior columns, spinocerebellar tracts, and corticospinal tracts
(D) an absence of high-density lipoproteins
(E) a slow virus disease related to immunizations

16. Atherosclerosis of the aorta in an elderly patient would most likely have which one of the following physiologic effects?

(A) Reduce systolic blood pressure
(B) Increase diastolic blood pressure
(C) Increase pulse pressure
(D) Increase capacitance in the aorta
(E) Have no effect on blood pressure

17. Which of the following factors would have the greatest effect on increasing total peripheral resistance in vessels?

(A) Increased hematocrit
(B) Increased vessel length
(C) Decreased hematocrit
(D) Decreased vessel radius
(E) Increased histamine concentration

18. A 45-year-old executive has marked elevation of triglyceride (hypertriglyceridemia). Chylomicrons are absent but a turbid infranate forms in the plasma after refrigeration at 4°C. Based on these findings, the mechanism most likely responsible for her hypertriglyceridemia is

(A) absent apolipoprotein B
(B) absent low-density lipoprotein (LDL) receptors
(C) defective apolipoprotein E
(D) decreased catabolism of very-low-density lipoproteins (VLDL)
(E) absent high-density lipoproteins (HDL)

19. A 40-year-old man with a history of two previous acute myocardial infarctions and a family history of premature coronary artery disease has a nodular mass in his Achilles tendon. The patient's lipid profile would be expected to exhibit

(A) an increase in chylomicrons
(B) high-density lipoprotein (HDL) level greater than 60 mg/dl
(C) triglyceride level greater than 1000 mg/dl
(D) low-density lipoprotein (LDL) level greater than 160 mg/dl
(E) plasma with a turbid supranate and infranate

20. A high cytoplasmic staining intensity for alkaline phosphatase in a peripheral blood smear indicates the presence of

(A) basophils
(B) monocytes
(C) neutrophils
(D) eosinophils
(E) promyelocytes

21. Which of the following sets of elements persist through the respiratory bronchioles?

(A) Glands and cilia
(B) Goblet cells and cilia
(C) Smooth muscle and cilia
(D) Smooth muscle and cartilage
(E) Elastic fibers and goblet cells

22. Fenestrated capillaries are typically found in which of the following structures or systems of the body?

(A) Lung
(B) Liver
(C) Muscle
(D) Endocrine glands
(E) Central nervous system

23. The accompanying photograph depicts palpable, erythematous lesions on the lower extremity of a 52-year-old man. The patient recently had a flu-like syndrome for which he took penicillin. These lesions would most likely be associated with which one of the following conditions?

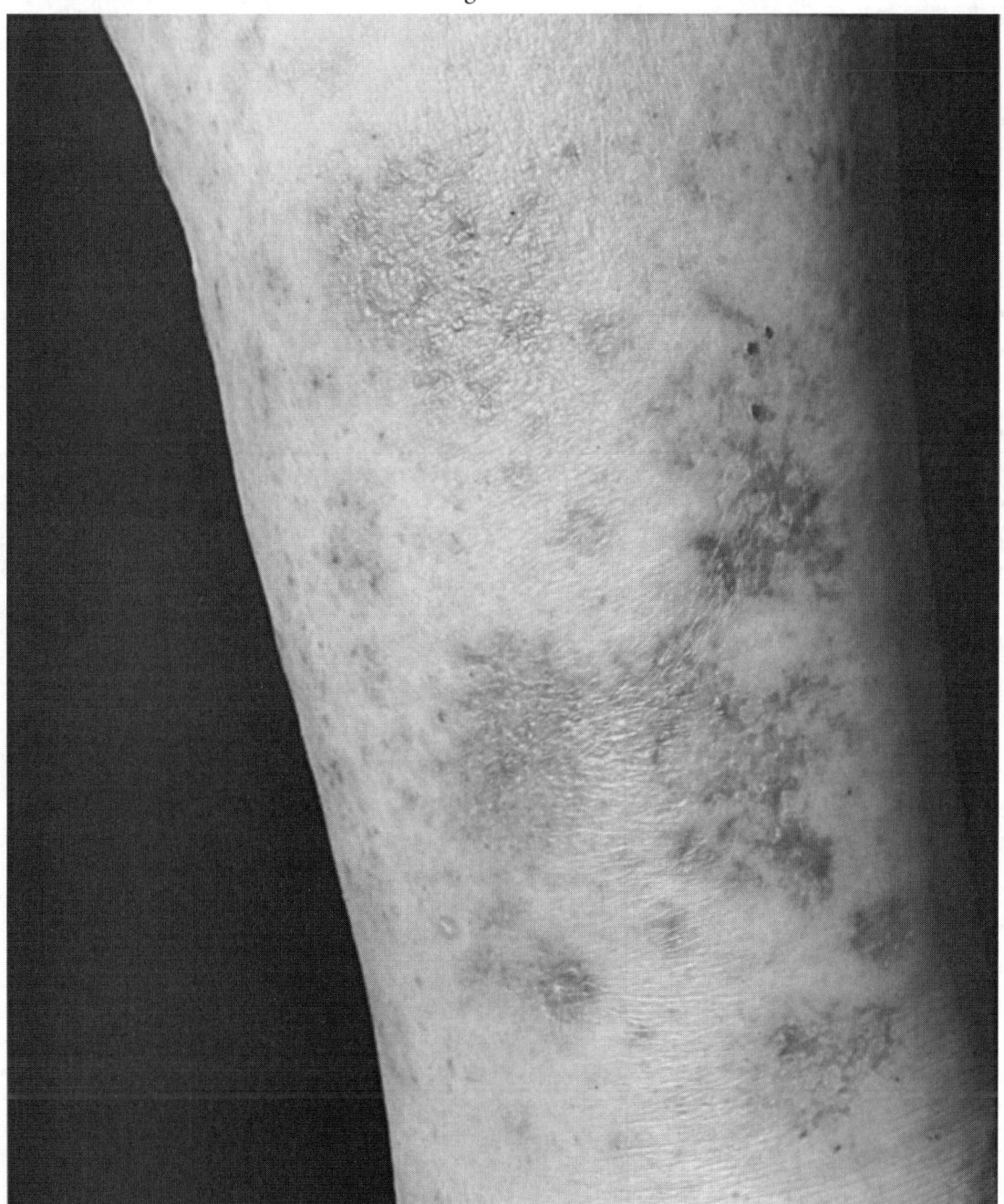

(A) Wegener granulomatosis
(B) Polyarteritis nodosa
(C) Giant cell arteritis
(D) Leukocytoclastic angiitis
(E) Kawasaki disease

24. The accompanying schematic depicts a carotid pulse and its relationship to inspiration. Which of the following conditions best correlates with the carotid pulse finding?

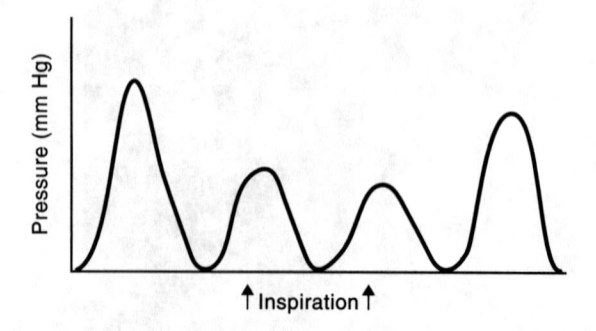

(A) Aortic stenosis
(B) Essential hypertension
(C) Pericardial effusion
(D) Coarctation of the aorta
(E) Mitral valve prolapse

25. The photograph depicts nonpruritic, reddish-brown truncal lesions on a 26-year-old homosexual man. The biopsy would be expected to show

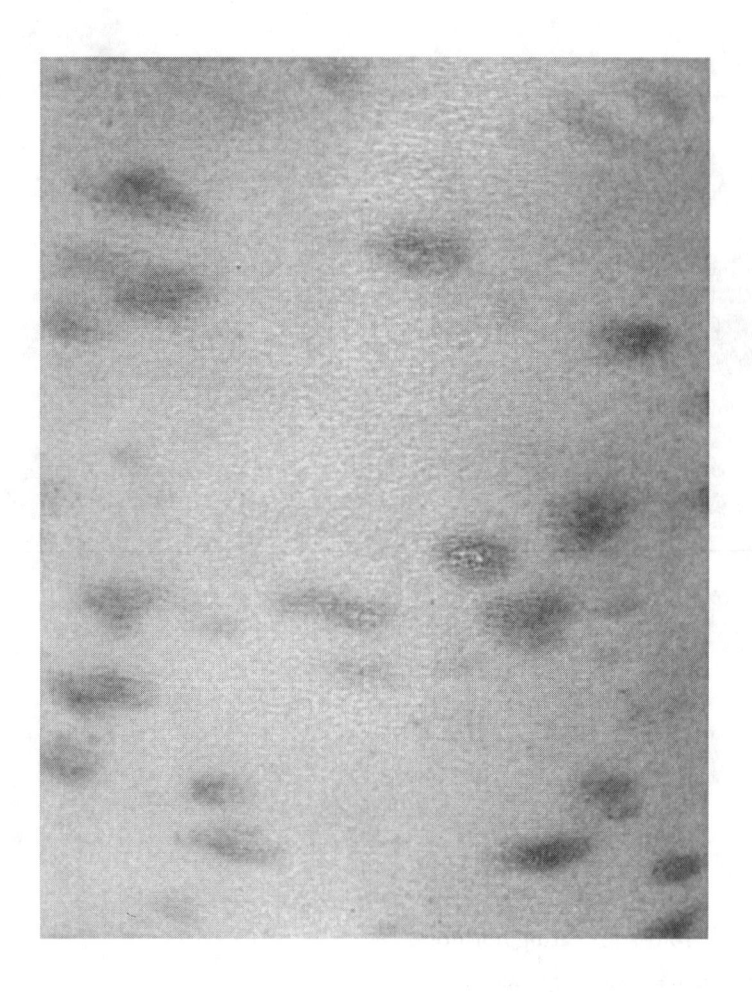

(A) a malignant B-cell proliferation in the epidermis and subcutaneous tissue
(B) granulation tissue with an atypical inflammatory infiltrate
(C) a superficial dermatophyte infestation limited to the stratum corneum
(D) a malignant vascular proliferation with extensive hemosiderin deposits
(E) metastatic foci of a highly vascular tumor of epithelial origin

88 Body Systems Review I: Hematopoietic/Lymphoreticular, Respiratory, Cardiovascular

26. This 62-year-old man most likely has which one of the following conditions?

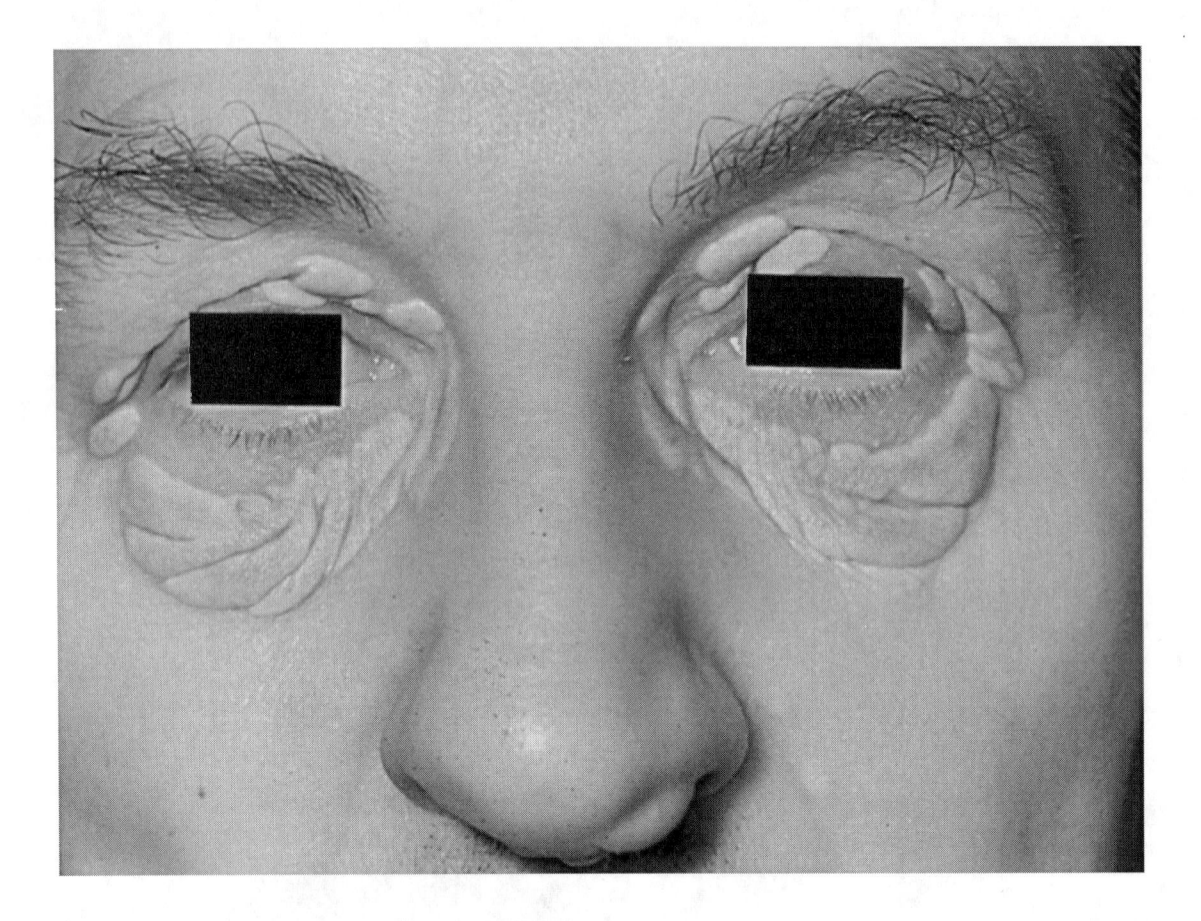

(A) Low level of serum thyroid-stimulating hormone
(B) Positive serum antinuclear antibody test
(C) Elevated low-density lipoprotein
(D) Elevated very-low-density lipoprotein
(E) Turbid plasma sample

27. The peripheral blood smear depicted in the micrograph exhibits

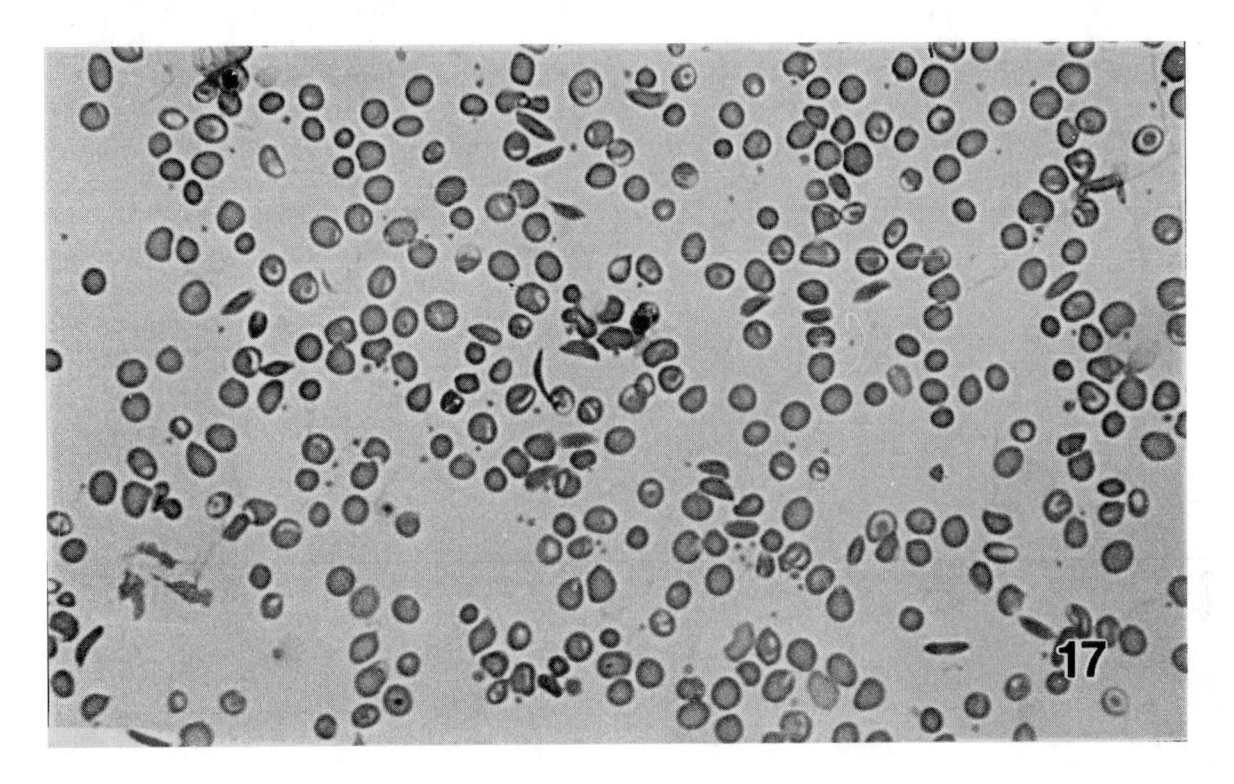

(A) reticulocytes
(B) poikilocytosis
(C) a shift to the left
(D) hypersegmentation
(E) Howell-Jolly bodies

28. A 48-year-old man has a lipid profile with the following results: serum cholesterol = 350 mg/dl; serum high-density lipoprotein = 20 mg/dl; serum triglyceride = 450 mg/dl. The calculated low-density lipoprotein level in this patient is

(A) 180 mg/dl
(B) 220 mg/dl
(C) 240 mg/dl
(D) 280 mg/dl
(E) 330 mg/dl

29. Following are two platelet aggregation curves recorded with a densitometer after the addition of adenosine diphosphate (ADP), a potent aggregating agent. Platelet aggregation curve A is a normal response, and curve B is an abnormal response from a patient who had a recent myocardial infarction. Phase I is normally due to the aggregating properties of ADP on the platelets when it is added to the test tube. Phase II is the endogenous response of the platelets after releasing their own ADP. Aggregation of platelets or the lack of same is recorded with a densitometer.

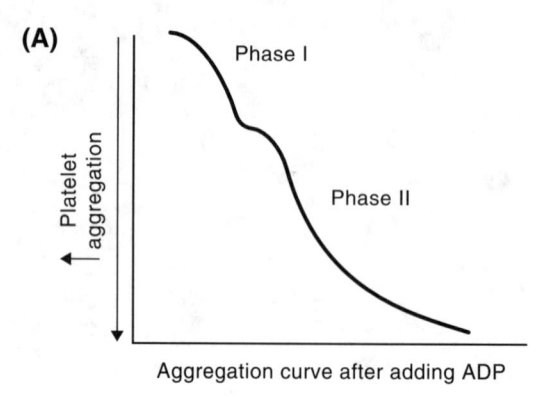

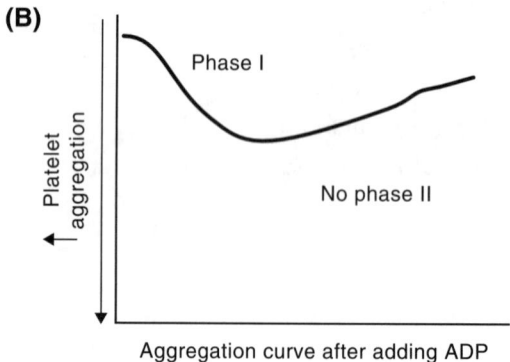

The abnormal ADP platelet aggregation (curve B) would be expected in a patient

(A) with von Willebrand disease
(B) with thrombocytopenia
(C) taking aspirin
(D) with absence of the von Willebrand factor receptor on platelets
(E) with absence of the fibrinogen receptors (IIb/IIIa) on platelets

30. A 35-year-old woman presents with fever, petechia, oliguria, and seizure activity. The results of her coagulation studies show normal prothrombin and partial thromboplastin times, a normal fibrinogen, absence of fibrinogen degradation products, and the presence of thrombocytopenia. The patient has a normocytic anemia with a corrected reticulocyte count greater than 3%. Schistocytes are noted in the peripheral blood. The serum blood urea nitrogen and creatinine levels are elevated. The patient most likely has

(A) disseminated intravascular coagulation
(B) idiopathic thrombocytopenic purpura
(C) autoimmune thrombocytopenia
(D) thrombotic thrombocytopenic purpura
(E) von Willebrand disease

31. A 17-year-old nonsmoking man presents with deep venous thrombosis and a pulmonary embolus. This is his second hospital admission for a similar presentation 1 year ago. The patient has a positive family history for recurrent pulmonary emboli. The partial thromboplastin time remains normal after initiation of heparin therapy. Another dose of heparin produces similar results. The patient most likely has

(A) von Willebrand disease
(B) antithrombin III deficiency
(C) antibodies against heparin
(D) a qualitative platelet disorder
(E) protein C deficiency

32. Vitamin K levels would most likely be normal in a

(A) 3-day-old infant delivered outside the hospital
(B) patient taking a broad spectrum antibiotic
(C) patient with celiac disease
(D) patient with bile salt deficiency
(E) woman taking birth control pills

33. The figure shows a peripheral blood finding from a 3-year-old child who has abdominal colic and growth retardation. His complete blood cell count exhibits a moderately severe microcytic anemia with a normal white cell and platelet count. Which of the following tests would be most useful in confirming the diagnosis in this patient?

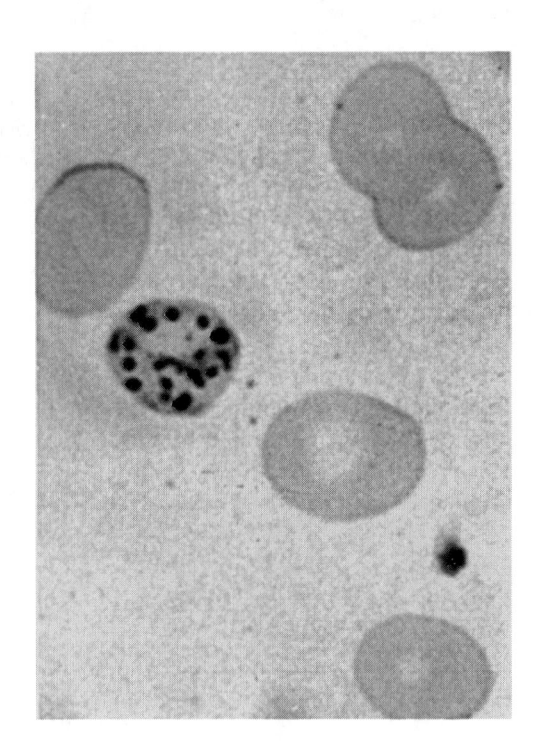

(A) Serum iron, total iron binding capacity, and percent saturation
(B) Bone marrow aspirate and biopsy
(C) Hemoglobin electrophoresis
(D) Blood lead level
(E) Heinz body preparation

34. A 25-year-old black woman has a moderately severe anemia. She would most likely have a history of

(A) an abnormal hemoglobin electrophoresis
(B) inflammatory bowel disease
(C) peptic ulcer disease
(D) hematuria
(E) menorrhagia

35. Reduced hemoglobin synthesis is the primary mechanism for anemia due to

(A) folate deficiency
(B) anemia of chronic disease
(C) congenital spherocytosis
(D) myeloproliferative disease
(E) leukemia

36. Which of the following ethnic groups would most likely have a proportionately greater number of normal hemoglobin electrophoreses than the other groups listed?

(A) Asians
(B) Africans
(C) Greeks
(D) Italians
(E) Hispanics

37. Which one of the following parasitic diseases would most likely be associated with a normal hemoglobin and hematocrit?

(A) Falciparum malaria
(B) Ancyclostomiases
(C) Diphyllobothriasis
(D) Babesiosis
(E) Ascariasis

38. The figure shows a peripheral smear from a 28-year-old white woman with symptomatic gallbladder disease and a mild normocytic anemia with an elevated corrected reticulocyte count. The patient has a family history of splenectomies for "anemia." The patient has mild splenomegaly and is not jaundiced. Further laboratory testing would be expected to show

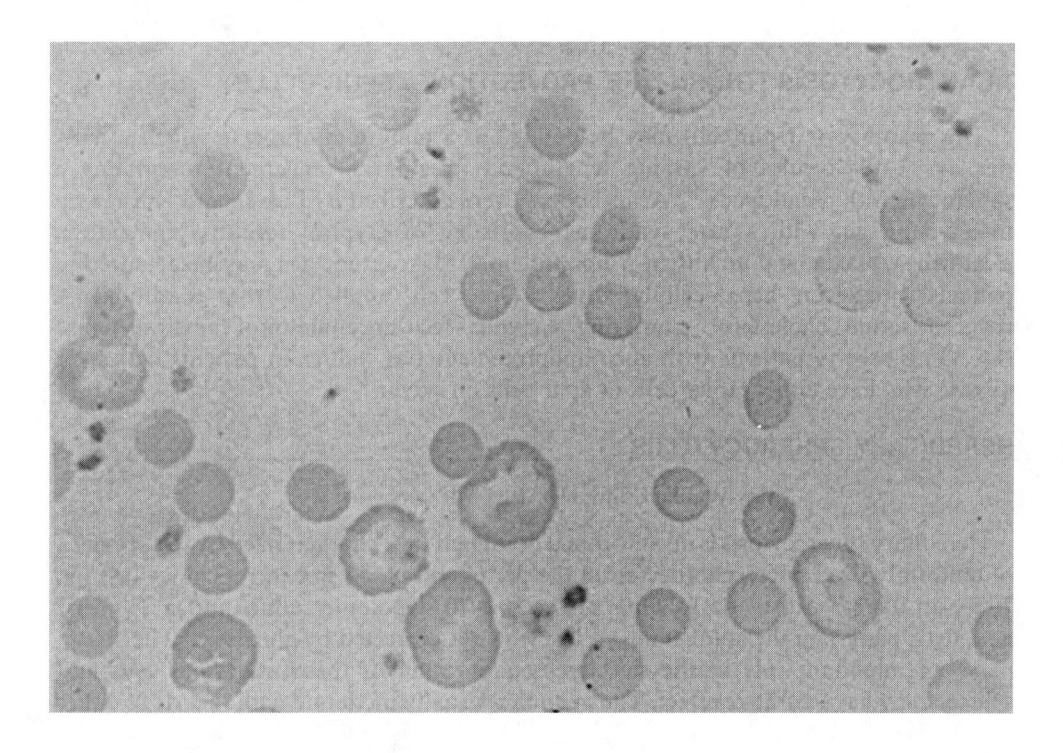

(A) an abnormal hemoglobin electrophoresis
(B) a positive direct Coombs test
(C) an abnormal Schilling test
(D) increased osmotic fragility of the red blood cells
(E) increased urine for hemosiderin

39. The figure shows a peripheral blood finding from a 35-year-old man with fever, generalized lymphadenopathy, and hepatosplenomegaly. The patient's complete blood cell count reveals a moderate normocytic anemia, thrombocytopenia, and an elevated white blood cell count with 20% of the cells similar to those noted in the slide. This patient most likely has an abnormality involving

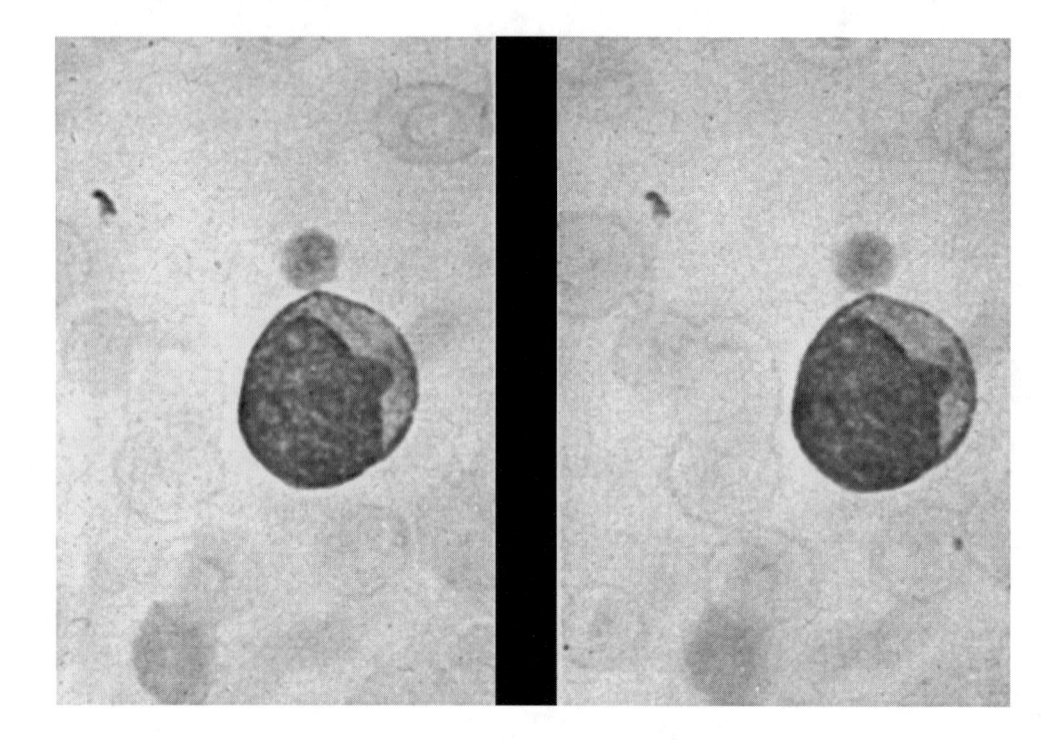

(A) lymphocytes
(B) monocytes
(C) neutrophils
(D) mast cells
(E) histiocytes

40. This figure shows a red blood cell finding in a 48-year-old woman with massive splenomegaly, left upper quadrant pain, and a small pleural effusion in the left pleural cavity. The patient's complete blood cell count shows a moderate normocytic anemia with thrombocytopenia, an elevated white blood cell count with both mature and immature cells in the granulocytic series, and numerous nucleated red blood cells. The leukocyte alkaline phosphatase score is elevated, and the result of the Philadelphia chromosome study is negative. The patient most likely has

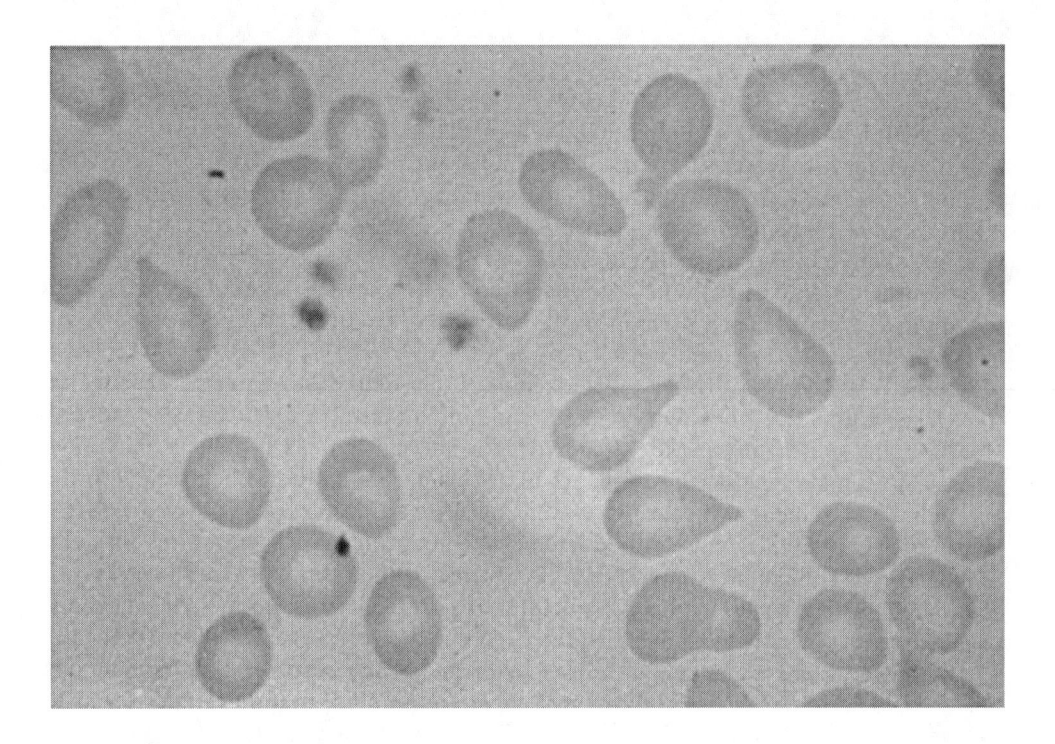

(A) polycythemia rubra vera
(B) agnogenic myeloid metaplasia
(C) hairy cell leukemia
(D) metastatic carcinoma to the bone marrow
(E) chronic myelogenous leukemia

96   Body Systems Review I: Hematopoietic/Lymphoreticular, Respiratory, Cardiovascular

41. The figures show peripheral blood findings from a 6-year-old black child with fever secondary to a *Streptococcal pneumoniae* septicemia. The conclusion that can be drawn from the peripheral blood findings and clinical history is that the

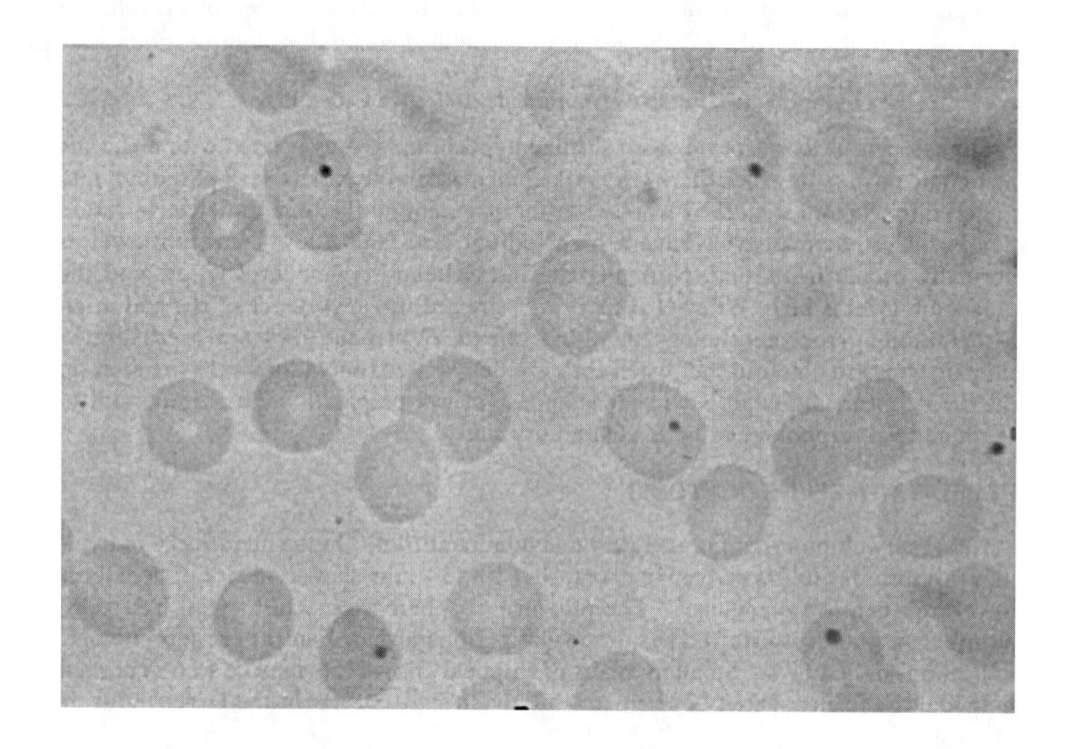

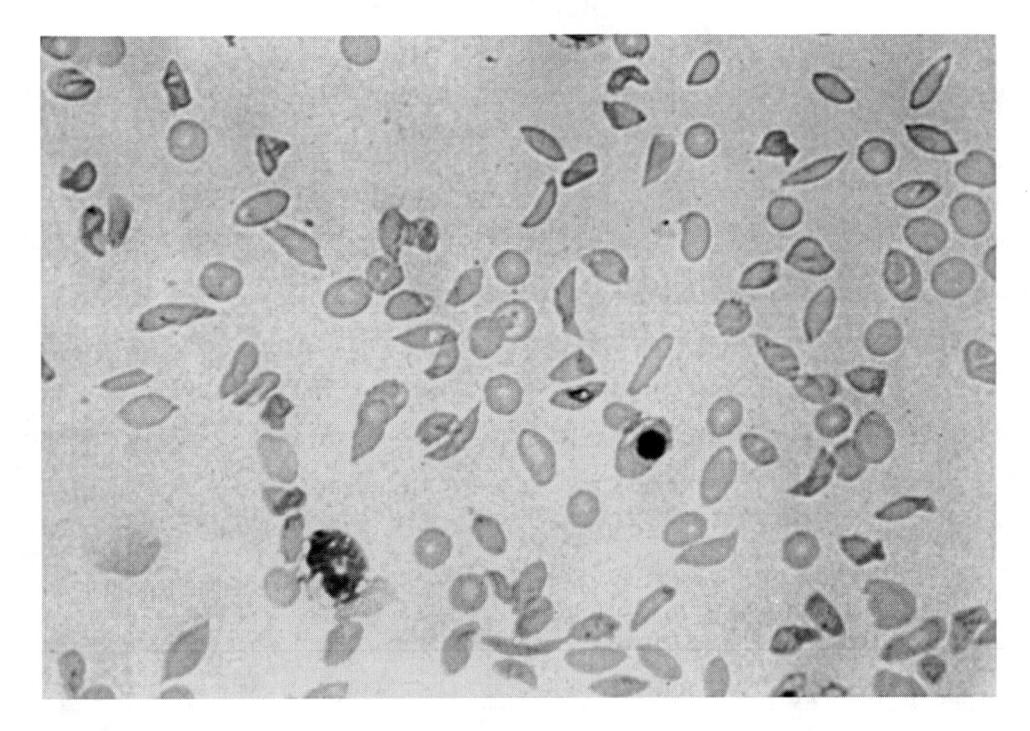

(A) hemoglobin electrophoresis would most likely show predominantly hemoglobin S, with a small amount of hemoglobin A and hemoglobin F
(B) child has a functional asplenia
(C) child received the Pneumovax immunization
(D) child has an autosomal dominant disease
(E) child has a normocytic anemia with a corrected reticulocyte count less than 2%

42. Which of the following is more likely associated with a platelet disorder rather than hemophilia A?

(A) Hemarthroses
(B) Epistaxis
(C) Late rebleeding
(D) Prolonged partial thromboplastin time
(E) Intramuscular hematomas

43. Which of the following is more likely to predispose to B$_{12}$ deficiency than to folate deficiency?

(A) Alcoholism
(B) Chronic pancreatitis
(C) Celiac disease involving the duodenum and jejunum
(D) Patient taking methotrexate
(E) Patient taking phenytoin

44. Which of the following normally serves a procoagulant role in the vessel endothelium?

(A) Tissue thromboplastin
(B) Prostacyclin (PGI$_2$)
(C) Heparin
(D) Protein C and S
(E) Tissue plasminogen activator

45. Study the following forward and back typing results on patients A through D. Forward typing is a blood bank procedure to determine the blood type of the patient. The patient's red blood cells are reacted against anti-A and anti-B test sera (forward type), and the serum is reacted against A and B tests red blood cells (back type).

	Forward Type Using		Back Type Using	
Patient	Anti-A	Anti-B	A Red Blood Cells	B Red Blood Cells
A	Positive	Negative	Negative	Positive
B	Negative	Negative	Positive	Positive
C	Negative	Positive	Positive	Negative
D	Positive	Positive	Negative	Negative

Which patient(s) could safely be transfused with group A packed red blood cells?

(A) Patient A only
(B) Patient B only
(C) Patient C only
(D) Patient D only
(E) Patients A and D

46. A 72-year-old man with myelodysplastic syndrome develops a temperature of 101°F during a transfusion of packed red blood cells. The results of his transfusion reaction work-up including an indirect and direct Coombs are negative. The mechanism of this patient's transfusion reaction relates most closely to a

(A) type I, IgE-mediated hypersensitivity reaction
(B) type II, cytotoxic antibody hypersensitivity reaction
(C) type III, immune complex hypersensitivity reaction
(D) type IV, cellular immunity hypersensitivity reaction
(E) bacterial contamination of the unit of blood

47. A woman in her second semester of her second pregnancy has laboratory evidence that her unborn child has hemolytic disease of the newborn due to anti-D antibodies. Which of the following is a correct statement concerning this disease?

(A) Intravascular hemolytic anemia is the mechanism of hemolysis in the D antigen–positive fetus
(B) The mother must be D antigen negative and was sensitized in her first pregnancy
(C) Rh immune globulin given to the mother either during during her pregnancy or at delivery will reduce the severity of hemolysis in the fetus
(D) Increased indirect bilirubin produced in the fetus as an end-product of hemolysis is primarily removed by the fetal liver
(E) The mother is probably group O and Rh negative, and her first child is either group A, AB, or B and Rh positive

48. A 32-year-old woman who received 8 units of packed cells in the hospital 1 week ago develops a sudden onset of fever and jaundice. Studies reveal a 2 g/dl decrease in her hemoglobin level since discharge from the hospital and a positive direct Coombs test on her red blood cells. The patient most likely has

(A) an expected response from any transfusion of RBCs from a donor
(B) ABO incompatibility with one of the donor units that was delayed in its presentation
(C) a leukocyte antibody directed against a leukocyte antigen from one of the donor units
(D) nonspecific antibody against a plasma protein from one of the donor units
(E) extravascular hemolytic anemia from an antibody in her serum directed against an RBC cell antigen from one of the donor units

49. The figure shows a peripheral smear from a 22-year-old woman with extreme fatigue lasting 2 weeks. The patient also has fever, exudative tonsillitis, tender posterior cervical lymph nodes, and tender hepatosplenomegaly. A complete blood cell count shows a normal hemoglobin and platelet count. The total white blood cell count is 8500 cell/μl, with 25% of the cells similar to those noted in the slide.

The most likely finding is

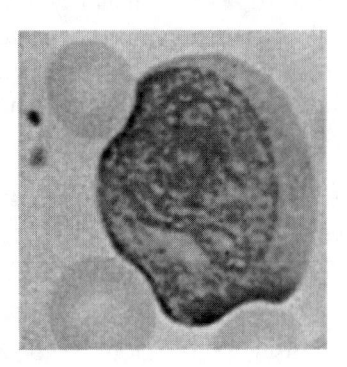

(A) group A streptococcus on culture of the tonsillar exudate
(B) heterophile antibodies in the serum
(C) normal serum transaminases
(D) direct and indirect hyperbilirubinemia
(E) anti-viral capsid antigen IgG antibodies in her serum

50. Malignant T cells in the peripheral blood, hypercalcemia, and lytic lesions in bone best describe a leukemia associated with

(A) Epstein-Barr virus
(B) HIV-1
(C) benzene exposure
(D) radiation exposure
(E) human leukemia lymphocyte virus type 1

51. A malignant B-cell lymphoma most commonly originates from the

(A) sinuses of lymph nodes
(B) germinal follicles of lymph nodes
(C) bone marrow lymphocytes
(D) paracortical tissue in the lymph nodes
(E) white pulp of the spleen

52. In the evaluation of the majority of white blood cell (WBC) disorders encountered in clinical practice, which of the following would be immediately unimportant in the evaluation of a patient with a leukocyte abnormality?

(A) To check whether mature or immature WBCs are present in the smear
(B) To check if there is normal or abnormal WBC morphology in the smear
(C) To check whether there is an accompanying anemia and/or thrombocytopenia
(D) To identify what specific WBCs are involved and whether there is an absolute or relative increase in these cells
(E) To perform a bone marrow aspirate and biopsy to evaluate the myeloid to erythroid ratio.

53. A 32-year-old woman with systemic lupus erythematosus is fatigued and jaundiced. Physical examination reveals mild scleral icterus, generalized lymphadenopathy, a low-grade fever, and hepatosplenomegaly. The patient has a hemoglobin level of 5 g/dl, a white blood count of 5000 cells/µl with an absolute monocytosis, and a normal platelet count. Shift cells (polychromasia), rare nucleated red blood cells, and spherocytes are noted in the peripheral smear. The patient's corrected reticulocyte count is 16%. Which of the following tests is most indicated in this patient?

(A) Osmotic fragility
(B) Heinz body preparation
(C) Schilling test
(D) Coombs test
(E) Bone marrow aspirate/biopsy

54. The figure shows a peripheral smear from a 72-year-old woman who has fatigue, dyspnea, and angina. In the course of the patient's work-up, she is noted to have pale conjunctiva and a grade II pulmonary flow murmur. A complete blood cell count reveals macrocytic indices and pancytopenia with white red blood cells and leukocytes similar to those noted in the slide. The patient's corrected reticulocyte count is less than 2%. She has decreased vibratory sensation in the lower extremities and glossitis on physical examination. The mechanism for her anemia most closely relates to

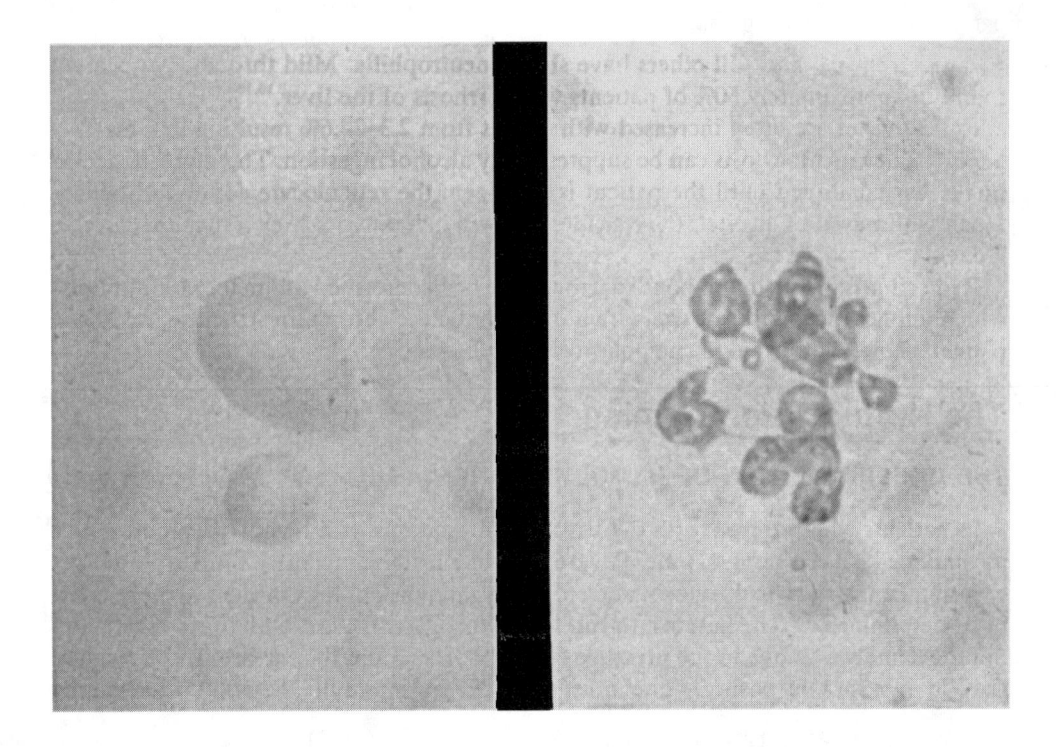

(A) nutritional deficiency
(B) autoimmune disease
(C) a hemolytic process
(D) blood loss
(E) a defect in hemoglobin synthesis

55. Which of the following conditions would most likely have a normal prothrombin and partial thromboplastin time?

(A) Long-standing alcoholic cirrhosis
(B) Celiac disease
(C) Disseminated intravascular coagulation
(D) Fulminant hepatic failure
(E) Thrombotic thrombocytopenic purpura

56. Generalized, nontender lymphadenopathy in a man over 65 years of age is most likely caused by

(A) acute myelogenous leukemia
(B) an autoimmune disease
(C) primary lung cancer with metastasis
(D) chronic lymphocytic leukemia
(E) a normal age-dependent process

57. Which of the following lymph node enlargements is most likely benign?

(A) Unilateral inguinal lymph nodes in an asymptomatic 55-year-old man
(B) Left supraclavicular lymph node in an adult with weight loss
(C) Lymph nodes in the neck of a smoker with hoarseness
(D) Cervical nodes in a man with a thyroid nodule
(E) Para-aortic lymph nodes in a man with a history of cryptorchidism

58. A 55-year-old black woman presents with a 2-cm left axillary mass. It is nontender and has been present for 3 to 4 months. The patient denies fever or night sweats. Her complete blood cell count and biochemistry profile (SMAC) are normal. Physical examination including breast, pelvic, and rectal examinations is otherwise unremarkable. Which of the following options should be recommended?

(A) Observation unless it increases in size
(B) Bone marrow aspiration and biopsy
(C) ELISA screen for HIV antibody
(D) Excisional biopsy of the lymph node
(E) Purified protein derivative intermediate skin test

59. A 28-year-old woman with postpartum bleeding and a pretransfusion hemoglobin level of 5 g/dl is still bleeding after being given 6 units of packed red blood cells. Her prothrombin time (PT) is normal; her partial thromboplastin time (PTT) is prolonged; and her platelet count and bleeding time are normal. Fibrin(ogen) degradation products are negative. Her factor VIII concentration is only 10%. One milliliter of her plasma mixed with 1 ml of normal plasma does not correct the PTT. The patient most likely has

(A) disseminated intravascular coagulation
(B) hemophilia A
(C) von Willebrand disease
(D) a factor VIII inhibitor
(E) a qualitative platelet defect

60. Which of the following is the best indicator that no significant coagulation abnormality is present in a patient?

(A) The patient is not taking birth control pills
(B) The patient has no history of epistaxis
(C) The patient has no excessive bleeding with superficial scratches
(D) The patient had a molar extraction without any bleeding problems
(E) The patient has no family history of bleeding

61. Which of the following would be expected in a patient with epistaxis, spontaneous bruising, and petechia?

(A) Prolonged prothrombin time
(B) Prolonged partial thromboplastin time
(C) Decreased antithrombin III concentration
(D) Prolonged bleeding time
(E) Increase in fibrinogen degradation products

62. A 23-year-old woman has menorrhagia with associated iron deficiency. Her prothrombin time (PT) is normal; the partial thromboplastin time (PTT) is prolonged; the bleeding time is prolonged; and the platelet count is normal. This patient's menorrhagia would most likely diminish with

(A) a packed red blood cell transfusion
(B) infusion of platelet concentrates
(C) administration of desmopressin
(D) treatment of her iron deficiency with iron
(E) an intramuscular injection of vitamin K

63. A major hemorrhage in a patient is most likely caused by which of the following hemostasis abnormalities?

(A) Coagulation factor deficiency
(B) Acquired platelet defect
(C) Vasculitis
(D) Primary fibrinolysis
(E) Thrombocytopenia

64. In a patient with severe anemia, which of the following would be expected?

(A) An increased alveolar-arterial gradient
(B) Hypoxemia
(C) A diastolic flow murmur
(D) A decreased oxygen saturation
(E) A wide pulse pressure

65. Which of the following statements concerning iron metabolism is correct?

(A) Vitamin C is the most important factor for enhancing iron absorption
(B) Patients who are pure vegetarians primarily ingest heme iron, which is easier to absorb than non-heme iron
(C) Gastric acid is responsible for reducing iron from ferric to ferrous
(D) A Billroth II procedure for peptic ulcer disease does not interfere with normal iron reabsorption
(E) The percentage of iron reabsorbed in the gastrointestinal tract decreases in patients with anemia

Questions 66-68

The following data were obtained from a skeletal muscle capillary and its surrounding tissue (interstitial) space

Plasma oncotic pressure	10 mm Hg
Tissue oncotic pressure	5 mm Hg
Capillary hydrostatic pressure	30 mm Hg
Tissue hydrostatic pressure	4 mm Hg

66. This capillary is

(A) in filtration equilibrium
(B) filtering small amounts of fluid
(C) reabsorbing small amounts of fluid
(D) filtering large amounts of fluid
(E) reabsorbing large amounts of fluid

67. The net force favoring the movement of fluid is

(A) -21 mm Hg
(B) -5 mm Hg
(C) +21 mm Hg
(D) -25 mm Hg
(E) 0 mm Hg

68. Which of the following conditions is most consistent with the data?

(A) Venodilator administration for angina pectoris
(B) Cardiogenic shock
(C) Anemia
(D) Polycythemia
(E) Cirrhosis

Questions 69-71

The information which follows was obtained from a 23-year-old patient during a complete work-up (direct spirometry; helium-washout technique) in a pulmonary function laboratory.

Volumes:	
Inspiratory reserve volume	3.50 L
Inspiratory capacity	4.00 L
Total lung capacity	7.00 L
Expiratory reserve volume	1.50 L

69. The functional residual capacity (FRC) is

(A) 1.5 L
(B) 2.5 L
(C) 3.0 L
(D) 4.0 L
(E) not measurable given the above data

70. The tidal volume (V_T) is

(A) 300 ml
(B) 350 ml
(C) 450 ml
(D) 500 ml
(E) cannot be determined given the above data

71. The vital capacity (VC) is

(A) 3.5 L
(B) 4.0 L
(C) 5.0 L
(D) 5.5 L
(E) 6.0 L

Questions 72-74

The following spirogram was obtained from a 23-year-old woman, who was asked to inspire maximally (A to B) and then exhale maximally (B to C).

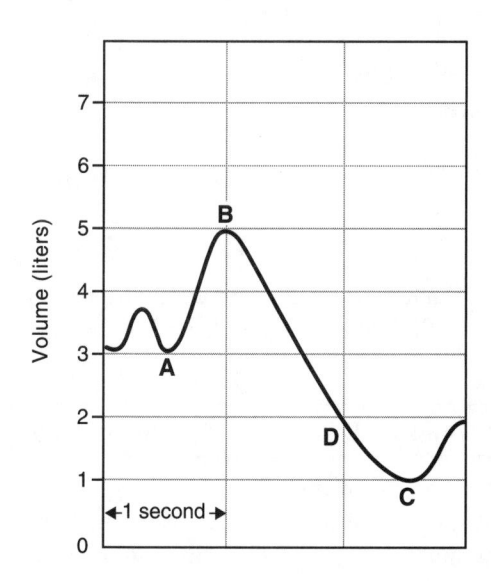

72. The forced vital capacity (FVC) is

(A) 500 ml
(B) 2000 ml
(C) 4000 ml
(D) 5500 ml
(E) cannot be determined

73. The forced expiratory volume (FEV) in 1 second is

(A) 500 ml
(B) 2200 ml
(C) 3000 ml
(D) 5500 ml
(E) cannot be determined

74. The residual volume (RV) is

(A) 1000 ml
(B) 1500 ml
(C) 1200 ml
(D) one half of the functional residual capacity
(E) cannot be determined

Questions 75-77

A middle-aged man, who is being treated with chemotherapy for lung cancer, has both serious respiratory problems and anemia. He is brought to the emergency room with a hemoglobin content of 10 gm%, a $P_{arterial}O_2$ of 50 mm Hg, and a $P_{venous}O_2$ of 25 mm Hg. Curve B describes his O_2 dissociation on admittance. For each of the curves below, the O_2-combining capacity is 1.34 ml/gm Hb.

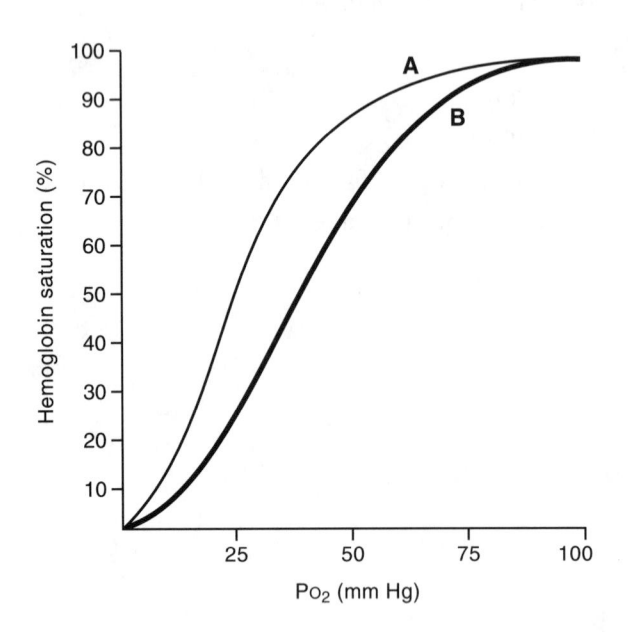

75. What is the arterial O_2 content on hemoglobin?

(A) 7.62 vol%
(B) 7.80 vol%
(C) 9.38 vol%
(D) 10.35 vol%
(E) 13.40 vol%

76. What is the A-V O_2 difference (vol%) across the tissues?

(A) 2.00 vol%
(B) 4.50 vol%
(C) 5.80 vol%
(D) 6.08 vol%
(E) 6.70 vol%

77. The patient is unknowingly transfused with outdated blood. If the partial pressure of his blood gases remains unchanged, the percentage of saturation of his venous blood will be

(A) 40%
(B) 50%
(C) 60%
(D) 70%
(E) 90%

78. Which of the following comparisons between a "pink puffer" and a "blue bloater" is correct?

	Pink puffer	Blue bloater
(A)	Chronic bronchitis	Emphysema
(B)	Respiratory acidosis common	Normal arterial blood gases to mild respiratory alkalosis
(C)	Normal to slightly increased anteroposterior diameter	Increased anteroposterior diameter
(D)	Minor ventilation/perfusion mismatch	Major ventilation/perfusion mismatch
(E)	Hypoxemia moderate to severe	Hypoxemia mild to absent

79. Which of the following patients has the greatest likelihood for progression into cor pulmonale? A patient with

(A) mitral stenosis
(B) a ventricular septal defect
(C) pulmonary stenosis
(D) multiple recurrent pulmonary emboli
(E) an atrial septal defect

Questions 80-81

The barometric pressure on the summit of Mt. Everest is 200 mm Hg, with the water vapor pressure already subtracted out. Answer the following questions assuming that a mountain climber has no supplemental oxygen.

80. If the mountain climber has an alveolar CO_2 of 40 mm Hg, and the respiratory quotient is 1 (normally 0.8 at sea level), what would the alveolar P_{O_2} be?

(A) 0 mm Hg
(B) 2 mm Hg
(C) 4 mm Hg
(D) 6 mm Hg
(E) 8 mm Hg

81. Which of the following actions would be the most efficient way of increasing the alveolar P_{O_2}?

(A) Having the patient breathe into a paper bag
(B) Administering acetazolamide
(C) Increasing the water intake
(D) Increasing the respiratory rate
(E) Administering a loop diuretic

82. Bronchial asthma is similar to chronic bronchitis caused by cigarette smoking in that both diseases

(A) are associated with immunoglobulin E (IgE)-mediated inflammation
(B) are characterized by episodic flare ups
(C) are likely to initially present with respiratory alkalosis
(D) have the site of obstruction to airflow in the small caliber airways
(E) are marked by a productive cough throughout most of the year

83. Which of the following types of bronchogenic carcinomas are most likely to develop within a residual area of peripheral scar tissue?

(A) Small-cell carcinoma
(B) Squamous-cell carcinoma
(C) Large-cell carcinoma
(D) Bronchial carcinoid
(E) Adenocarcinoma

84. A lobar consolidation in the lung of a middle-aged, nonsmoking man did not respond to antibiotics. Sputum cytology is negative; however, a fine needle aspiration of the lesion is reported to contain neoplastic cells. The patient most likely has a

(A) primary small-cell carcinoma
(B) metastatic squamous-cell carcinoma
(C) primary bronchioalveolar carcinoma
(D) bronchial carcinoid
(E) scar carcinoma

85. Inspiratory stridor characterizes which one of the following groups of disease?

(A) Chronic bronchitis and foreign body obstruction in the upper airway
(B) Respiratory distress syndrome and adult respiratory distress syndrome
(C) Bronchial asthma and bronchiolitis
(D) Bronchopneumonia and laryngeal carcinoma
(E) Acute epiglottitis and croup

86. A decrease in the pulmonary diffusion capacity with carbon monoxide (D_LCO) would most likely be associated with

(A) removal of an entire lung
(B) chronic bronchitis
(C) a mesothelioma
(D) bronchial asthma
(E) a primary small-cell carcinoma

87. A 30-year-old, nonsmoking woman who is 2 days postpartum has a history of a sudden onset of tachypnea, dyspnea, and left lower lobe chest pain on inspiration. Scattered expiratory wheezes, dullness to percussion, and absent tactile fremitus are noted in the left lower lobe. Her left calf is slightly swollen and tender. Dorsiflexion of the foot and anteroposterior compression of the calf are painful. The arterial blood gas drawn with the patient breathing room air is depicted below.

pH	7.48 (7.35–7.45)
Pa_{CO_2}	24 mm Hg (33–44 mm Hg)
Pa_{O_2}	70 mm Hg (75–105 mm Hg)
HCO_3	19 mEq/L (22–28 mEq/L)

Which of the following statements is correct?

(A) Hypoxemia is primarily related to a diffusion abnormality
(B) The alveolar-arterial gradient is greater than 30 mm Hg
(C) The arterial blood gases exhibit an uncompensated respiratory alkalosis
(D) Left lower lobe findings are consistent with a consolidation without a pleural effusion
(E) A ventilation scan would most likely reveal a defect in the left lower lobe

88. A 65-year-old man with urinary retention from prostate hyperplasia presents with fever, dyspnea, tachypnea, and cyanosis involving his skin and mucous membranes. His skin feels warm. Inspiratory crackles are noted in all lung fields. An arterial blood gas reveals a Pa_{O_2} of 51 mm Hg while breathing 40% oxygen. The Pa_{O_2} does not significantly increase after breathing 100% oxygen for 20 minutes. The patient most likely has

(A) a massive pulmonary embolus
(B) lobar pneumonia
(C) hypovolemic shock
(D) adult respiratory distress syndrome
(E) congestive heart failure

89. Reactivation tuberculosis and primary squamous-cell carcinoma of the lung are similar in that they both are commonly associated with

(A) cavitation
(B) scar carcinomas
(C) silicosis
(D) ectopic secretion of a parathormone-like peptide
(E) spread to the same most common extrapulmonary site

Questions 90-91

The photograph is of a Gram stain of sputum from a 4-year-old boy with cystic fibrosis, who has fever and a productive cough. He has a sudden onset of dyspnea and left-sided chest pain. Physical examination reveals widening of the intercostal muscles and decreased breath sounds on the left side. The trachea in the sternal notch is deviated to the right.

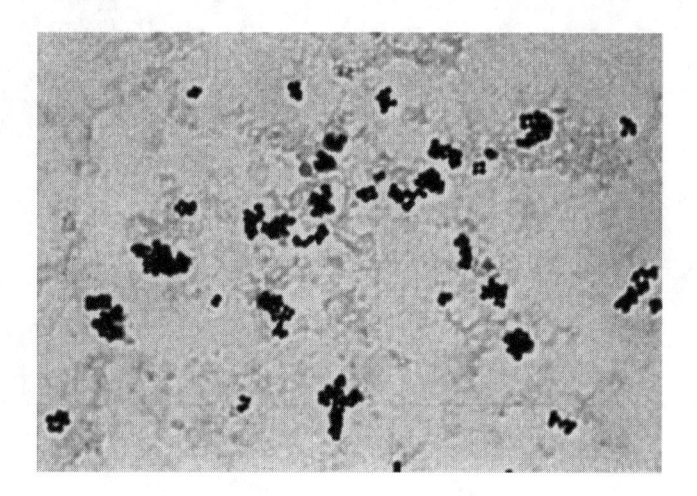

90. The Gram stain is most consistent with

(A) *Streptococcus pneumoniae*
(B) *Haemophilus influenza*
(C) group A streptococcus
(D) *Staphylococcus aureus*
(E) *Pseudomonas aeruginosa*

91. The most likely cause for the sudden onset of dyspnea in this patient is a

(A) build-up of empyema fluid in his left chest
(B) spontaneous pneumothorax involving his left lung
(C) lung abscess secondary to the bronchopneumonia
(D) pulmonary infarction in the left lung
(E) tension pneumothorax involving the left lung because of rupture of a pneumatocyst

92. The radiograph and Gram stain of sputum are from a 26-year-old man, who presented with a sudden onset of 103°F temperature, dyspnea, cough productive of rusty-colored sputum, and pleuritic chest pain in the right upper lung. If the patient has no drug allergies, the treatment of choice would be

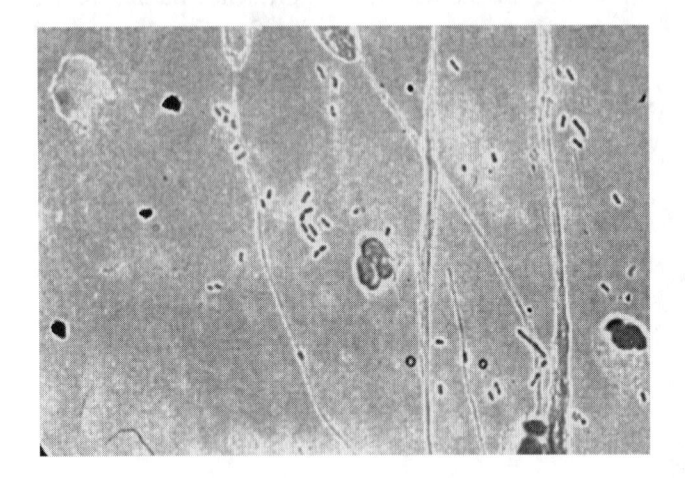

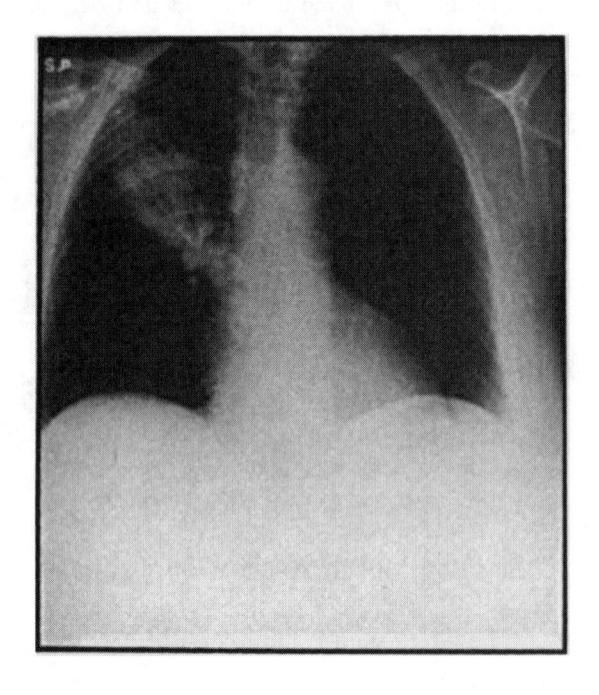

(A) penicillin V potassium
(B) tetracycline
(C) ceftriaxone
(D) erythromycin
(E) trimethoprim/sulfamethoxazole

93. The following photograph of the lungs was taken during an autopsy from a 62-year-old woman who had severe dyspnea. The gross features are most compatible with

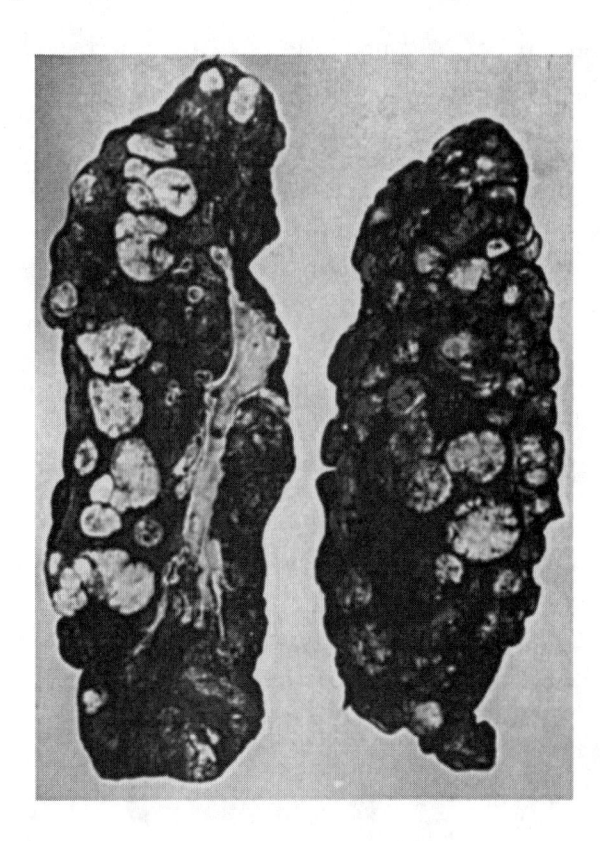

(A) granulomatous disease
(B) primary cancer
(C) metastatic cancer
(D) multiple areas of liquefactive necrosis
(E) multiple areas of coagulation necrosis

94. Which drug inhibits the release of histamine, prostaglandins, and other mediators from inflammatory cells?

(A) Diphenhydramine
(B) Cromolyn sodium
(C) Phenylephrine
(D) Albuterol
(E) Ipratropium

95. Which of the following bronchodilators is most likely to cause tachycardia and cardiac arrhythmias?

(A) Albuterol
(B) Isoproterenol
(C) Terbutaline
(D) Salmeterol
(E) Bitolterol

DIRECTIONS:

Each of the numbered items or incomplete statements in this section is negatively phrased, as indicated by a capitalized word such as NOT, LEAST, or EXCEPT. Select the ONE lettered answer or completion that is BEST in each case.

96. The superior mediastinum contains all of the following structures EXCEPT

(A) the trachea
(B) the esophagus
(C) the left recurrent laryngeal nerve
(D) the right recurrent laryngeal nerve
(E) the right phrenic nerve

97. All of the following are true of the vocal folds of the larynx EXCEPT

(A) they are covered with a mucosa that is innervated by branches of the vagus nerve
(B) they are abducted during respiration by contraction of the posterior cricoarytenoid muscles
(C) they are adducted during swallowing by contraction of the lateral cricoarytenoid muscles
(D) they are made more tense by contraction of the thyroarytenoid muscles
(E) they are inferior to the vestibular folds of the larynx

98. All of the following statements regarding the circulatory system are true EXCEPT

(A) a functional end artery may be occluded without causing the death of any cells, whereas occlusion of an anatomical end artery leads to some cell death
(B) the elasticity of the elastic arteries is responsible for the maintenance of blood pressure between contractions of the heart
(C) anastomoses may occur between arteries or between veins
(D) heat may be transferred from blood without the blood passing though a capillary bed
(E) blood cells that leave from the left ventricle may return to the heart without passing through a capillary

99. Tissue derived from the respiratory diverticulum include all of the following EXCEPT

(A) thyroid epithelium
(B) alveolar epithelium
(C) bronchial epithelium
(D) laryngeal epithelium
(E) tracheal epithelium

100. Which of the following vascular lesions has the LEAST clinical significance?

(A) Mönckeberg's medial calcification
(B) Hyaline arteriolosclerosis
(C) Hyperplastic arteriolosclerosis
(D) Glomus tumor
(E) Port wine nevus flammeus in the trigeminal distribution

101. Which of the following statements regarding lipoproteins is NOT correct?

(A) Lipoprotein (a) [Lp(a)] reduces the risk for developing an acute myocardial infarction
(B) Apolipoprotein A is associated with high-density lipoproteins
(C) Apolipoprotein B-48 is associated with chylomicrons
(D) Apolipoprotein B-100 is associated with very-low-density lipoprotein (VLDL) and low-density lipoprotein fractions
(E) Apolipoprotein C-II enhances capillary lipoprotein lipase, causing the release of fatty acids and glycerol from chylomicrons and VLDL

102. Hypercholesterolemia would LEAST likely be associated with a patient who has

(A) minimal change disease
(B) Hashimoto thyroiditis
(C) diabetes mellitus
(D) primary biliary cirrhosis
(E) fatty change in the liver resulting from alcoholism

103. Which of the following statements concerning lipids and heart disease is LEAST accurate?

(A) Saturated fat should comprise no more than 30% of the total calories consumed each day in order to lower the risk for heart disease
(B) Coconut oil contains lauric, myristic, and palmitic saturated fatty acids, and is considered the "worst" saturated fat for increasing serum cholesterol
(C) *Trans* fatty acids in hydrogenated vegetable oils raise low-density lipoprotein (LDL) and decrease high-density lipoprotein (HDL) levels
(D) The American Heart Association and the American Cancer Society both recommend at least 25 grams of fiber per day to help lower serum cholesterol levels and lower the risk for colon cancer
(E) Antioxidants, such as vitamin C, vitamin E, and beta-carotene, do not consistently lower serum cholesterol levels but do lower oxidized LDL levels

104. The blood–air barrier consists of all of the following EXCEPT

(A) alveolar pores
(B) a layer of surfactants
(C) type I alveolar cell cytoplasm
(D) fused basal lamina of alveolar and endothelial cells
(E) endothelial cell cytoplasm

105. Endothelial cells may have all the following features or functions EXCEPT

(A) gap junctions
(B) keratin filaments
(C) they are part of the blood–brain barrier
(D) they contain coagulating factor
(E) they convert angiotensin I to angiotensin II

106. The photograph depicts the right leg of a 58-year-old man. Which of the following statements concerning this case is NOT correct?

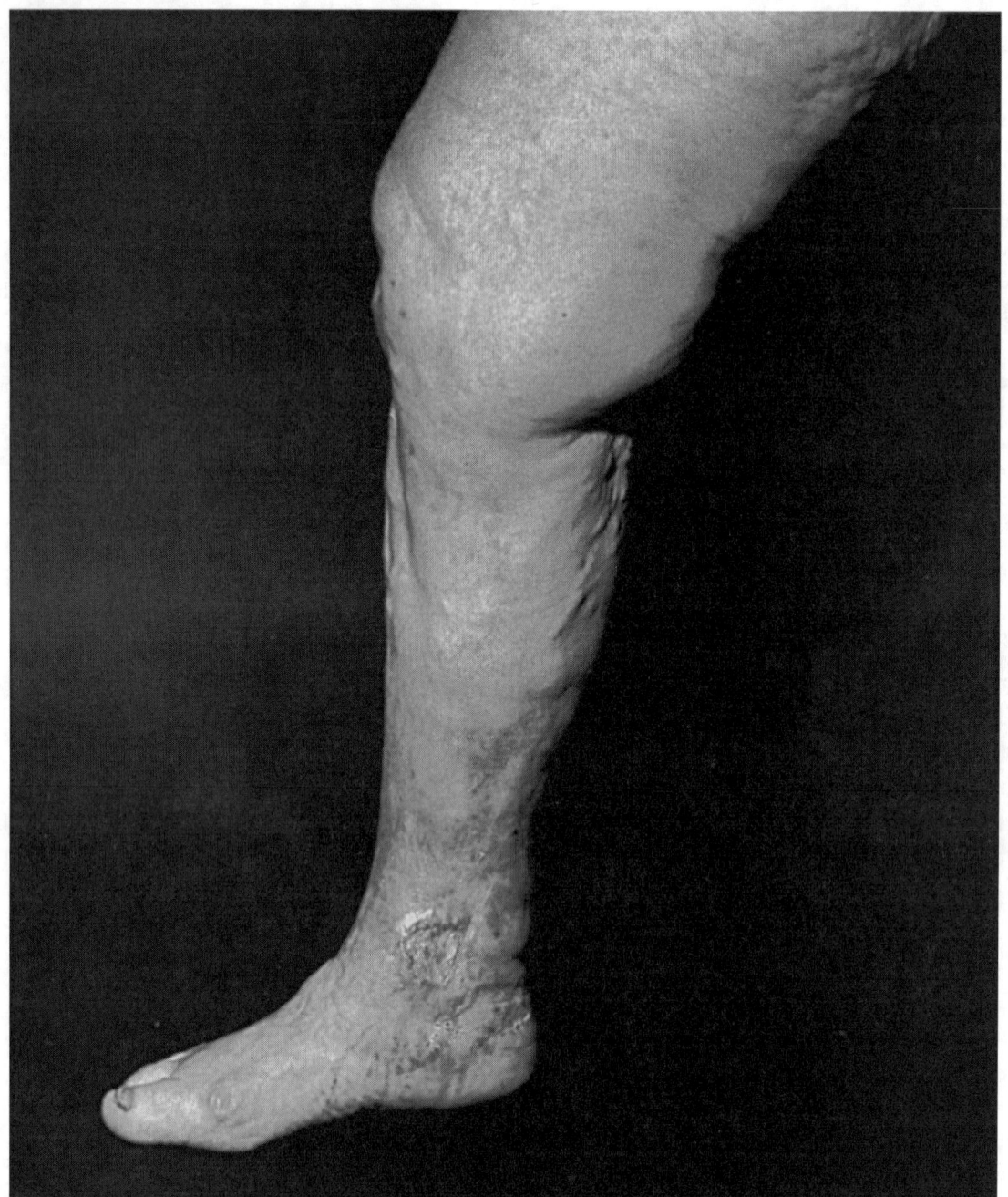

(A) Discoloration in lower leg is related to hemosiderin deposition in subcutaneous tissue
(B) Venous dilatation is due primarily to thrombosis of the greater saphenous vein with chronic stasis of venous blood below the area of obstruction
(C) The malleolar ulcer is secondary to ischemia from chronic venous stasis
(D) Flow of venous blood in the ankle is most likely from the deep to the superficial saphenous system via incompetent penetrating branches around the ankle
(E) An area of acute vessel inflammation (thrombophlebitis) is present along the inner right knee

107. These are the hands of a 35-year-old non-smoking woman. Which of the following conditions would LEAST likely be associated with this finding?

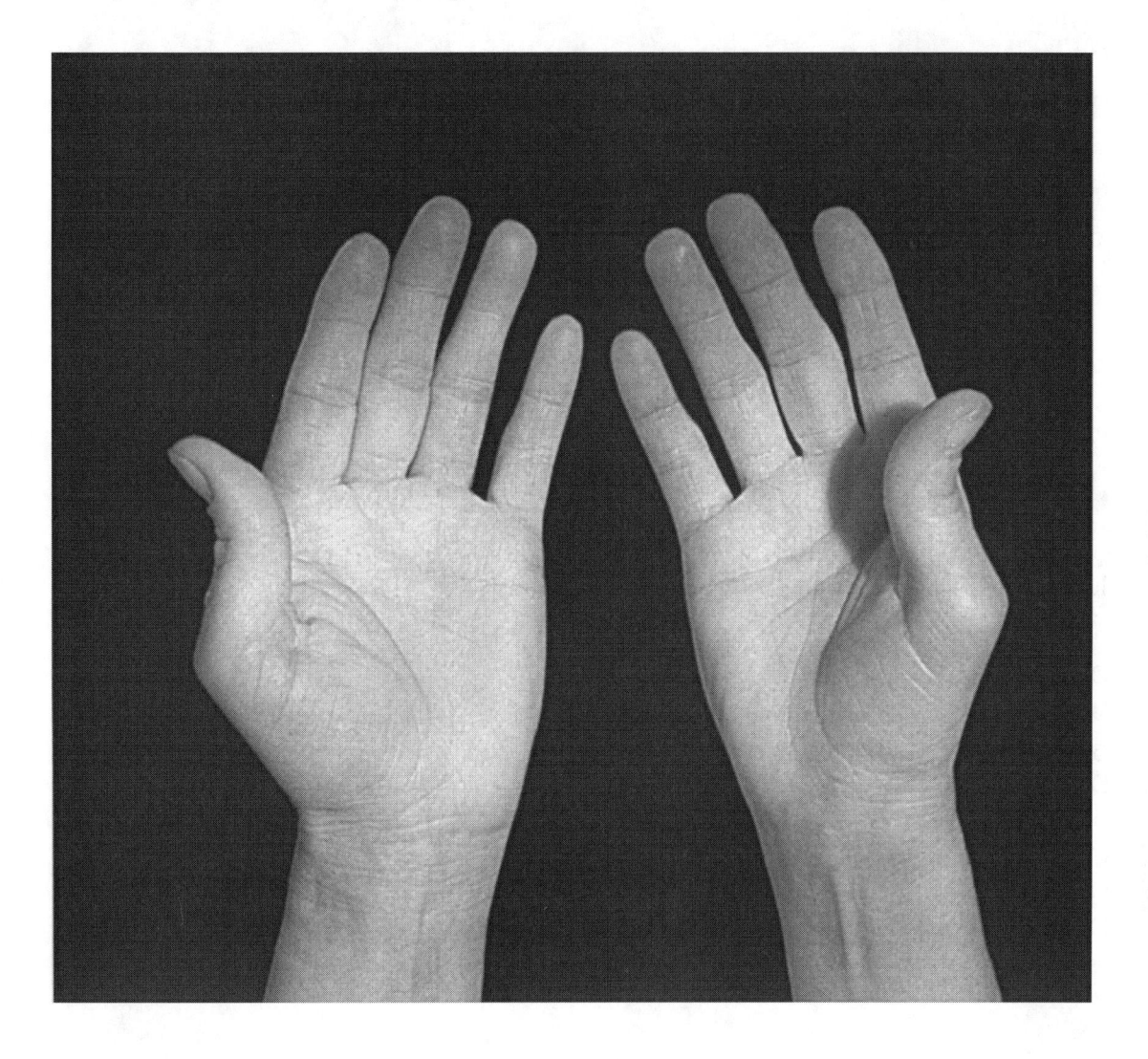

(A) Takayasu disease
(B) CREST syndrome
(C) Cryoglobulinemia
(D) Progressive systemic sclerosis
(E) Thromboangiitis obliterans

108. In the diagram depicting the metabolism of folate, letters represent compounds, and numbers represent enzymes.

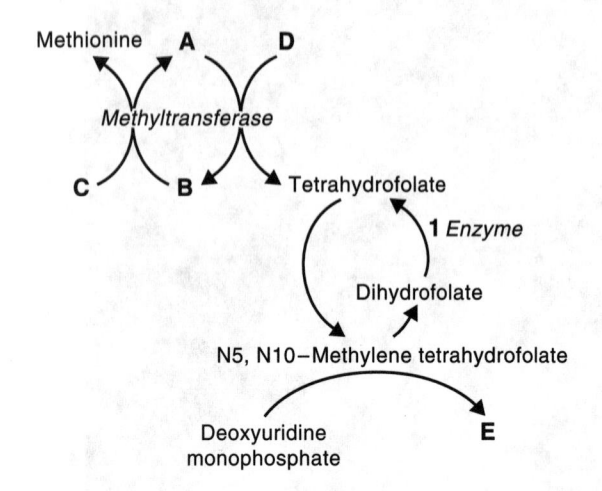

Methionine ⇌ A   D
Methyltransferase
C   B   Tetrahydrofolate
         ↑ 1 Enzyme
         Dihydrofolate
N5, N10–Methylene tetrahydrofolate
Deoxyuridine monophosphate → E

Which of the following statements is NOT true?

(A) Deficiency of either compound A or D decreases DNA synthesis
(B) Methotrexate and trimethoprim inhibit enzyme 1
(C) Deficiency of compound A or D or blockage of enzyme 1 results in a megaloblastic anemia
(D) Compound D is involved in propionate metabolism
(E) Deficiency of compound A or D increases compound C, which predisposes to vessel thrombosis

109. The figure shows a peripheral smear from a patient who has a microcytic anemia with low serum ferritin levels. Which of the following is the LEAST likely patient profile?

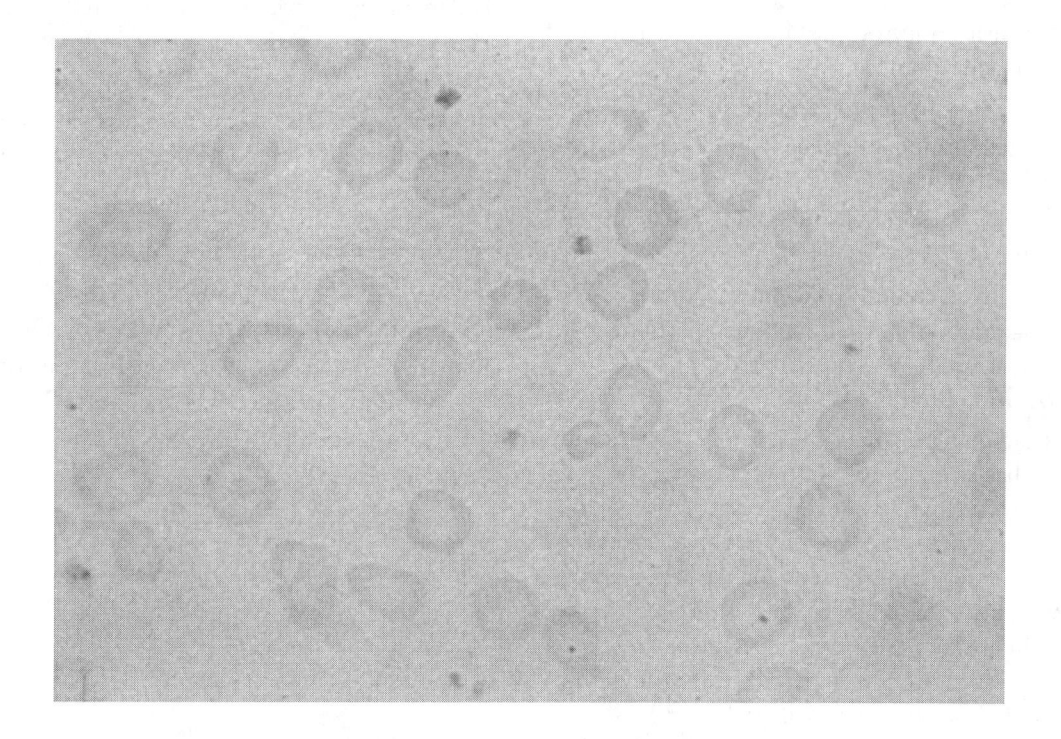

(A) Premature infant without iron supplementation
(B) 23-year-old woman with menorrhagia
(C) 26-year-old woman with three previous pregnancies without prenatal vitamins
(D) 14-month-old child who primarily drinks milk
(E) 15-year-old boy who eats junk food

110. Which of the following patients would LEAST likely require a bone marrow examination?

(A) Adult with pancytopenia and a normal mean corpuscular volume
(B) Adult with a myeloproliferative disease
(C) Adult with a monoclonal spike on a serum protein electrophoresis
(D) Adult with fever and a white blood cell count of 20,000 cells/µl with a left shift
(E) Child with fever, bone pain, anemia, thrombocytopenia, and an elevated lymphocyte count

Questions 111-114

A 58-year-old man presents with a history of a chronic cough that has lasted 9 out of 12 months for the last 7 years. He states that his cough is worse in the morning and is productive of opaque white to greenish sputum, the latter being present for the last 3 days. In addition, he states that he feels short of breath when climbing stairs. He denies chest pain.

His significant history includes 40 years of smoking two packs of cigarettes a day (80 pack years).

Physical examination reveals a temperature of 38°C, respirations of 14/min with pursed lips on expiration, pulse of 108 with a regular rhythm, and blood pressure of 120/76 mm Hg. The anteroposterior diameter of his chest is increased. He is actively using accessory muscles of the neck and shoulders for breathing. The percussion note is diffusely hyperresonant. There are scattered inspiratory and expiratory wheezes and sibilant rhonchi. Expiration is prolonged. Heart sounds are distant. The point of maximal impulse cannot be felt. The abdominal exam is unremarkable. There is no ankle edema.

His complete blood count reveals a hemoglobin of 19 g/dl (13.5–17.5 g/dl) and a hematocrit of 57% (41%–53%). The white blood cell count is 18,000 cells/ul (4500–11,000 cells/ul) with a 15% band neutrophil count and toxic granulation present. Arterial blood gases while on low flow oxygen (30%) are:

pH	7.24 (7.35–7.45)
Pa_{O_2}	46 (75–105 mm Hg)
Pa_{CO_2}	76 (33–44 mm Hg)
HCO_3	32 (22–28 mEq/L)

The radiograph reveals flattening of the diaphragms and an increased anteroposterior diameter. The bronchial markings are prominent. A Gram stain of sputum exhibits gram-negative diplococci intermixed with neutrophils.

111. Regarding this patient's productive cough, which of the following pathologic changes in the airways is LEAST likely to occur?

(A) Bronchial smooth muscle hypertrophy
(B) Goblet-cell hyperplasia of the major bronchi
(C) Squamous metaplasia with loss of cilia
(D) An increased ratio of bronchial mucous gland width to total mucosal thickness (Reid index)
(E) Destruction of elastic tissue and abnormal dilatation of the bronchi

112. The reason for the patient's most recent acute exacerbation of coughing most closely relates to

(A) acute bronchitis secondary to *Klebsiella pneumoniae*
(B) acute bronchitis secondary to *Haemophilus influenzae*
(C) bronchiectasis superimposed on primary lung disease
(D) a viral pneumonia with a secondary staphylococcal pneumonia
(E) acute bronchitis secondary to *Moraxella (Branhamella) catarrhalis*

Pulmonary function studies in this patient are as follows:

Total lung capacity	7.10 liters (130% of predicted)
Residual volume	3.90 liters (170% of predicted)
Forced vital capacity (FVC)	3.0 liters (74% of predicted)
Forced expiratory volume 1 second (FEV$_1$)	1.0 liter (20% of predicted)
FEV$_1$/FVC ratio	33% (75% of predicted)
Carbon monoxide diffusion capacity (D$_L$CO)	20 ml/min/mm Hg (60% of predicted)

113. Which of the following statements concerning the pulmonary function studies and other laboratory findings in this patient is NOT correct?

(A) The pulmonary function studies primarily exhibit an obstructive disease pattern
(B) The arterial blood gases reveal a respiratory acidosis with a partially compensated metabolic alkalosis as well as an increased alveolar-arterial P$_{O_2}$ gradient
(C) There is a secondary polycythemia primarily owing to hypoxemia
(D) The decreased diffusion capacity is caused by destruction of the alveolar capillary units throughout both lung fields
(E) The Gram stain of sputum is consistent with normal flora

114. Three years later, the patient returns with neck vein distention, a cannon a-wave of the jugular venous pulses, pitting edema around the ankles, and accentuation of P$_2$. His radiograph now reveals prominence of the right border of the heart and irregular pulmonary vascular obliteration. The primary mechanism for these new findings is

(A) right heart failure secondary to chronic ischemic heart disease
(B) pulmonic stenosis secondary to chronic obstructive pulmonary disease
(C) tricuspid stenosis secondary to pulmonary hypertension
(D) pulmonary hypertension secondary to chronic obstructive pulmonary disease
(E) acute cor pulmonale secondary to thromboembolism

115. A 42-year-old man presents to the emergency room with a 3-week history of a dry, nonproductive cough associated with shortness of breath at rest and with exertion. He has bilateral chest pain with inspiration, but no radiation of pain into his arms or jaw. He has felt tired and has had low grade fever for the last week.

Significant history includes the fact that he received 20 units of blood from blood loss in a motorcycle accident 11 years ago.

He is unmarried and admits to having unprotected anal intercourse with a male on only one occasion, approximately 1 year ago.

Physical examination reveals a temperature of 101°F (37.8°C), a pulse of 110/min, a respiratory rate of 26/min, and a blood pressure of 124/82 mm Hg. He has generalized, tender lymphadenopathy. Examination of the lungs reveals bilateral inspiratory rales at both lung bases. There is no calf tenderness or swelling. The remainder of the examination is unremarkable.

A radiograph reveals an extensive bilateral interstitial and alveolar infiltrate. A complete blood count exhibits a mildly decreased hemoglobin and hematocrit, a normal mean corpuscular volume (MCV), a normal white blood cell count with a relative increase in neutrophils, and an absolute decrease in lymphocytes. His corrected reticulocyte count is decreased and the serum ferritin is increased. An arterial blood gas on 30% oxygen reveals the following:

pH	7.52 (7.35–7.45)
P_{O_2}	50 (75–105 mm Hg)
P_{CO_2}	20 (33–44 mm Hg)
HCO_3	16 (22–28 mEq/L)

He is admitted to the hospital and a bronchoscopy is ordered after an induced sputum was reported as negative for pathogens. A transbronchial biopsy and bronchoalveolar lavage are performed. The transbronchial biopsy and a silver stain of the specimen are illustrated in the photographs. After receiving the results of this procedure, a serologic test is ordered.

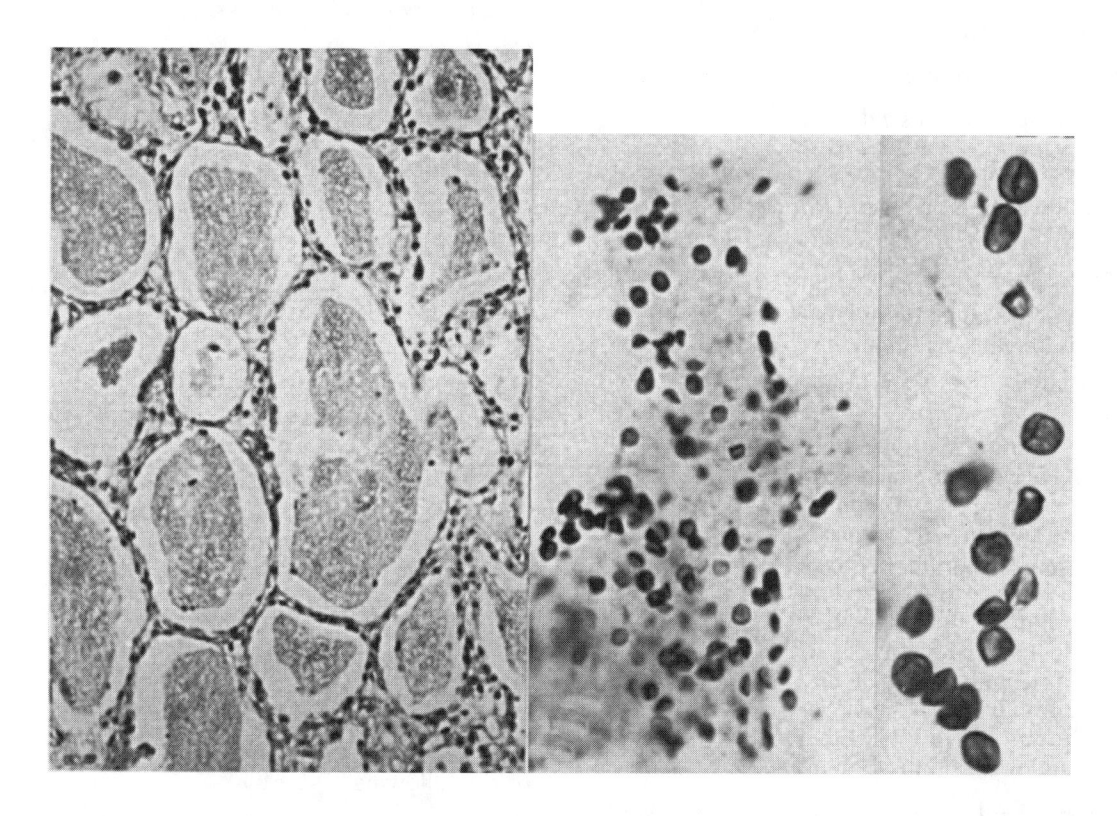

Which of the following statements is NOT correct?

(A) He has a partially compensated respiratory alkalosis with an abnormal alveolar-arterial gradient
(B) Trimethoprim-sulfamethoxazole is the drug of choice for his pulmonary disease
(C) The special test is an ELISA screen for human immunodeficiency antibody
(D) He has the anemia of inflammation
(E) He most likely contracted the disease from anal intercourse

116. Wheezing is LEAST likely to be associated with

(A) left heart failure
(B) pulmonary embolus
(C) respiratory syncytial virus infection
(D) carcinoid syndrome
(E) respiratory distress syndrome

117. A 29-year-old black woman presents with dyspnea, uveitis, and bilateral parotid gland enlargement. Fine, dry inspiratory crackles are heard at both lung bases. A radiograph reveals bilateral hilar adenopathy and bibasilar nodular infiltrates. Which of the following clinical findings would the patient be LEAST likely to have?

(A) An increase in angiotensin-converting enzyme
(B) Anergy to skin testing
(C) An increased peripheral blood T-helper cell count
(D) Noncaseating granulomas in a scalene lymph node biopsy
(E) A low carbon monoxide diffusion capacity (D_LCO)

118. Gross blood or rusty-colored sputum would LEAST likely be present in a patient with

(A) primary lung cancer
(B) bronchiectasis
(C) mitral stenosis
(D) an aspergilloma
(E) metastatic lung disease

119. In which of the following lung diseases is obstruction LEAST likely to be a primary, pathophysiologic factor?

(A) Bronchiectasis
(B) Pneumoconioses
(C) Primary lung cancer
(D) Endogenous lipoid pneumonia
(E) Atelectasis

120. Which of the following patients is LEAST at risk for developing a bronchogenic carcinoma?

(A) A 70-year-old patient who smokes, with a past history of tuberculosis
(B) A 52-year-old nonsmoking patient with progressive massive fibrosis
(C) A 56-year-old patient who smokes, with a history of asbestos exposure 20 years ago
(D) A 48-year-old woman who works around beryllium
(E) A 51-year-old man who works in a uranium mine

121. Which of the following conditions is LEAST likely to be a risk factor for contracting tuberculosis?

(A) Treatment with corticosteroid therapy
(B) Native American heritage
(C) Low socioeconomic status
(D) Residence in a long-term care facility
(E) Exposure to asbestos

122. Benefits of the cessation of smoking include all of the following EXCEPT

(A) coronary heart disease risk is halved in 1 year when compared with risk for those patients who continue smoking, and is halved in 15 years when compared with risk for patients who do not smoke
(B) lung cancer risk is halved in 1 year when compared with risk for those patients who continue smoking
(C) stroke risk is reduced to that of nonsmoking patients 5 to 15 years after quitting
(D) pancreatic cancer risk is reduced in 10 years when compared with risk for patients who continue smoking
(E) risk of death from chronic obstructive lung disease is reduced after a prolonged time when compared with risk for patients who continue smoking

123. Which of the following statements concerning the Mantoux test using intradermal injection of purified protein derivative (PPD) is NOT correct?

(A) The area of induration, rather than the area of erythema, is measured after 48 to 72 hours
(B) A negative PPD does not exclude active or inactive disease
(C) A positive PPD does not distinguish active from inactive disease
(D) Atypical mycobacteria often produce a weakly positive PPD
(E) A positive PPD offers no immunity against later reinfection by tubercle bacilli

124. Adverse effects of diphenhydramine can include all of the following EXCEPT

(A) dry mouth
(B) blurred vision
(C) sedation
(D) excitement
(E) bradycardia

125. As a nasal decongestant, pseudoephedrine should be used cautiously in patients with any of the following conditions EXCEPT

(A) hypertension
(B) hyperthyroidism
(C) ischemic heart disease
(D) urinary incontinence
(E) prostatic hyperplasia

126. All of the following effects are true of guaifenesin EXCEPT that it

(A) facilitates expectoration of sputum
(B) increases hydration of mucous membranes
(C) converts a nonproductive to a productive cough
(D) inhibits the cough reflex
(E) decreases mucous viscosity

124 Body Systems Review I: Hematopoietic/Lymphoreticular, Respiratory, Cardiovascular

DIRECTIONS:
Each set of matching questions in this section consists of a list of four to twenty-six lettered options (some of which may be in figures) followed by several numbered items. For each numbered item, select the ONE lettered option that is most closely associated with it. To avoid spending too much time on matching sets with a large number of options, it is generally advisable to begin each set by reading the list of options. Then for each item in the set, try to generate the correct answer and locate it in the option list, rather than evaluating each option individually. Each lettered option may be selected once, more than once, or not at all.

Questions 127-130

Phonocardiograms of various heart murmurs are depicted in the accompanying schematics. Match the following clinical descriptions with the correct schematic.

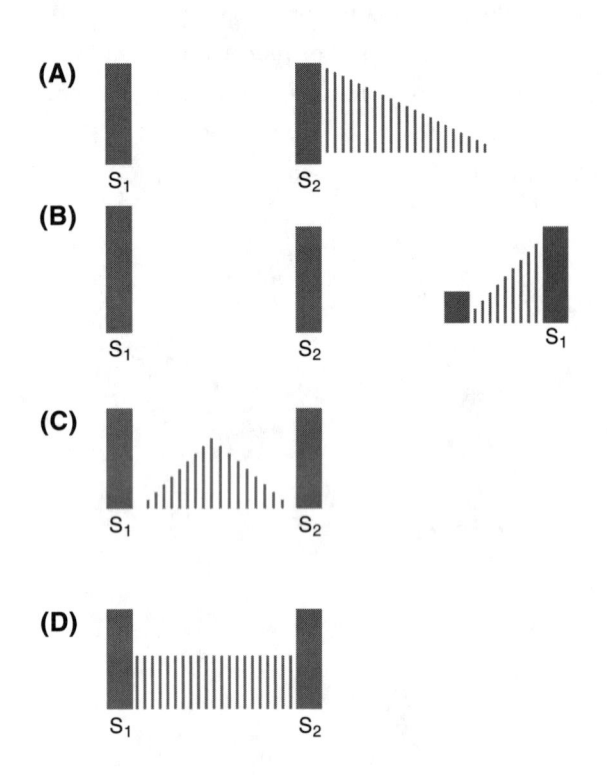

(A) S₁ ▮ S₂ ▮▁▁▁
(B) S₁ ▮ S₂ ▮   ▁▁▮ S₁
(C) S₁ ▮ ▁▮▁ S₂ ▮
(D) S₁ ▮ ▮▮▮▮▮▮▮▮ S₂ ▮

127. A 65-year-old man with diminished pulses and a history of angina and syncope with exercise has a murmur radiating into the carotid arteries

128. A 46-year-old woman has a malar flush, dysphagia for solids, hoarseness, an accentuated P₂ heart sound, and a productive cough with rust-colored sputum

129. A 62-year-old woman with chronic ischemic heart disease has bibasilar wet rales, a point of maximal impulse beyond the midclavicular line, a prominent S₃ heart sound, and radiation of a murmur into the axilla

130. A 73-year-old man has bounding pulses, bobbing of the head, and a prominent pulsating mass on his anterior chest. Serologic testing of the patient is positive for syphilis

Questions 131-133

The accompanying diagram is a schematic of the jugular venous pulse and heart sounds. Match the following descriptions with the appropriate letter in the diagram.

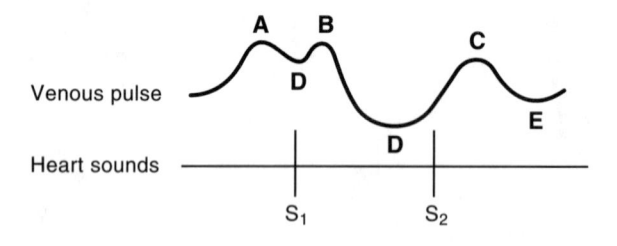

131. Upward displacement of this wave occurs in tricuspid stenosis

132. An accentuation of this wave occurs in tricuspid regurgitation

133. This wave is absent in a patient with an irregularly irregular beat

Questions 134-140

The figure depicts electrocardiogram (ECG) results from a normal individual. Match the letter to the statement.

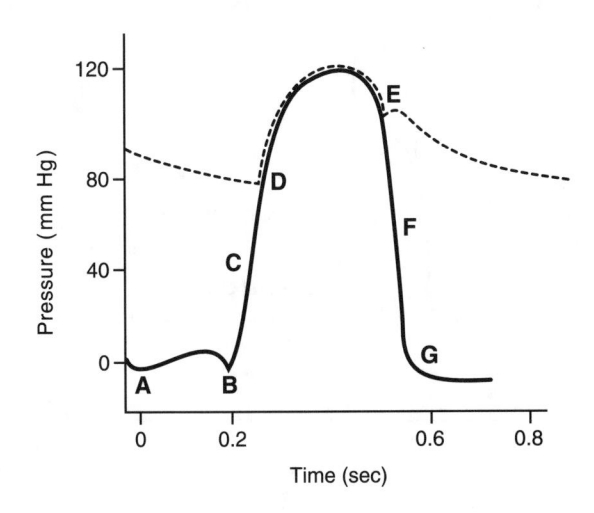

134. The first heart sound is heard at [ ]?

135. The P wave of the ECG occurs at [ ]?

136. The point that best represents preload [ ]?

137. Ejection begins at [ ]?

138. Isovolumic relaxation occurs during [ ]?

139. The second heart sound occurs at [ ]?

140. The end-systolic volume would be at [ ]?

Questions 141-143

Match the following lettered iron studies with the numbered clinical scenario that best represents the findings.

	Serum Iron	TIBC	% Saturation	Serum Ferritin
(A)	Low	High	Low	Low
(B)	Low	Low	Low	Normal to high
(C)	High	Normal	High	High
(D)	Normal	Normal	Normal	Normal
(E)	High	High	High	High

TIBC = total iron binding capacity.
% Saturation = serum iron/TIBC × 100.

141. A 28-year-old Asian woman with a mild microcytic anemia and a normal hemoglobin electrophoresis

142. A patient with long-standing rheumatoid arthritis with a guaiac-negative stool

143. A patient who has been transfused over 100 times

Body Systems Review I: Hematopoietic/Lymphoreticular, Respiratory, Cardiovascular

Questions 144-150

The results shown in the figures were obtained from individuals with various valve disorders. Match the statements to the appropriate diagram.

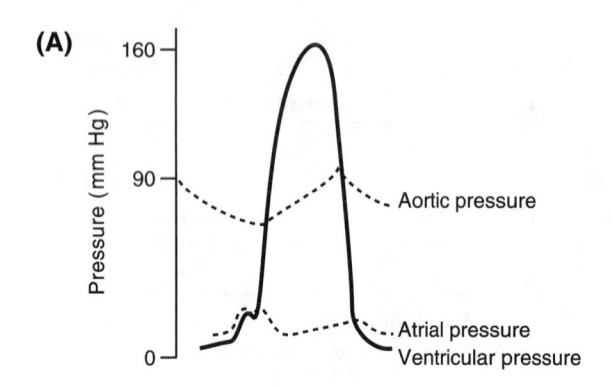

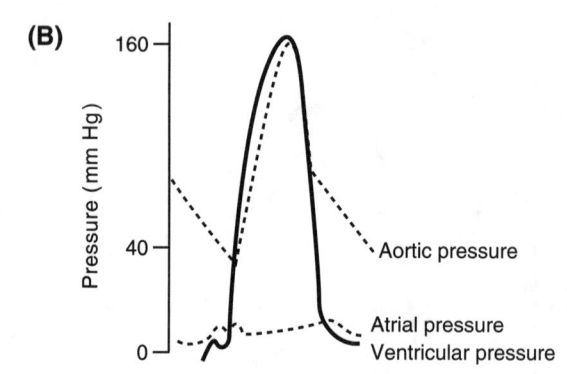

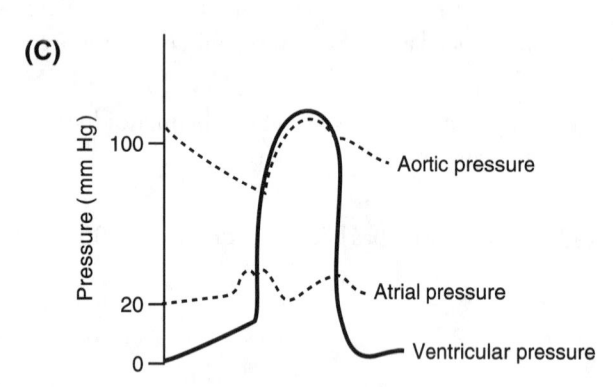

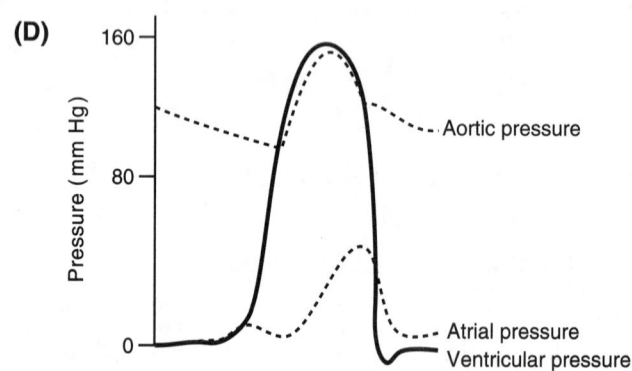

144. Aortic stenosis

145. Aortic regurgitation

146. Associated with a systolic crescendo–decrescendo murmur

147. Mitral stenosis

148. Mitral regurgitation

149. Associated with a presystolic murmur

150. Associated with concentric ventricular hypertrophy

Questions 151-157

The pressure–volume loop shown in the figure provides data from a normal heart. Match the statement to the appropriate point on the loop.

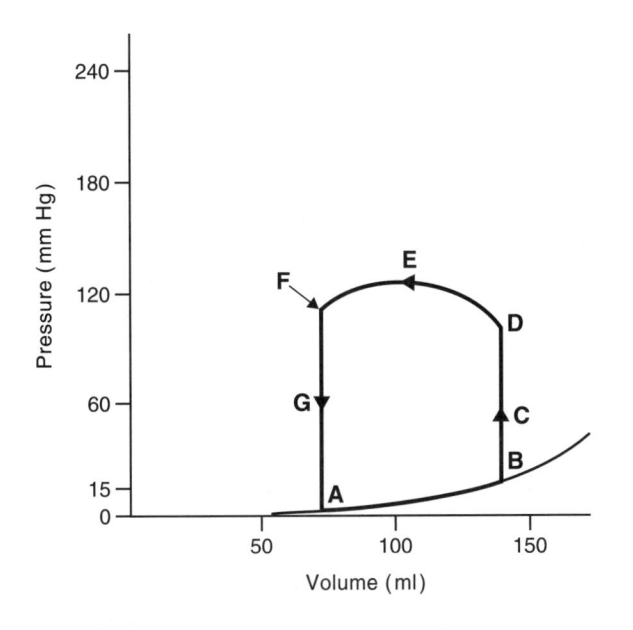

151. The second heart sound would be heard at [ ]?

152. The QRS complex of the electrocardiogram (ECG) occurs just before [ ]?

153. Aortic diastolic blood pressure is closest to [ ]?

154. Isovolumic relaxation occurs at [ ]?

155. End-diastolic volume is at [ ]?

156. Ventricular filling begins at [ ]?

157. The first heart sound would be heard at [ ]?

Questions 158-160

The figure below illustrates changes in theophylline clearance in a patient with asthma who was first treated at age 5. Select the most likely cause of the changes in clearance during the designated intervals from the lettered choices.

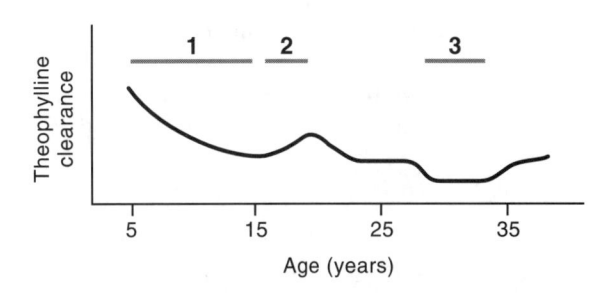

(A) Addition of inhaled cromolyn to asthma regimen
(B) Treatment of peptic ulcer with cimetidine
(C) Age-related change in theophylline clearance
(D) Addition of ipratropium to treatment asthma regimen
(E) Use of alcohol, tobacco, or marijuana

158. Interval 1

159. Interval 2

160. Interval 3

Questions 161-163

Identify the drugs that act at the lettered stages in the pathogenesis of bronchial asthma indicated in the figure below.

$$\text{Antigen + antibody}$$
$$\downarrow$$
$$\text{Mast cells}$$
$$\text{Stage } \mathbf{A} \downarrow$$
$$\text{Leukotrienes, PAF, histamine}$$
$$\swarrow \qquad \searrow$$
$$\text{Stage } \mathbf{B} \qquad \text{Stage } \mathbf{C}$$
$$\text{Bronchoconstriction} \qquad \text{Inflammation}$$

PAF = platelet-activating factor

(A) Diphenhydramine and cromolyn sodium
(B) Albuterol, theophylline, and ipratropium
(C) Beclomethasone and cromolyn sodium
(D) Cromolyn sodium only
(E) Beclomethasone only

161. Stage A

162. Stage B

163. Stage C

ANSWER KEY

1. E	29. C	56. D	83. E	110. D	137. D
2. B	30. D	57. A	84. C	111. E	138. F
3. B	31. B	58. D	85. E	112. E	139. E
4. E	32. E	59. D	86. A	113. E	140. E
5. C	33. D	60. D	87. B	114. E	141. D
6. B	34. E	61. D	88. D	115. E	142. B
7. B	35. B	62. C	89. A	116. E	143. C
8. C	36. E	63. A	90. D	117. C	144. A
9. E	37. E	64. E	91. E	118. E	145. B
10. A	38. D	65. A	92. A	119. B	146. A
11. A	39. C	66. D	93. C	120. B	147. C
12. D	40. B	67. C	94. B	121. E	148. D
13. A	41. B	68. E	95. B	122. B	149. C
14. D	42. B	69. C	96. D	123. E	150. A
15. B	43. B	70. D	97. D	124. E	151. F
16. C	44. A	71. D	98. A	125. D	152. B
17. D	45. E	72. C	99. A	126. D	153. D
18. D	46. B	73. C	100. A	127. C	154. G
19. D	47. B	74. E	101. A	128. B	155. B
20. C	48. E	75. C	102. E	129. D	156. A
21. C	49. D	76. E	103. A	130. A	157. B
22. D	50. B	77. B	104. A	131. A	158. C
23. D	51. B	78. D	105. B	132. C	159. E
24. C	52. E	79. D	106. B	133. A	160. B
25. D	53. D	80. B	107. E	134. B	161. C
26. C	54. B	81. D	108. D	135. A	162. B
27. B	55. E	82. D	109. E	136. B	163. E
28. C					

ANSWERS AND EXPLANATIONS

1. The answer is E. *(Gross anatomy; Cardiovascular)* The septomarginal trabecula, or moderator band, is a trabecula carneae that extends from the right surface of the interventricular septum to the anterior papillary muscle of the right ventricle. It contains conducting fibers (Purkinje fibers) from the right bundle branch of the atrioventricular bundle and conducts depolarization to the papillary muscle. The atrioventricular bundle carries the depolarization signal from the atrioventricular node into the interventricular septum. The bundle then divides into right and left bundle branches, which then distribute to the ventricular muscle. Disruption of this conducting pathway results in heart block. The conducting fibers in the moderator band ensure that the papillary muscle is contracting at the very beginning of systole.

2. The answer is B. *(Gross anatomy; Cardiovascular)* The pericardial space is between the two layers of the serous pericardium: the parietal layer and the visceral layer. The fibrous pericardium is fused to the outer surface of the parietal pericardium. The epicardium is synonymous with the visceral pericardium. This layer is fused to the myocardium. Accumulation of fluid in the pericardial space (e.g., blood, exudate) compresses the heart and restricts cardiac filling during diastole (cardiac tamponade).

3. The answer is B. *(Gross anatomy; Cardiovascular)* The right ventricle is mostly behind the sternum but extends to the left of the sternum. The right atrium is to the right of the sternum. The left ventricle is several centimeters to the left of the sternum. The left atrium is almost entirely on the posterior side of the heart. The left lung is displaced well to the left of the sternum at the level of the fifth intercostal space by the presence of the heart.

4. The answer is E. *(Gross anatomy; Cardiovascular)* Tetralogy of Fallot is the most common cyanotic congenital heart defect. Neither atrial septal defect, ventricular septal defect, nor patent ductus arteriosus are cyanotic defects. Septal defects and patent ductus arteriosus result in left-to-right shunts that do not cause cyanosis. Tetralogy of Fallot results in a right-to-left shunt that does cause cyanosis. The elements of tetralogy of Fallot are: pulmonary stenosis, ventricular septal defect, overriding aorta, and right ventricular hypertrophy. It results from abnormal positioning of the aorticopulmonary septum.

5. The answer is C. *(Gross anatomy; Cardiovascular)* The septum secundum forms on the right side of the septum primum and covers the ostium secundum. In this manner the foramen ovale is formed. The most common type of atrial septal defect involves faulty development of the septum secundum. As the septum primum grows downward, the ostium primum gets smaller. Before the ostium primum closes, the ostium secundum develops within the septum primum to allow for continuous communication between the atria.

6. The answer is B. *(Gross anatomy; Pulmonary/respiratory)* Elevation of the ribs increases the diameter of the chest and thus causes inspiration. The ability of the scalene muscles to elevate the ribs makes them accessory muscles of respiration. Similarly, an asthmatic patient may fix an upper limb in order to use the pectoralis muscle as a muscle of inspiration because it will elevate the ribs.

The diaphragm is innervated by the phrenic nerve. The pleural cavity is a potential space and normally has no air in it. Contraction of the diaphragm lowers the diaphragm, thereby increasing pressure in the abdomen. Contraction of the abdominal wall muscles increases abdominal pressure, thereby pushing the diaphragm up to assist with forced expiration.

7. The answer is B. *(Gross anatomy; Pulmonary/respiratory)* The aortic arch lies against the anterolateral wall of the esophagus on the left side. Pulsations of this artery are often visible from within the esophagus. This region of the esophagus should be avoided when doing an esophageal biopsy. The ascending aorta and the brachiocephalic trunk are separated from the esophagus by the trachea.

8. The answer is C. *(Gross anatomy; Pulmonary/respiratory)*
The fifth or sixth tracheal cartilage is below the isthmus of the thyroid gland. The thyroidea ima artery, when present, approaches the isthmus in the midline from below. All of the other vessels listed approach the thyroid gland from more lateral directions. The inferior thyroid arteries cross the recurrent laryngeal nerves. The superior thyroid arteries are closely associated with the superior laryngeal nerves. Care must be taken to avoid these nerves when ligating these arteries during surgery on the thyroid gland.

9. The answer is E. *(Pathology; Endocrine)*
The patient most likely has a type V hyperlipoproteinemia, which is associated with eruptive xanthomas secondary to increased triacylglycerol. Type V hyperlipoproteinemia is a combination of type I, characterized by elevated chylomicrons, and type IV, characterized by increased very-low-density lipoproteins (VLDL). Diabetic ketoacidosis and alcoholism are two of the most common factors that promote type V hyperlipoproteinemia. The absence of insulin characteristic of diabetic ketoacidosis decreases capillary lipoprotein lipase activity, so neither chylomicrons nor VLDL are properly metabolized in the liver. The increased triacylglycerol concentration that results from the accumulation of chylomicrons and VLDL in the blood produces a turbid specimen. Because chylomicrons have the lowest density, they form a turbid supranate in plasma left at 4°C overnight. The density of VLDL is slightly higher than that of chylomicrons, so it forms a turbid infranate. Presence of a supranate and infranate, as in this case, indicates type V hyperlipoproteinemia. A supranate without an infranate occurs in type I hyperlipoproteinemia, and an infranate without a supranate occurs in type IV hyperlipoproteinemia. Another feature of type V disease is hyperchylomicronemia. Clinical features include eruptive xanthomas caused by the deposition of triacylglycerol in the subcutaneous tissue. Once the triacylglycerol concentration is reduced, the xanthomas disappear.

10. The answer is A. *(Behavioral science; Psychosocial, cultural, and environmental influences)*
There are currently more female than male smokers in the United States. Women are less likely than men to abuse alcohol, have spinal cord injuries, take illegal drugs, or engage in risk-taking behavior.

11. The answer is A. *(Pathology; Cardiovascular)*
A stress electrocardiogram is most useful in the work-up of chest pain. Positive results for angina are ST depression, which indicates exertional or unstable angina, and ST elevation, which indicates Printzmetal angina. A stress test is also useful in evaluating congenital and acquired heart disease, and in the detection of arrhythmias. It is not used to document the presence of an acute myocardial infarction.

12. The answer is D. *(Pathology; Cardiovascular)*
Echocardiography is a noninvasive test that utilizes sound waves reflected from the heart. The ejection fraction represents the stroke volume divided by the left ventricular end-diastolic volume. It is measured in all patients with a myocardial infarction in order to determine prognosis. A low ejection fraction indicates poor ventricular contractility and carries a poor prognosis.

Echocardiography is the primary test that distinguishes systolic from diastolic heart failure. Systolic heart failure results from poor contractility of the ventricle, and diastolic heart failure is associated with reduced compliance of the ventricle. In systolic heart failure, the ejection fraction is low. In diastolic heart failure, the ejection fraction is normal to slightly decreased, and left atrial pressure is increased.

Pericardial effusions are life threatening. Echocardiography is a rapid means of securing a diagnosis so that a pericardiocentesis (withdrawal of fluid with a needle) can relieve the intrapericardial pressure that inhibits filling of the cardiac chambers.

13. The answer is A. *(Pathology; Cardiovascular)*
Myocardial dysfunction is manifested within seconds after ischemia. Compliance (i.e., the ability to stretch) of the affected tissue decreases, which reduces filling of the heart in diastole. Contractility is impaired, which has a negative impact on systolic function. This reduces stroke volume and cardiac output. After 20 to 40 minutes, irreversible injury is likely to occur unless perfusion is re-established. Pain results in the release of catecholamines, but they are unable to increase contractility and maintain stroke volume.

14. The answer is D. *(Physiology; Cardiovascular)*
The Valsalva maneuver is performed by expiring against a closed glottis. This increases intrathoracic pressure, resulting in the following:

- Increased jugular venous pressure (central venous pressure)
- Decreased venous return of blood to the right heart
- Decreased cardiac output
- Increased heart rate as compensation for the decreased cardiac output
- Increased cerebrospinal fluid pressure

15. The answer is B. *(Pathology; Gastrointestinal)*
The patient has abetalipoproteinemia, which is a rare autosomal recessive disease characterized by an absence of apolipoprotein B. Apolipoprotein B is an integral component of chylomicrons, very-low-density lipoproteins (VLDL), and low-density lipoproteins (LDL). Without the apoprotein, triglyceride and cholesterol cannot be packaged into chylomicrons and VLDL. The disease has its onset in childhood. Malabsorption occurs because dietary fat cannot be packaged into chylomicrons, and intestinal cells fill with lipid. This blocks the reabsorption of fat and other nutrients. Loss of essential fatty acids results in membrane abnormalities in red blood cells with production of acanthocytes (thorny-appearing cells). Neurologic findings, including ataxia, nystagmus, and sensory abnormalities, also occur.

16. The answer is C. *(Pathology; Cardiovascular)*
Capacitance (ml/mm Hg) = volume (ml)/pressure (mm Hg). Capacitance is related to the distensibility of a blood vessel. Atherosclerosis of the aorta decreases distensibility, thereby decreasing capacitance and increasing systolic pressure. Pulse pressure is the difference between systolic and diastolic pressure. Systolic pressure is synonymous with stroke volume, and diastolic pressure represents arterial pressure during diastole. The most important determinant of pulse pressure is stroke volume, which is the amount of blood ejected from the left ventricle into the aorta. When capacitance of the aorta is reduced, the blood ejected from the left ventricle into the aorta causes the systolic pressure to increase, because the vessel is no longer able to distend and accommodate the blood as it normally should. This increases the pulse pressure. Because diastolic pressure should remain the same during ventricular systole, the rise in the pulse pressure should correlate with the rise in the systolic pressure.

17. The answer is D. *(Physiology; Cardiovascular)*
Poiseuille's equation involves all those factors that change resistance of blood vessels. It states that $R = 8nl/\pi r^4$, where R = resistance, n = viscosity, l = length of the vessel, and r^4 = radius of the vessel to the fourth power. Peripheral resistance is reduced by decreasing the viscosity of blood (e.g., in anemia) or the length of the vessel, or by increasing the radius of the vessel. Histamine is a potent vasodilator and would be expected to increase the radius of the resistance arterioles, thus decreasing the peripheral resistance. Peripheral resistance is increased by increasing viscosity (e.g., increased hematocrit) or vessel length, or by decreasing the radius of the vessel. Because resistance is inversely proportional to the fourth power of the radius of the vessel, altering the radius has a much greater significance than altering viscosity or the length of the vessel. This underscores why the arterioles are the most important vessels controlling the total peripheral resistance in the systemic circulation.

18. The answer is D. *(Pathology; Gastrointestinal)*
This patient with a turbid infranate representing very-low-density lipoprotein (VLDL) and absent supranate representing chylomicrons has type IV hyperlipoproteinemia. This is the most common type of hyperlipoproteinemia. It is thought to result from decreased catabolism of VLDL synthesized in the liver. Like all hyperlipoproteinemias, acquired causes are more common than genetic disease. Familial hypertriglyceridemia is an autosomal dominant disease characterized by triglyceride levels that begin to increase at puberty. Patients have an increased incidence of atherosclerosis involving coronary arteries and peripheral vessels. It can be exacerbated by concomitant diabetes, birth control pills, and alcohol ingestion. These factors often result in an increase of both VLDL and chylomicrons, which is called type V hyperlipoproteinemia. Tiny yellow papules, representing eruptive xanthomas, develop on the skin.

19. The answer is D. *(Pathology; Gastrointestinal)*
A nodular mass in the Achilles tendon most likely represents a tendon xanthoma, which is pathognomonic for the genotype of familial hypercholesterolemia. According to the Freidrickson's phenotype classification, the biochemical phenotype is type II hyperlipoproteinemia. Type II-a hyperlipoproteinemia is characterized by increased low-density lipoproteins (LDL) [often > 260 mg/dl], increased serum cholesterol, and normal triglyceride level. Type II-b shows increased triglyceride as well as increased cholesterol and LDL. The mechanisms for the increase in LDL are absent or defective LDL receptors and defective internalization of LDL.

There are three different genotypic causes of hypercholesterolemia: polygenic hypercholesterolemia (85% of cases), familial combined hypercholesterolemia (10%), and familial hypercholesterolemia (5%). Familial hypercholesterolemia is an autosomal dominant disease. Heterozygotes have a 2–3-fold increase in serum cholesterol and an increase in LDL. Homozygotes (extremely rare) have a 6–8-fold increase in these lipids. Tendon xanthomas are common (75%). Polygenic hypercholesterolemia, the most common type, is not typically associated with a clear-cut family history or tendon xanthomas. Most patients have a type II-a phenotype. Familial combined hypercholesterolemia is an autosomal dominant disease characterized by normal serum cholesterol and triglyceride levels at birth that begin to increase at puberty. These patients also have an increased frequency of obesity, hyperuricemia, and glucose intolerance. They do not have tendon xanthomas. They have either a type II-a or II-b phenotype.

20. The answer is C. *(Cell biology; Hematopoietic/lymphoreticular)*
Alkaline phosphatase is the marker enzyme for the mature neutrophil. Alkaline phosphatase is a lysosomal enzyme that appears in specific granules (lysosomes). Specific granules are more abundant than azurophilic granules in neutrophils.

The specific granules of basophils contain eosinophilic chemotactic factor, heparin, histamine, and peroxidase. Monocytes contain some azurophilic granules that contain digestive enzymes, but no alkaline phosphatase. The granules of eosinophils contain several enzymes, including acid phosphatase, arylsulfatase, cathepsin, phospholipase, and RNAase; but no alkaline phosphatase. The promyelocyte is an early cell in the differentiation pathway of granulocytes. It is characterized by basophilic cytoplasm and large azurophilic granules. These granules contain lysosomal enzymes and myeloperoxidase, thus myeloperoxidase is the marker enzyme of the immature neutrophil. Promylocytes give rise to the three known granulocytes.

21. The answer is C. *(Histology; Pulmonary/respiratory)*
Respiratory bronchioles are similar to terminal bronchioles except that alveoli interrupt the walls of respiratory bronchioles. Smooth muscle is present beneath the epithelium and cilia are present on simple cuboidal cells. Glands and cartilage are present in bronchi, but absent in bronchioles. Goblet cells persist through the earlier bronchioles but disappear in the respiratory bronchioles. Cilia remain, allowing mucus to be swept clear of the respiratory portion. Elastic fibers extend through the bronchioles and also encircle the openings of the alveoli. This allows for expansion during inspiration.

22. The answer is D. *(Histology; Cardiovascular)*
Fenestrated capillaries are characterized by the presence of fenestrae that provide channels across capillary walls. Pinocytotic vesicles may also be present. Endocrine glands produce hormone that quickly accesses the blood stream via these fenestrated capillaries.

Lung contains continuous capillaries joined by tight junctions. Liver contains discontinuous capillaries, with gaps between endothelial cells. These capillaries are also called sinusoidal capillaries, or sinusoids. Muscle and the central nervous system contain continuous capillaries.

23. The answer is D. *(Pathology, Immunology; Skin/connective tissue)*
The patient most likely has a hypersensitivity vasculitis syndrome (HVS) related to penicillin. HVS encompasses a group of diseases that produce an immune-complex vasculitis involving small vessels such as arterioles, capillaries, and postcapillary venules (the most common site). Small-vessel vasculitis is recognized by

the presence of palpable purpura. When immune complexes deposit in the vessel, they activate the complement system, which leads to the generation of C5a, a chemotactic agent that attracts neutrophils to the area. Neutrophils infiltrate and disrupt the integrity of the vessel, resulting in thrombosis and rupture. Antineutrophil cytoplasmic antibodies with a perinuclear immunofluorescent pattern are noted in most cases. Histologically, vessels exhibit leukocytoclastic angiitis consisting of a neutrophilic infiltrate, nuclear dust, and fibrinoid necrosis. Unlike polyarteritis nodosa, in which lesions are present in varying stages of development, HSV lesions share the same stage of development. Diseases associated with HVS include:

- Serum sickness
- Drug reactions to penicillins, sulfonamides, or allopurinol
- Henoch-Schönlein purpura (the most common hypersensitivity vasculitis in children)
- Connective tissue diseases (e.g., systemic lupus erythematosus, rheumatoid arthritis)
- Malignancy

24. The answer is C. *(Pathology; Cardiovascular)*
The schematic represents pulsus paradoxus, in which the amplitude of a pulse diminishes with inspiration. It correlates with a blood pressure drop greater than 10 mm Hg during inspiration, and may be caused by pericardial effusion. On inspiration, blood is drawn into the right side of the heart by a diminishing intrathoracic negative pressure as the diaphragms descend like a giant bellows. Normally, the excess return of blood to the right ventricle is easily handled by the pulmonary vascular bed, resulting in a net reduction of blood flow to the left side of the heart. This should result in a systolic pressure drop on inspiration of less than 10 mm Hg. Pericardial effusion, however, markedly impairs the return of blood to the right side of the heart, causing reflux of blood into the jugular vein on inspiration (Kussmaul sign) and pulsus paradoxicus with a systolic drop in pressure greater than 10 mm Hg, and often greater than 20 mm Hg.

Left-sided heart diseases such as aortic stenosis, essential hypertension producing left ventricular hypertension, coarctation of the aorta, and mitral valve prolapse do not significantly affect venous return to the right side of the heart, so pulsus paradoxicus is not seen in these settings.

25. The answer is D. *(Pathology; Skin/connective tissue)*
The lesions are most consistent with Kaposi sarcoma. The patient would most likely test positive for the HIV antibody. These lesions frequently progress from patches, to plaques, to nodules, which commonly ulcerate. The lesions usually appear on the skin and on mucocutaneous and visceral tissue, such as the lung and gastrointestinal tract. Histologic sections reveal slit-like spaces with tufts of malignant cells projecting into the lumen. Red blood cells are present in the vessel lumens and hemosiderin is present in surrounding tissue. Kaposi sarcoma is the most common cancer associated with AIDS. It affects 25% of AIDS patients, and predisposes to secondary malignancies, such as malignant lymphoma, acute leukemia, and multiple myeloma.

The pathogenesis of Kaposi sarcoma is still speculative. HIV has been implicated as a cofactor. Presumably, endothelial cells are the cell of origin for the neoplastic proliferation. However, smooth muscle cells, pericytes, and, most recently, primitive mesenchymal cells activated either by various cytokines or by the HIV virus itself have been added to the list of suspects.

26. The answer is C. *(Pathology; Skin/connective tissue, Cardiovascular)*
The patient has xanthelasmas, which are yellowish plaques located around the eyes. Sections through these lesions reveal foamy macrophages with increased cholesterol. Approximately 50% of patients with xanthelasmas have elevated low-density lipoprotein (LDL), or type II hyperlipoproteinemia. Xanthelasmas are also associated with primary biliary cirrhosis, in which obstruction of bile flow resulting from autoimmune destruction of the bile ducts in the portal triads interferes with the excretion of cholesterol.

27. The answer is B. *(Hematology; Hematopoietic/lymphoreticular)*
Poikilocytosis is variability of erythrocyte shape in blood. Sickle cells, teardrop cells, and elongated erythrocytes are all evident in this smear, which consists

totally of erythrocytes. Some nucleate erythrocytes are present.

Reticulocytes are young erythrocytes (recently released from the bone marrow) with ribosomal RNA present in a reticular pattern. A supravital stain is necessary to view this pattern. Reticulocytes make up about 1% of circulating erythrocytes. Neutrophils with more than five nuclear lobes are hypersegmented. Typically these are older cells, but in some pathologic conditions, young cells may have five or more lobes. Howell-Jolly bodies are round, solid-staining, dark-blue to purple inclusions in erythrocytes. They are predominately composed of DNA and thus may represent fragments of the nucleus. An increase in the proportion of immature neutrophils is called a shift to the left.

28. The answer is C. *(Biochemistry; Gastrointestinal)*
The concentration of low-density lipoprotein (LDL) is usually calculated rather than directly measured. The following calculation is used:

LDL = cholesterol − high-density lipoprotein − triglyceride/5

In this case, the calculation is as follows:

350 − 20 − 450/5 = 240 mg/dl

29. The answer is C. *(Hematology; Hematopoietic/lymphoreticular)*
Platelet aggregation studies evaluate the in vitro response of platelet-rich plasma to aggregating agents such as adenosine diphosphate (ADP), epinephrine, collagen, and ristocetin. When ADP is added to a test tube filled with platelet-rich plasma, ADP normally causes the platelets to aggregate (phase I in diagram A). The first phase of aggregation is followed by the platelets releasing their own ADP, which causes further aggregation of the platelets (phase II in diagram A). These changes are recorded on a densitometer tracing of the reactions.

If a patient is taking aspirin, which inhibits thromboxane A_2 formation, the patient's platelets will have a normal response to exogenously added ADP (phase I) but no response on phase II, because platelets cannot generate their own ADP (diagram B). This patient was most likely taking aspirin to prevent further coronary artery thrombosis.

Ristocetin causes normal platelets that contain von Willebrand factor and receptors for von Willebrand factor to aggregate. The platelets of patients with von Willebrand disease, however, do not aggregate in the presence of ristocetin because they are missing von Willebrand factor. However, because platelets do aggregate in the presence of ADP, a phase II would be present.

Patients with thrombocytopenia should have a normal aggregation study, because platelet function is normal. Absence of the fibrinogen receptors (IIb/IIIa) on platelets interferes with platelet aggregation, because fibrinogen links the platelets together to form a loose clot. Glanzmann disease (thrombasthenia) is characterized by absent IIb/IIIa fibrinogen receptors. Platelet aggregation studies demonstrate no phase I or II response.

30. The answer is D. *(Pathology; Tissue biology and associated response to disease)*
The patient has thrombotic thrombocytopenic purpura (TTP). In this female-dominant disease, the endothelium in the microvasculature is damaged by a circulating factor in plasma, causing platelet thrombi to develop over the areas of endothelial injury. The platelet thrombi damage red blood cells, resulting in intravascular hemolysis. There is widespread small vessel thrombosis and organ injury. The classic constellation of findings includes (1) a microangiopathic hemolytic anemia (platelet thrombi) with schistocytes (fragmented red blood cells), (2) thrombocytopenia, (3) renal failure, (4) neurologic disease, and (5) fever. Plasma pheresis is extremely useful in removing the endothelial damaging agent. The prognosis is poor.

TTP is not a variant of disseminated intravascular coagulation (DIC), because the clotting factors (fibrinogen, V, and VIII) are not consumed. The prothrombin time and partial thromboplastin time are normal in TTP, but are prolonged in DIC.

Idiopathic thrombocytopenic purpura and autoimmune thrombocytopenia are not associated with the constellation of findings listed for TTP.

Von Willebrand disease is not associated with anemia and renal failure. It is a disease with a qualitative platelet and deficiency of the factor VIII complex.

31. The answer is B. *(Pathology; Tissue biology and associated response to disease)*

The patient has antithrombin III (AT III) deficiency. Factors suggesting a hereditary thrombotic disorder include (1) a family history of deep venous thrombosis with or without pulmonary embolus, (2) deep venous thrombosis and/or pulmonary embolus at a young age, and (3) thrombosis in unusual sites, such as the axilla or dural sinuses. AT III normally inhibits 75% of thrombin and factor X, but also inhibits other serine proteases. Heparin enhances its activity at least 2000-fold. AT III is synthesized in the liver. Acquired causes of AT III deficiency include birth control pills and liver disease. Hereditary AT III deficiency is autosomal dominant. When initially treating the patient with heparin, there is a failure to prolong the partial thromboplastin time, since heparin's function as an anticoagulant is through AT III. Treatment involves the use of very high doses of heparin when acute thrombosis is present, with possible use of fresh frozen plasma (contains AT III) or AT III concentrates as adjuncts. Warfarin is primarily reserved for prophylaxis.

Von Willebrand disease, antibodies against heparin, and a qualitative platelet defect are not associated with signs of hypercoagulability, which have been found in this patient.

Protein C deficiency is another cause of a hypercoagulable state, but it is not resistant to heparin therapy.

32. The answer is E. *(Pathology; Tissue biology and associated response to disease)*
Vitamin K levels would most likely be normal in a woman on birth control pills, which have no adverse effect on vitamin K.

The vitamin K-dependent factors are prothrombin (II), VII, IX, X, protein C, and protein S. The liver produces these factors in a nonfunctional form. Vitamin K_1 renders them functional by γ carboxylation of their terminal amino acid residues. Patients with vitamin K deficiency have a hemorrhagic diathesis.

Causes of vitamin K deficiency include:

- Fat malabsorption because vitamin K is a fat-soluble vitamin. Malabsorption may be due to (1) bile salt deficiency (e.g., liver disease, obstruction to bile flow, bacterial overgrowth, terminal ileal disease), (2) intestinal diseases (e.g., celiac disease, Whipple disease), or (3) pancreatic disease (e.g., deficient lipase).
- Broad spectrum antibiotics, which destroy colonic bacteria that normally synthesize vitamin K
- Warfarin derivatives, which inhibit epoxide reductase, an enzyme that keeps vitamin K in an active form

Hemorrhagic disease of the newborn is due to the normal drop of vitamin K in the blood 2 to 5 days after birth. It is caused by the lack of bacterial colonization of the newborn colon that is necessary for vitamin K synthesis. This is the rationale for giving intramuscular vitamin K to newborns in the hospital. However, infants born outside the hospital who do not receive the injection may be at risk for developing serious bleeds, the most disastrous of which is bleeding in the central nervous system.

Vitamin K deficiency resulting in severe bleeding may be corrected immediately by giving the patient fresh frozen plasma plus 10 mg of subcutaneous vitamin K. Less significant bleeding may be controlled with intramuscular injection of vitamin K.

33. The answer is D. *(Pathology; Biochemistry and molecular biology)*
The patient has lead poisoning. The history of growth retardation, abdominal colic, and the peripheral blood findings of a microcytic anemia associated with coarse basophilic stippling in the red blood cells is virtually pathognomonic for this diagnosis. The blood lead level is the screening and confirmatory test of choice.

Lead poisoning is most prevalent in the inner city, where the most common source of poisoning in children is eating paint containing lead. In adults, it is primarily an occupational hazard in battery factories. Less common causes include spray paints containing lead, automobile exhaust, and lead mining. Lead is readily absorbed in the lungs (inhalation) and the gastrointestinal tract. It denatures the sulfhydryl groups in ferrochelatase in the mitochondria that normally binds iron to protoporphyrin to form heme. It also denatures aminolevulinic acid (ALA) dehydrase, an enzyme that converts δ ALA into porphobilinogen. Because ferrochelatase is denatured, iron cannot combine with protoporphyrin to form heme, thus interfering with hemoglobin synthesis in the developing normoblast. Unbound iron accumulates in the mitochondria. Because mitochondria are located around the nucleus, the excess iron produces a ringed sideroblast (normoblast with iron). Lead poisoning is one of the many causes of sideroblastic anemia characterized by the presence of ringed sideroblasts. Lead also denatures ribonuclease and the breakdown of ribosomes resulting in coarse basophilic stippling. In the brain, lead interferes with enzymes resulting in central nervous system disturbances (cerebral edema) and demyelination. Demyelination of peripheral nerves produces neuropathies such as wrist and foot drop. In the gastrointestinal tract of patients with lead poisoning, there is frequently a lead line along the gums and abdominal colic. Like most heavy metals (e.g., arsenic, mercury), lead produces a proximal nephrotoxic tubular necrosis. However, unlike the other heavy metals, it deposits in the epiphyses and is easily visualized on radiograph.

Laboratory findings other than those previously discussed include increased levels of free erythrocyte protoporphyrin, which builds up behind the ferrochelatase block, and increased δ ALA in the urine, resulting from the block of ALA dehydrase. Patients are usually treated with a combination of BAL (British anti-lewisite) and ethylenediamine tetraacetic acid. EDTA Serum iron, total iron binding capacity, and percent saturation are usually normal in lead poisoning.

A bone marrow aspirate/biopsy is not necessary in diagnosing lead poisoning.

Hemoglobin electrophoresis primarily is used in the diagnosis of hemoglobinopathies, like the thalassemias and sickle cell variants.

Heinz body preparations are performed to diagnose glucose-6-phosphate dehydrogenase deficiency in the acute phase of hemolysis.

34. The answer is E. *(Clinical medicine; Reproductive)*
Regardless of race, iron deficiency is still the most common cause of anemia in women. Menorrhagia is the primary cause of iron deficiency in women under 50 years of age. Sickle cell disease is the most common genetic anemia in the black population, but not the most common overall anemia. Peptic ulcer disease is the most frequent cause of iron deficiency in men less than 50 years of age. Hematuria is rarely a cause of anemia. Inflammatory bowel disease (ulcerative colitis, Crohn disease) is not a common cause of severe anemia.

35. The answer is B. *(Pathology; Biochemistry and molecular biology)*
Reduced hemoglobin (Hgb) synthesis is the primary mechanism for the microcytic anemias. These include (1) iron deficiency, the most common microcytic anemia, (2) anemia of chronic disease (ACD), (3) the thalassemias (α and β), and (4) the sideroblastic anemias, which include lead poisoning. Hgb is composed of heme (iron + protoporphyrin) and globin chains (α, β, δ, and γ). A decrease in Hgb concentration in a developing normoblast in the bone marrow results in additional mitoses, which produces a microcytosis. The mean corpuscular volume is a red blood cell (RBC) index of cell size. In both iron deficiency and ACD, iron delivery to the developing RBCs is decreased. In iron deficiency, there is an actual decrease in iron stores, most commonly from blood loss. In ACD, the iron is blockaded in macrophages and is not available for Hgb synthesis. Problems in forming heme occur in sideroblastic anemias due to (1) reduced synthesis of protoporphyrin (e.g., pyridoxine deficiency, lead poisonng), (2) inability to combine iron with protoporphyrin via ferrochelatase (e.g., lead poisoning), or (3) generalized damage of the mitochondria by alcohol, a mitochondrial poison. These anemias are called sideroblastic because the iron accumulates in the mitochondria. The thalassemias are autosomal recessive diseases in which problems with globin chain synthesis are a hallmark. α-Thalassemia is characterized by a decrease in α-chain synthesis, whereas β-thalassemia is characterized by a decrease in β-chain synthesis. In both types, there is a normal production of the uninvolved chains. It is possible to have a microcytosis (decreased mean corpuscular volume) without anemia in very mild α- and β-thalassemia and in patients with polycythemia rubra vera, a myeloproliferative disease. In polycythemia rubra vera, patients are phlebotomized to decrease RBC mass and to render them iron deficient so there is less RBC proliferation.

Folate deficiency produces a defect in DNA synthesis resulting in a macrocytic anemia.

Congenital spherocytosis is an autosomal dominant disease with a defect in spectrin in the cell membrane resulting in a hemolytic anemia.

Myeloproliferative diseases are disorders involving the stem cells in the bone marrow.

Leukemias are malignancies originating in the bone marrow.

36. The answer is E. *(Hematology; Hematopoietic/lymphoreticular)*
Hemoglobinopathies refer to abnormalities in globin structure or globin synthesis. Examples include sickle cell disease and the thalassemias, in which there is a decrease in the production of globin chains. Hispanics are least likely to have a hemoglobinopathy when compared with the other ethnic groups listed in the question. Blacks have an increased incidence of sickle cell trait/disease and α- and β-thalassemias. Asians have an increased incidence of α-thalassemia and hemoglobin E disease. The Greek and Italian populations have an increased incidence of β-thalassemia.

37. The answer is E. *(Pathology; Microbial biology and infection)*
Ascariasis would most likely have a normal hemoglobin and hematocrit, because the adult worms live free in the lumen of the bowel and do not attach to the mucosal surface. Falciparum malaria produces an intravascular hemolysis with hemoglobinuria, the latter assuming a black color in the presence of an acid pH (blackwater fever). Ancyclostomiases, or hookworm disease, is associated with iron deficiency, because the adult worms attach to the tips of the villi and feed on the blood. Diphyllobothriasis, or fish tapeworm disease, produces a B_{12} deficiency. Babesiosis, contracted by the bite of *Ioxides dammini*, is an intraerythrocytic parasite that produces intravascular hemolysis.

38. The answer is D. *(Genetics, Pathology; Biochemistry and molecular biology)*
The patient has congenital spherocytosis, which is an autosomal dominant disease with an increased prevalence in people of Northern European extraction. Congenital spherocytosis is an example of an intrinsic hemolytic anemia with extravascular hemolysis. The intrinsic abnormality is a membrane defect in spectrin, which decreases the amount of red blood cell membrane resulting in a low surface-to-volume ratio and spherocyte formation. Spherocytes are trapped in the splenic cords because they are not flexible enough to enter through the narrow openings into the sinusoids. Stagnation in the cords results in the red blood cell release of lactic acid, which decreases pH and decreases glycolysis. Reduced glycolysis results in less adenosine triphosphate formed; therefore, the spherocytes have difficulty in extruding sodium and using glucose. Macrophages in the splenic cords remove the spherocytes, resulting in an extravascular hemolytic anemia. Anemia, splenomegaly, and jaundice (indirect hyperbilirubinemia) are a common triad. A family history of splenectomy or gallstones at an early age from increased bilirubin metabolism (calcium bilirubinate stones) is strongly suggestive of spherocytosis. Patients usually have a normocytic anemia with a corrected reticulocyte count greater than 3%. Spherocytes are small, round cells with no central pallor. Because they have less membrane, the hemoglobin concentration is increased, which is reflected by an increased mean corpuscular hemoglobin concentration. The osmotic fragility test is the confirmatory test for spherocytosis. This test documents increased susceptibility of the spherocytes to hemolysis in hypotonic saline solutions when compared with normal cells. For example, spherocytes begin hemolyzing at 0.65%, whereas normal red blood cells begin hemolyzing at 0.50%. Splenectomy is the treatment of choice. Patients should receive Pneumovax before splenectomy to protect the patient against *Streptococcus pneumoniae* sepsis. Spherocytes remain in the peripheral blood after splenectomy, but do not hemolyze in the absence of a spleen.

An abnormal hemoglobin electrophoresis with a patient history of gallbladder disease may occur in sickle cell disease. However, the patient is white and the spleen is enlarged rather than autosplenectomized.

A positive direct Coombs test indicates the presence of an autoimmune hemolytic anemia. Spherocytes are present in autoimmune hemolytic anemia, but most patients have a history of another autoimmune disease (e.g., systemic lupus erythematosus).

An abnormal Schilling test is noted in B_{12} deficiency. It is a defect in DNA synthesis, which is not normally associated with splenomegaly and an increased incidence of gallstones.

An increased urine for hemosiderin is noted in chronic intravascular hemolysis. Spherocytosis is an extravascular hemolysis.

39. The answer is C. *(Pathology; Tissue biology and associated response to disease)*
The cells are myeloblasts with auer rods in the cytoplasm. This finding is pathognomonic for acute myelogenous leukemia, the most common leukemia in

individuals aged 15 to 39 years. Acute myelogenous leukemia is tied with chronic myelogenous leukemia for commonality in individuals aged 40 to 59 years. Possible causes of acute myelogenous leukemia include radiation, alkylating agents, benzene, immunodeficiency states, and Down syndrome. The classification of the acute nonlymphocytic leukemias, using the French-American-British classification is as follows.

- MO: minimally differentiated acute myelogenous leukemia; incidence of 2%–3%; no auer rods
- M1: acute myelogenous leukemia without differentiation; incidence of ~20%; rare auer rods
- M2: acute myelogenous leukemia with maturation; incidence of 30%–40%; most common acute myelogenous leukemia; auer rods are easy to find; a t(8;21) translocation is a favorable sign
- M4: acute myelomonocytic leukemia; incidence of 15%–20%; auer rods are not usually present; chromosome 6 abnormalities (inversion) associated with increased eosinophils, which indicates a better prognosis
- M5: acute monocytic leukemia; incidence ~10%; no auer rods present; gum infiltration is common
- M6: acute erythroleukemia (Di Guglielmo); incidence ~5%; bizarre, multinucleated and often megaloblastoid appearing erythroblasts that are periodic acid-Schiff positive
- M7: acute megakaryocytic leukemia; incidence ~1%; myelofibrosis in the marrow; the platelet peroxidase stain is positive on electron microscope findings

Lymphocytic and monocytic leukemias do not have auer rods. Systemic mastocytosis and histiocytosis X also are not associated with auer rods.

40. The answer is B. *(Pathology; Tissue biology and associated response to disease)*
The patient has agnogenic myeloid metaplasia, a myeloproliferative disease. The red blood cell (RBC) finding shows a teardrop shape, which is due to membrane damage of the cell when exiting the sinusoids in the fibrosed bone marrow. Agnogenic myeloid metaplasia is uncommon in patients under 60 years of age. It is due to a proliferation of neoplastic stem cells that begin dividing in the bone marrow and later find residence in the spleen, where the primary process continues as a trilineage production of RBCs, granulocytes, and platelets, called extramedullary hematopoiesis. Marrow fibrosis is a reactive phenomenon and is not related to neoplastic fibroblast proliferation. Megakaryocytes in the bone marrow secrete platelet-derived growth factor, which stimulates marrow fibrosis. Marrow fibrosis must be differentiated from metastatic disease to the marrow with reactive myelofibrosis, which is called myelophthisic anemia. Patients have massive splenomegaly from the extramedullary hematopoiesis, which is commonly associated with left upper quadrant pain due to infarctions, often resulting in friction rubs and a left-sided pleural effusion, as in this patient. Laboratory findings include a normocytic anemia, a white blood cell (WBC) count between (10,000 and 50,000 cells/μl, and a leukoerythroblastic smear, with immature WBCs and nucleated RBCs. Teardrop RBCs are prominent.

Polycythemia rubra vera is manifested by an increase in the hemoglobin rather than a normocytic anemia.

Hairy cell leukemia is more common in older men.

Chronic myelogenous leukemia often has marrow fibrosis, but the result of the Philadelphia chromosome study is positive, and the leukocyte alkaline phosphatase score is low.

41. The answer is B. *(Genetics, Pathology, Hematology; Hematopoietic/lymphoreticular)*
The peripheral smears exhibit numerous sickle cells; scattered target cells, which are excellent markers for hemoglobinopathies; and Howell-Jolly bodies, representing remnants of nuclear material. If the spleen was functional in this patient, Howell-Jolly bodies would have been removed, therefore, their presence indicates a functional asplenia.

Sickle cell disease is an intrinsic hemolytic anemia with predominantly extravascular hemolysis. It is an autosomal recessive disease; therefore, both parents must have either the trait [hemoglobin SA (Hgb SA)] or the disease (Hgb SS). The prevalence of Hgb SS disease is 1 in 600 blacks, making it the most common hemoglobinopathy in this population. Valine is substituted for glutamic acid in the 6th position of the β chain. Deoxygenation of the Hgb S molecule causes it to aggregate and polymerize. Sickling is initially reversible, but with repeated sickling and unsickling there is membrane damage, which causes irreversible damage and permanent sickling.

The two main problems in Hgb SS disease are a chronic hemolytic anemia and occlusion of the microvasculature by sickle cells with ischemic damage to multiple organs (vaso-oclusive disease). Factors that induce sickling are (1) reduced oxygen tension, such as high altitude, (2) a concentration of Hgb S greater than 60% in the red blood cells, which is the most important factor, (3) the presence of other Hgbs, like Hgb C, (4) dehydration, which increases the Hgb S concentration, and (5) acidosis, which right shifts the oxygen dissociation curve, decreases oxygen affinity for Hgb, and increases the amount of deoxygenated Hgb resulting in sickling. Due to its high oxygen affinity, Hgb F inhibits sickling, thus preventing deoxygenation. Newborns with Hgb SS disease do not sickle until the Hgb F has been replaced by red blood cells with Hgb A in a few months.

Sickle cells become sequestered in the splenic cords, rendering them susceptible to extravascular removal by macrophages. Vaso-oclusive crisis with organ damage is the most common clinical manifestation. Dactylitis involving the hands and feet in a 4- to 6-month-old infant is the first manifestation. Bone infarcts produce swelling of the hands and feet. Vaso-oclusive crises occur in other organs such as the brain, leading to strokes at a young age; lungs, resulting in hypoxemia; liver, producing liver cell necrosis; spleen, leading to a possible sequestration crisis with severe anemia; and penis, producing priapism (permanent, painful erection).

There is a predisposition for salmonella osteomyelitis, since the spleen is nonfunctional and is unable to filter out the organisms. Aseptic necrosis of the femoral head is common. Children initially have splenomegaly due to sequestration of sickle cells in the cords of Billroth. However, by 3 to 4 years of age, it is nonfunctional, and by adulthood, it is autosplenectomized. Patients have a susceptibility to *Streptococcus pneumoniae* and *Haemophilus influenzae*, the former representing the most common cause of death in children. Microhematuria is seen in both sickle trait and disease, because the low oxygen tension in the renal medulla may induce sickling in the peritubular capillaries, producing microinfarctions, hematuria, dilution and concentration defects, and, possibly, renal papillary necrosis. All blacks with microhematuria should have a sickle cell screen.

Laboratory findings include sickle cells in the peripheral blood only in Hgb SS or Hgb S/thalassemia, but not sickle cell trait (Hgb SA). There is a normocytic anemia with a corrected reticulocyte count greater than 3%. The erythrocyte sedimentation rate is zero, because the sickled cells are unable to aggregate and settle properly in the tube. Sickle cell screens include the solubility test (Sickledex) and the metabisulfite test, which reduces the oxygen tension and induces sickling. A positive sickle screen is always followed by a Hgb electrophoresis, which is the gold standard test. Patients with Hgb SS disease have 80% to 95% Hgb S, small amounts of Hgb A_2 (more if β-thalassemia is present), 5% to 15% Hgb F, and no Hgb A. Sickle cell trait has 35% to 45% Hgb S, 60% Hgb A, and small amounts of Hgb A_2 and F. Recently, hydroxyurea has been approved for the treatment of sickle cell disease, because it increases the synthesis of Hgb F, which reduces the number of vaso-oclusive crises. Pneumovax and *Haemophilus influenzae* vaccine are recommended in all patients with sickle cell disease to prevent septicemia, which occurred in this patient.

42. The answer is B. *(Pathology; Tissue biology and associated response to disease)*
Platelet disorders are more likely associated with epistaxis rather than a coagulation disorder such as hemophilia A. Other findings include symmetrical distribution of petechia and ecchymoses, bleeding of the mucosal membranes, bleeding from superficial scratches, a history of spontaneous bruising with no trauma, and gingival bleeding associated with brushing of the teeth. Platelet defects do not interfere with the prothrombin or partial thromboplastin times, which are tests that evaluate coagulation deficiencies.

Pure coagulation deficiencies, such as hemophilia A, present with hematuria, gastrointestinal bleeds, large solitary ecchymoses, hemarthrosis (bleeding into joints), bleeding into potential spaces (e.g., retropharynx, muscle), continuous oozing after tooth extractions, and late rebleeding, since the vessels only have a temporary platelet plug that keeps them from bleeding.

43. The answer is B. *(Hematology; Hematopoietic/lymphoreticular)*
Chronic pancreatitis is more likely to be associated with B_{12} than folate deficiency, because the R factor attached to the B_{12} molecule must be enzymatically cleaved off in the duodenum for intrinsic factor to

combine with the B_{12} to form a complex that is reabsorbed in the terminal ileum.

Alcoholism is more likely to predispose to folate deficiency, because of poor diet (only 3–4 month supply of folate in the liver).

Celiac disease involving the duodenum and jejunum would have a deleterious effect on reabsorbing iron from the duodenum and folate from the jejunum. Because B_{12} is reabsorbed in the terminal ileum, B_{12} deficiency should not occur.

A patient taking methotrexate may develop folate deficiency because methotrexate blocks dihydrofolate reductase, which converts dihydrofolate back into the active tetrahydrofolate.

A patient taking phenytoin may have problems with the reabsorption of folate because phenytoin inhibits intestinal conjugase, which is necessary to convert polyglutamates into monoglutamates for absorption in the jejunum.

44. The answer is A. *(Hematology; Hematopoietic/lymphoreticular)*
Tissue thromboplastin serves a procoagulant role by activating factor VII in the extrinsic coagulation system to generate small amounts of thrombin for clot formation.

Normal hemostasis requires the interaction of vessels, platelets, the coagulation system, and the fibrinolytic system. Factors inhibiting clot formation on the vessel endothelium (anticoagulant activities) include:

- The generation of prostacyclin (PGI_2) by vessel endothelium. Prostacyclin inhibits platelet aggregation and is a vasodilator, which increases blood flow and decreases the chances for platelets to stick to the endothelium.
- The release of heparan sulfate, which enhances antithrombin III (AT III) activity. AT III neutralizes those coagulation factors that are serine proteases, such as plasmin, thrombin, factors XII (Hageman factor), XI, X, IX, and II (prothrombin).
- Thrombomodulin, which is involved in the activation of proteins C and S. Proteins C and S inactivate factors V and VIII in the coagulation system and enhance fibrinolytic activity, thus preventing clot formation.
- The release of tissue plasminogen activator activates plasminogen to form plasmin, which breaks up clots.

Procoagulants other than tissue thromboplastin that help form a clot in an injured vessel are:

- Von Willebrand factor (VIII:vWF), which is synthesized by the endothelial cells. When VIII:vWF is exposed after endothelial injury, it binds to platelet receptors, which causes the platelets to adhere to the endothelium. This factor is absent in classic von Willebrand disease.
- Platelet adhesion, which is a stimulus for the generation of thromboxane A_2 (TXA_2) by the platelet. The platelet synthesis of TXA_2 is inhibited by nonsteroidal drugs via their inhibition of platelet cyclooxygenase activity. TXA_2 stimulates the platelet release reaction, which involves the release of numerous chemical mediators from both the dense bodies and α granules within the platelets. TXA_2 is also a potent vasoconstrictor, which enhances the chances for platelet adhesion. Platelet aggregation occurs due to the presence of adenosine diphosphate and thrombin from the release reaction. The platelets are held loosely together by fibrinogen connecting IIb/IIIa receptors on the platelet membranes, thus forming a temporary hemostatic plug, which stops bleeding. The temporary hemostatic plug is unstable. It undergoes further alterations to become a stable clot, which requires the participation of the coagulation system. Tissue thromboplastin is released the moment vessel injury occurs. It activates factor VII in the extrinsic coagulation system, which generates thrombin. The exposed collagen on the vessel activates the intrinsic coagulation system vis factor XII, or Hageman factor. Thrombin from both systems converts the fibrinogen strands loosely connecting the platelets together into fibrin monomers, thus forming a stable clot. Deficiencies of certain of the clotting factors thwart the ability of the coagulation system to generate thrombin, thus inhibiting the formation of a stable clot. In patients with these deficiencies, only temporary hemostatic plugs are restricting blood flow. Fibrinolysis of the clot occurs by activation of the fibrinolytic system, thus restoring blood flow. Factor XII, or Hageman factor, has the additional function of activating the fibrinolytic system. Tissue plasminogen activator is also released into the circulation with vessel injury; therefore, it cooperates with factor XII in stimulating fibrinolysis.

45. The answer is E. *(Hematology; Hematopoietic/lymphoreticular)*
The blood groups for patients A through D are as follows:

Patient	Forward Type Using Anti-A	Forward Type Using Anti-B	Back Type Using A RBCs	Back Type Using B RBCs	Interpretation Blood Group
A	Positive	Negative	Negative	Positive	A with anti-B
B	Negative	Negative	Positive	Positive	O with anti-A and anti-B
C	Negative	Positive	Positive	Negative	B with anti-A
D	Positive	Positive	Negative	Negative	AB with no antibodies

To safely transfuse group A packed red blood cells, the patient would have to be either blood group A (patient A), since the patient's anti-B would not react with donor A cells, or blood group AB (patient D), since AB patients do not have anti-A or -B antibodies to react with donor A cells. Patient B, who is blood group O, has anti-A and -B antibodies that would destroy group A donor cells, and patient C, who is blood group B. Patients who are blood group O may only receive O blood, but they may donate blood to any blood group (universal donor), because the O cells lack A and B antigen. Patients with blood group AB may be transfused with any blood group (universal recipients), because they lack antibodies against A and B antigens.

46. The answer is B. *(Hematology, Pathology; Hematopoietic/lymphoreticular)*
The patient has a febrile transfusion reaction. The three types of transfusion reactions are (1) febrile reactions (most common), (2) allergic reactions, and (3) hemolytic transfusion reactions. Fever is seen in all three types of reactions; therefore, a transfusion reaction work-up is always indicated to distinguish which type is present. The pathogenesis of febrile reactions involves the reaction of cytotoxic antibodies directed against the donor leukocytes. The leukocytes are destroyed and release pyrogens that initiate a febrile reaction. It is a type II hypersensitivity reaction. Patients with these antibodies must have had previous exposure to blood components from a transfusion or pregnancy, since they are not normally present in the serum. Symptoms and signs include fever and chills approximately 1 hour after infusion accompanied by headache, flushing, and tachycardia. Patients are treated with antipyretics. These reactions are prevented by using microaggregate blood filters to filter out the leukocytes while the blood is being infused into the patient.

Allergic reactions are due to IgE antibodies (type I hypersensitivity) directed against plasma proteins in the donor plasma. Signs and symptoms range from urticaria to possible anaphylactoid reactions. Fever, tachycardia, wheezing, dyspnea, and cyanosis may also be present. Nearly all reactions respond to oral or intramuscular antihistamines. Severe anaphylactoid reactions occur in IgA-deficient patients who have antibodies against IgA in their plasma. IgA-deficient patients should never receive any blood component.

Type III hypersensitivity reactions (immune complexes) and type IV hypersensitivity (cellular immunity) are not associated with transfusion reactions.

Bacterial contamination of the unit of blood is a rare complication and is associated with shock and a high rate of mortality.

47. The answer is B. *(Hematology; Hematopoietic/lymphoreticular)*
For this woman to already have anti-D antibodies, she must be Rh negative and have been sensitized to D antigen during her first pregnancy. If she was O negative and her first child either blood group A, AB, or B- and D-antigen positive, she would not have become sensitized. A group O Rh-negative woman

with an A, AB, or B fetus would be ABO incompatible with her fetus. If at delivery, fetal A, AB, or B red blood cells (RBCs) that are D-antigen positive entered her blood stream by a fetal–maternal bleed, the mother's anti-A IgM and anti-B IgM antibodies would immediately destroy the fetal cells intravascularly. IgM antibodies are potent activators of the classical complement pathway, and all of the fetal cells would be subject to complement destruction within seconds without any chance for maternal sensitization against the D antigen. Therefore, ABO incompatibility protects women against Rh sensitization. Because this woman has already been sensitized, her first pregnancy was not likely complicated by ABO incompatibility.

During a normal pregnancy, fetal RBCs often gain access to the mother's circulation, usually during delivery. If the fetus's RBCs are D-antigen positive and the mother is D-antigen negative, the mother may become sensitized (develop anti-D antibodies) unless she received Rh immune globulin during her pregnancy or after her delivery. The newborn is not affected in the first Rh incompatible pregnancy. However, there is an increased risk for Rh hemolytic disease of the newborn in subsequent pregnancies if the woman develops anti-D antibodies. Assuming that she has another D-antigen positive fetus, her anti-D IgG antibodies will cross the placenta, attach to the fetal RBCs, and be extravascularly destroyed by macrophages in the fetal spleen and liver. Severe anemia may develop in the fetus with subsequent enlargement of the spleen and liver secondary to extramedullary hematopoiesis as compensation for the anemia. The indirect bilirubin, which is the end-product of extravascular hemolysis, is primarily handled by the mother's liver and to a lesser extent by the fetus.

48. The answer is E. *(Hematology; Hematopoietic/lymphoreticular)*
The patient most likely has a delayed hemolytic transfusion reaction, whereby antibodies in her serum are destroying donor red blood cells (RBCs).

Hemolytic transfusion reactions are subdivided into intravascular and extravascular hemolysis. Clinical findings in either type include hypotension, burning at the site of infusion, fever, chills, headache, pain in the lower back or chest, bleeding from intravenous sites and/or surgical wounds (disseminated intravascular coagulation), and evidence of hemoglobinuria and oliguria (acute tubular necrosis). Intravascular hemolysis is the most serious of the two reactions and is most often the result of ABO mismatched blood being infused into the patient. The majority of cases are due to human error on the part of the nurse, medical student, or physician. For example, a group A patient may receive B blood if the identification number of the unit of blood is not matched with the number of that unit previously placed on the patient's arm bracelet by the blood bank. In this unfortunate situation, anti-B IgM antibodies in the patient (people with blood group A have anti-B IgM antibodies) will attach to all the infused donor B RBCs, activate the classical complement system, and produce a massive intravascular hemolysis as C9 punches holes through the RBC membranes. This occurrence often precipitates hypovolemic shock, disseminated intravascular coagulation, and renal failure. ABO incompatibility is a type II cytotoxic antibody hypersensitivity reaction.

The patient in this case most likely has an extravascular hemolysis due to an undetected atypical antibody in her plasma against a donor RBC antigen that she does not have. Although these antibodies are usually detected in the pretransfusion work-up with an antibody screen of the patient's serum, the titers of antibody are sometimes too low for detection. For discussion purposes, let us assume that the patient is Kell antigen negative and has an anti-Kell antibody from previous exposure to the antigen. Reexposure of her memory B cells to Kell antigen on RBCs from the infused donor unit will result in either a rapid or, as in this case, a delayed synthesis of IgG anti-Kell antibodies. These IgG antibodies will eventually coat all the Kell antigen–positive donor RBCs, which will be removed extravascularly by the macrophages in the patient's spleen, liver, and bone marrow. The removal of macrophages produces fever, jaundice from indirect hyperbilirubinemia, a drop in the hemoglobin, and a positive Coombs test, which detects the anti-Kell antibodies on the surface of the donor RBCs. This type of hemolytic transfusion reaction is also a type II hypersensitivity reaction, but the mechanism of hemolysis is extravascular rather than intravascular. Treatment involves immediate cessation of the transfusion and delivery of the unit and a sample of patient blood to the blood bank for a STAT transfusion reaction work-up.

Fever, a drop in hemoglobin, jaundice, and a positive direct Coombs test are never expected responses from any transfusion of donor RBCs.

ABO incompatibility with one of the donor units would not be delayed in its presentation, as previously noted in the discussion.

Leukocyte antibodies directed against a leukocyte antigen from one of the donor units are the cause of febrile reactions. This patient's reaction is a hemolytic transfusion reaction.

Nonspecific antibodies against a plasma protein from one of the donor units would produce an allergic reaction, not a hemolytic transfusion reaction.

49. The answer is D. *(Hematology; Hematopoietic/lymphoreticular)*
The patient has infectious mononucleosis (IM). The peripheral smear shows a large lymphocyte with ample cytoplasm and a coarse chromatin pattern.

IM is caused by the Epstein-Barr virus (EBV). Transmission is through kissing, because the virus is in high concentration in the saliva. EBV attaches to EBV receptors on the B cells and remains within the B cells for an indefinite period of time. Atypical lymphocytes, like the one in the figure, are T cells reacting against the infected B cells.

IM is primarily a disease of adolescents and young adults. The hallmarks of the disease are fever (seldom > 104°F), fatigue, painful anterior and/or posterior cervical lymphadenopathy, and an exudative tonsillitis, mainly due to EBV and less frequently group A streptococcus. Petechia on the soft palate is a characteristic finding. Anicteric hepatitis is the rule. Painful hepatosplenomegaly is commonly present. Patients who are placed on ampicillin often develop a pruritic, maculopapular rash within 24 hours that is not an allergic reaction to the drug.

The majority of patients have a positive result on heterophile antibody test (Monospot). The heterophile antibody unique to IM is an IgM antibody against sheep and horse red blood cells. This is the basis for the Monospot test. The test is positive in the first 2 to 3 weeks of the disease in 90% of patients. False-negatives more commonly occur in children and in the elderly. The test becomes nonreactive over the ensuing 8 to 12 weeks. In the peripheral blood, there is an initial leukopenia followed by atypical lymphocytosis. If the heterophile test is negative, EBV serologies are useful, albeit expensive. The anti VCA (viral capsid antigen)-IgM is the first marker of the disease and has a sensitivity of 100%. Anti-EA (early antigen) has a sensitivity of 70% in acute IM, whereas anti-EBNA (Epstein-Barr nuclear antigen) occurs late in the disease. Almost all patients have hepatitis with an increase in transaminases, which are markers of liver cell necrosis, and an increase in alkaline phosphatase from a mild intrahepatic cholestasis. The total bilirubin is normal in the majority of cases. The hepatitis invariably resolves and is not associated with a chronic state.

There is no specific treatment for IM. Bed rest is recommended in severe cases. Patients should be advised to avoid contact sports for 6 weeks to prevent splenic rupture.

Other relationships with EBV include (1) polyclonal malignant lymphoma, (2) hairy leukoplakia of the tongue in HIV infections, (3) sex-linked lymphoproliferative disease, (4) Burkitt lymphoma, and (5) nasopharyngeal carcinoma.

50. The answer is E. *(Pathology; Hematopoietic/lymphoreticular)*
Malignant T cells in the peripheral blood, hypercalcemia, and lytic lesions in bone best describe adult T-cell leukemia associated with the human T-cell leukemia virus (HTLV) type 1 retrovirus. This type of leukemia is common in Japan and is only sporadic in the United States. The majority of patients are middle-aged men. Clinical findings include lymphadenopathy, hepatosplenomegaly, skin infiltration (which is a common with any T-cell neoplasm), and lytic lesions in bone often associated with hypercalcemia (90%). Malignant T cells are noted in the peripheral blood, lymph nodes, and bone marrow. Hypercalcemia is caused by the secretion of osteoclast-activating factor by the T lymphoblasts in the bone marrow. The majority of patients die in 1 year despite chemotherapy. There is no association with the Epstein-Barr virus, HIV-1, benzene, or radiation exposure.

51. The answer is B. *(Pathology; Tissue biology and associated response to disease)*
Most malignant B-cell lymphomas derive from the germinal follicles in lymph nodes, where B cells are normally located. Approximately 35% of adult and 50% of childhood lymphomas derive from follicular

cells. In the Luke-Collins classification, these are designated follicular lymphomas. The sinuses of the lymph nodes are the primary site of origin for histiocytic disorders (e.g., malignant histiocytosis), which are rare diseases. Bone marrow lymphocytes are the primary source for acute and chronic lymphocytic leukemias. The paracortical tissue of lymph nodes is populated by T cells and are the primary site of origin for malignant T cell lymphomas. The white pulp of the spleen has both B and T lymphocytes. Primary lymphomas of the spleen are extremely rare, and most malignant lymphomas represent metastatic lesions.

52. The answer is E. *(Pathology, Hematology; Hematopoietic/lymphoreticular)*
The majority of white blood cell (WBC) disorders do not require a bone marrow examination to evaluate the myeloid to erythroid ratio. Normally, this ratio is 3:1. It is altered by quantitative changes in the WBCs and red blood cell precursors in the bone marrow. For example, the ratio is decreased in erythroid hyperplasia and increased in leukemia. Evaluation of the peripheral smear morphology, the differential WBC count, and the presence or absence of anemia and thrombocytopenia are the most important initial steps in evaluation of a WBC disorder. Factors that may prompt a clinician to order a bone marrow study include pancytopenia or the presence of immature WBCs in the peripheral blood.

53. The answer is D. *(Hematology, Pathology; Hematopoietic/lymphoreticular)*
The patient has an autoimmune hemolytic anemia (AIHA), most likely the warm type, which involves IgG antibodies. Fever, generalized lymphadenopathy, jaundice, and hepatosplenomegaly are commonly present. A Coombs test is the test of choice for documenting AIHA. The warm type of AIHA has IgG and/or C3b on the surface of the red blood cells (RBCs), which renders them susceptible to extravascular removal by macrophages in the spleen and/or liver. Causes of warm AIHA include autoimmune disease, particularly systemic lupus erythematosus, chronic lymphocytic leukemia, malignant lymphoma (particularly Hodgkin disease), and drugs. The direct Coombs test detects antibody and/or complement on the patient's RBCs. Other laboratory findings in warm AIHA include a normocytic (sometimes macrocytic) anemia with an increased corrected reticulocyte count, indirect hyperbilirubinemia from extravascular hemolysis, and spherocytes in the peripheral blood. The shift cells (polychromasia) and nucleated RBCs in the patient's peripheral blood are the result of accelerated erythropoiesis in her bone marrow, which is trying to keep pace with the extravascular removal of her RBCs. Shift cells are marrow reticulocytes that take at least 2 to 3 days to become mature RBCs.

Osmotic fragility is a test to document congenital spherocytosis. Although spherocytes are present in the patient's blood, her history of systemic lupus erythematosus and severe anemia is more compatible with AIHA than congenital spherocytosis, which would have been a life-long problem. The Coombs test is the best test to distinguish AIHA from congenital spherocytosis.

A Heinz body preparation is used to document the hemolytic anemia associated with glucose-6-phosphate dehydrogenase deficiency, a sex-linked recessive disease. The sex of the patient and lack of association of the hemolysis with infection or drugs rule out this diagnosis.

The Schilling test is used to locate the cause of B_{12} deficiency, which is not a hemolytic anemia.

A bone marrow aspirate/biopsy is not recommended in the work-up of the majority of hemolytic anemias.

54. The answer is B. *(Hematology; Hematopoietic/lymphoreticular)*
The patient has B_{12} deficiency, most likely caused by pernicious anemia, an autoimmune disease characterized by destruction of the parietal cells in the body and fundus of the stomach. The peripheral blood reveals egg-shaped macro-ovalocytes and a hypersegmented neutrophil. Because the liver normally has a 6- to 10-year supply of B_{12}, the likelihood of a nutritional deficiency is remote. The patient does not have folate deficiency, because the physical examination demonstrates decreased vibratory sensation in the lower extremities, consistent with subacute combined degeneration from B_{12} deficiency. A hemolytic process is not present because the corrected reticulocyte count is less than 2%. Defects in hemoglobin synthesis (e.g., iron deficiency, anemia of inflammation, thalassemia, sideroblastic anemia) are usually microcytic rather than

macrocytic and do not have hypersegmented neutrophils in the peripheral blood.

55. The answer is E. *(Hematology; Hematopoietic/lymphoreticular)*
Patients with thrombotic thrombocytopenic purpura have a circulating factor in plasma that damages endothelial cells causing platelet thrombi to develop in the microcirculation. The platelet thrombi damage red blood cells and produce an intravascular hemolytic anemia. Unlike in disseminated intravascular coagulation, there is no consumption of clotting factors; therefore, the prothrombin and partial thromboplastin times (PT and PTT, respectively) are normal. In alcoholic cirrhosis and fulminant hepatic failure, there are multiple coagulation deficiencies because the liver is the primary site for coagulation factor synthesis. The PT and PTT are prolonged in these conditions. Celiac disease is a cause of malabsorption; therefore, vitamin K, a fat-soluble vitamin, is often depleted, resulting in multiple coagulation deficiencies, which prolong the PT and PTT.

56. The answer is D. *(Pathology; Tissue biology and associated response to disease)*
Generalized, nontender lymphadenopathy in a patient over 65 years of age is most commonly caused by chronic lymphocytic leukemia, which is also the most common leukemia in this age bracket. Generalized lymphadenopathy is associated with disseminated infections, malignant lymphomas, or leukemia. Acute myelogenous leukemia is the most common leukemia in individuals 15 to 39 years of age and shares the lead with chronic myelogenous leukemia in the 40- to 59-year-old age bracket. Autoimmune disease is associated with generalized lymphadenopathy but is not as prevalent as chronic lymphocytic leukemia in the elderly population. Primary lung cancer, although the most common cause of cancer death among men and women, does not produce generalized lymphadenopathy secondary to lymph node metastases. Generalized lymphadenopathy is not an age-dependent process.

57. The answer is A. *(Pathology; Tissue biology and associated response to disease)*
Inguinal lymph nodes in an asymptomatic 55-year-old man are most likely benign. Both anterior cervical and inguinal lymph nodes are common sites for drainage of infected material, making them the prime site for lymphadenopathy.

A left supraclavicular lymph node (Virchow node) in an adult with weight loss most likely is metastatic cancer from either primary lung cancer or stomach cancer.

Lymph nodes in the neck of a smoker with hoarseness most likely implicates metastasis from a primary squamous-cell carcinoma of the true vocal cords.

Cervical nodes in a man with a thyroid nodule is highly predictive of metastatic papillary adenocarcinoma of the thyroid.

Para-aortic lymph nodes in a man with a history of cryptorchidism is highly predictive of metastatic seminoma from a primary malignancy in the cryptorchid testis.

58. The answer is D. *(Pathology; Tissue biology and associated response to disease)*
Any woman over 50 years of age who has an axillary mass that is nontender should have an excisional biopsy, since it is most likely metastatic primary breast cancer. As a general rule, lymph nodes that are tender most likely have an inflammatory etiology. Observation, bone marrow examination, screening for the HIV antibody, and purified protein derivative skin testing would not be high priorities.

59. The answer is D. *(Pathology, Hematology; Hematopoietic/lymphoreticular)*
The patient has a circulating anticoagulant (inhibitor) most likely against factor VIII. Circulating anticoagulants are antibodies against certain coagulation factors. This condition is often confused with a coagulation deficiency because the factor assays are low. The prothrombin time (PT) and/or the partial thromboplastin time (PTT) are prolonged depending on the type of circulating anticoagulant. The distinction between an antibody versus a true coagulation deficiency is made by performing mixing studies. When patient plasma containing a circulating anticoagulant is mixed with normal plasma, the PT and/or PTT is not corrected

because the antibodies inactivate the coagulation factors in the normal plasma as well. However, in a true factor deficiency, the PT and/or PTT is corrected when normal plasma is mixed with patient plasma, because the normal plasma provides enough of the missing factor to correct the prolonged coagulation test.

Classic anticoagulant relationships include (1) factor VIII inhibitors (most common) associated with birth, chlorpromazine therapy, and treating hemophilia A with factor VIII, (2) factor V inhibitor, associated with streptomycin and aminoglycosides, and (3) factor XIII inhibitor, associated with isoniazid therapy.

Another anticoagulant is the lupus anticoagulant (LA), which appears in 10% of patients with systemic lupus erythematosus. LA may also be seen in AIDS and in patients taking procainamide. LA is an antibody against the phospholipid added to the test system, resulting in a prolonged PTT. Unlike other circulating anticoagulants, LA is not associated with bleeding. It may be part of the antiphospholipid syndrome, which has both LA and anticardiolipin antibodies. These latter antibodies react with the cardiolipin of the RPR/VDRL test system and produce a false-positive syphilis serology. They are also associated with a high incidence of vessel thrombosis often manifested by habitual late trimester abortions.

Disseminated intravascular coagulation is an unlikely diagnosis because the PT, bleeding time, and platelet count are normal.

Hemophilia A is an unlikely diagnosis because the mixing studies did not correct the prolonged PTT.

Von Willebrand disease is an unlikely diagnosis because the mixing studies did not correct the PTT and the bleeding time is normal.

A qualitative platelet defect is not present because the PTT is prolonged and the bleeding time is normal. Platelet abnormalities do not prolong either the PT or the PTT.

60. The answer is D. *(Hematology; Hematopoietic/lymphoreticular)*
A molar extraction without any bleeding problems provides the greatest assurance that there is no coagulation deficiency in a patient. Patients with a coagulation deficiency are unable to form a stable clot. A temporary hemostatic plug, which is composed of platelets held together by strands of fibrinogen, initially stops bleeding. Thrombin that is generated from the extrinsic and intrinsic pathways converts the fibrinogen into fibrin, forming a stable clot. Patients with a coagulation deficiency, therefore, have only a temporary hemostatic plug preventing bleeding and are prone to oozing from molar extractions and surgical wounds.

That a patient is not taking birth control pills offers no assurance that she will not bleed, since birth control pills do not interfere with platelet function or the coagulation system. Birth control pills may interfere with antithrombin III activity, which predisposes the patient to thrombus formation.

A negative history for epistaxis rules out a platelet abnormality.

No excessive bleeding with superficial scratches rules out a platelet abnormality, but this is not as significant a test of a patient's hemostatic function as is a molar extraction.

No family history of bleeding is an important finding, but does not carry the same significance as adequate control of bleeding after a tooth extraction.

61. The answer is D. *(Hematology; Hematopoietic/lymphoreticular)*
In a patient with epistaxis, spontaneous bruising, and petechia, a prolonged bleeding time (BT) would be expected because the constellation of findings suggests a platelet abnormality. The BT is performed as follows: a blood pressure cuff is inflated to 40 torr on the upper arm. An area on the volar aspect of the forearm is prepped with alcohol. A commercial template with sharp cutting blades is then applied, and a free flow of blood is established from a minor wound inflicted by the blades. A stopwatch is activated the moment blood flows from the wound, and filter paper is dabbed on the wound site every 30 seconds until there is no blood staining on the paper, signaling the end of the bleeding time (normally less than 10 minutes). The BT evaluates the integrity of the blood vessels and platelet function. Thrombocytopenia or functional defects in platelets will prolong the BT. Functional platelet defects include (1) absence of the von Willebrand factor in the endothelial tissue (von Willebrand disease) or its receptor on the platelet (Bernard-Soulier disease), which are both necessary for platelet adhesion; (2) inhibition of platelet cyclooxygenase by nonsteroidal drugs (most common), which block the synthesis

of thromboxane A_2, a platelet aggregator; (3) storage depot diseases in the platelet (e.g., deficiency of adenosine phosphate); and (4) problems with platelet fibrinogen receptors (Glanzmann thrombasthenia) that allow fibrinogen to hold the platelet plug together.

A prolonged prothrombin and/or partial thromboplastin time is associated with a coagulation deficiency. Coagulation deficiencies are not associated with epistaxis, spontaneous bruising, or petechia.

Decreased antithrombin III concentration predisposes to vessel thrombosis rather than to bleeding.

An increase in fibrinogen degradation products indicates the presence of excessive fibrinolytic activity. Although these products may interfere with platelet function, they are not the primary test ordered for a patient with this history.

62. The answer is C. *(Hematology; Hematopoietic/lymphoreticular)*
This patient most likely has von Willebrand disease, since the partial thromboplastin time (PTT) is prolonged, the bleeding time is prolonged, and the platelet count is normal. Classic von Willebrand disease (vWD) is inherited as an autosomal dominant trait. It is considered the most common congenital qualitative platelet defect and coagulation deficiency. Patients with vWD lack (1) von Willebrand factor, which is important in platelet adhesion, (2) VIII antigen, which carries VIII coagulant, and (3) VIII coagulant, which is important in the intrinsic coagulation pathway. Absence of the von Willebrand factor results in a prolonged bleeding time, whereas low VIII coagulant levels are responsible for the prolonged PTT. Epistaxis, easy bruising, and menorrhagia are common findings in vWD. Mild cases of vWD may be treated with desmopressin, which initiates the synthesis of all the factor VIII components, including von Willebrand factor, antigen, and coagulant. Cryoprecipitate is also used, but there is a danger of transmitting hepatitis C.

A packed red blood cell transfusion, infusion of platelets, or an intramuscular injection of vitamin K would not stop the bleeding in vWD, because there is a combined platelet adhesion defect and factor VIII deficiency. Factor VIII is not a vitamin K–dependent factor.

Treatment of the patient's iron deficiency with iron would correct the anemia but not vWD.

63. The answer is A. *(Hematology; Hematopoietic/lymphoreticular)*
Major hemorrhages are most likely associated with a coagulation deficiency rather than problems with platelets (e.g., defective function, thrombocytopenia); primary fibrinolysis, which is extremely rare; or vasculitis, which is the most frequently immune complex in origin. Coagulation deficiencies result in the formation of unstable clots that are only composed of aggregated platelets and fibrinogen strands. Reduction in the generation of thrombin from either the intrinsic or extrinsic coagulation pathways hampers the conversion of the fibrinogen into fibrin, which produces a stable clot. If these unstable plugs are dislodged after surgery, for example, significant bleeding may occur. In addition, severe factor VIII or IX deficiencies with less than 1% factor activity are associated with spontaneous bleeding into joints (hemarthrosis) or into potential spaces (e.g., retroperitoneum, fascial planes in muscle, retropharynx).

64. The answer is E. *(Physiology; Hematopoietic/lymphoreticular)*
Anemia refers to a decrease in hemoglobin (Hgb) or hematocrit (Hct) concentration. In general, an Hgb less than 14 g/dl in a man or less than 12 g/dl in a woman is considered anemia. Anemia is a sign of underlying disease rather than a specific diagnosis. It produces tissue hypoxia, since Hgb is the primary vehicle for carrying oxygen. The partial pressure of oxygen in arterial blood (PaO_2) and oxygen saturation (number of heme groups occupied by oxygen) are normal, since there are no problems with gas exchange in the lungs. Characteristic findings include dyspnea, weakness, fatigue, anorexia (intestinal hypoxia), insomnia, inability to concentrate, syncope, dizziness (central nervous system hypoxia), and possible exacerbation of angina/claudication (tissue hypoxia). In the heart, there may be a systolic flow murmur over the pulmonic area due to decreased blood viscosity. Decreased blood viscosity also reduces the peripheral resistance, which increases the cardiac output and stroke volume. The pulse pressure, or the difference between the systolic and diastolic pressure, widens due to the increase in stroke volume (systolic pressure) and decrease in diastolic pressure (decreased peripheral resistance). The alveolar-arterial gradient is normal because

there is no intrapulmonary or intracardiac shunting of blood.

65. The answer is A. *(Biochemistry; Gastrointestinal)*
Most people have an intake of approximately 10 to 20 mg of dietary iron a day, but only 10% of this amount is normally absorbed (1–2 mg/day) in the duodenum, the primary site for iron absorption. A Billroth II procedure connects the distal stomach to the jejunum, so iron absorption is decreased.

Most iron in the adult male and female is in hemoglobin (Hgb). The total amount of iron in men is approximately 3500 mg, with 2100 mg in hemoglobin, 1000 mg in iron stores, and the rest in myoglobin and enzymes. Women have approximately 2500 mg of iron with 1800 mg in Hgb and 400 mg in iron stores. The lower amount of iron in women is primarily the result of menstrual loss. Women lose approximately 30 ml of blood per period. Since 1 ml of blood contains 0.5 mg of iron, there is a loss of 15 mg of iron per period.

Iron must be in the ferrous state (+2) for reabsorption in the duodenum. Vitamin C is the most important factor for reducing iron to the ferrous state. Gastric acid is responsible for freeing up elemental iron from heme and nonheme products, but does not reduce iron to the ferrous state. Heme iron, which is primarily in meats, is already in the ferrous state and is directly absorbed. It is released from apoproteins in food by gastric acid and is directly taken up by mucosal cells in the duodenum. In the mucosal cell, heme is degraded to release iron, which is either stored as ferritin for later release, or directly delivered to transferrin for delivery to the liver or the bone marrow. Nonheme iron (plants) is in the ferric state and must be reduced by ascorbic acid to be reabsorbed. Luminal mucins bind iron at the acid pH of the stomach and keep it solubilized for mucosal absorption. Integrins at the surface of the mucosal cells bind iron and facilitate its passage into the cell. Another protein accepts the iron and delivers it to ferritin for storage or to transferrin for delivery to the liver or the macrophages in the bone marrow. Iron balance in the body is maintained by regulating the amount of iron reabsorbed (mucosal block therapy). The amount of ferritin in the mucosal cell seems to be an important regulator of how much is absorbed. Normally, there are fixed daily losses of iron ranging between 1 to 1.5 mg, which are maintained by the 1 to 2 mg of iron normally reabsorbed per day. Women must reabsorb more iron each day than men to get the required 1 to 2 mg of iron a day as replacement for the fixed losses. Many women do not ingest enough iron for their greater iron requirements. The percentage of iron reabsorbed increases (20%–25%) in anemia regardless of whether the body needs the iron or not. This may result in iron overload, particularly in patients with severe thalassemia, who are already iron overloaded due to a heavy transfusion requirement. The percentage of iron reabsorbed normally increases in pregnancy and lactation.

66-68. The answers are: 66-D, 67-C, 68-E. *(Physiology; Tissue biology and associated response to disease)*
The direction in which the bulk flow of fluid across a capillary occurs depends on a balance between net filtration forces and net reabsorption forces. The hydrostatic pressure inside a capillary favors the filtration of fluid from the inside of the vessel to the outside, whereas the hydrostatic pressure outside the capillary opposes it. Therefore,

$$\text{Net filtration} = \text{Hydrostatic pressure}_{\text{inside capillary}} - \text{Hydrostatic pressure}_{\text{outside capillary}}$$

The oncotic pressure inside a capillary favors the reabsorption of fluid from the outside of the capillary to the inside, whereas the oncotic pressure outside the capillary opposes it. Therefore,

$$\text{Net reabsorption} = \text{Oncotic pressure}_{\text{inside capillary}} - \text{Oncotic pressure}_{\text{outside capillary}}$$

Thus,

Net fluid movement = Net filtration − Net reabsorption
= (30 − 4) − (10 − 5)
= + 21 mm Hg

A positive result indicates that the forces favoring filtration exceed those favoring reabsorption. A negative result indicates that the forces favoring reabsorption are greater than those favoring filtration. The oncotic pressure is due to the presence of impermeable substances and is a measure of the concentration of plasma proteins. A normal oncotic pressure is 25 mm Hg. In cirrhosis, the plasma protein concentration is low, and the forces that favor filtration far exceed those that favor reabsorption, causing edema. The net movement of fluid across a capillary is not affected by the hematocrit, so that neither anemia nor polycythemia are important determinants of the direction in which fluid moves. Capillary hydrostatic pressure decreases in shock and with the use of venodilators, and this favors the reabsorption of fluid.

69-71. The answers are: 69-C, 70-D, 71-D. *(Physiology; Pulmonary/respiratory)*

The physiologic subdivisions of lung volume are listed as follows. Most can be measured directly by simple spirometry, others require special methods, such as the helium-dilution or N_2-washout techniques. A spirometer can only measure the volume of air exhaled into or inhaled from it. Even after a maximum expiration, a considerable volume of air, called the residual volume, remains in the lung and cannot be measured directly. Thus, the residual volume and any capacity of which it is a part (i.e., the functional residual capacity [FRC] and the total lung capacity [TLC]) require the use of the techniques previously mentioned. Lung volumes are determined by many variables, including age, sex, and body surface area. The spirogram that follows is typical only for a 70-kg man.

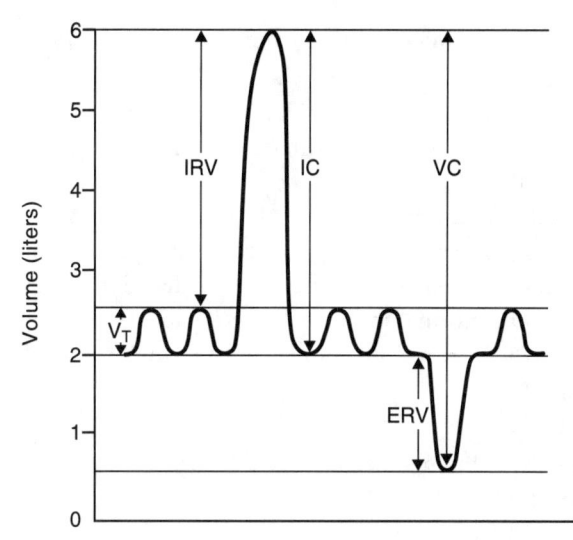

Lung volumes capable of being determined directly by spirometry:
- Tidal volume (V_T)—volume of gas inspired during quiet breathing
- Inspiratory reserve volume (IRV)—additional volume of gas that can be inspired from end-tidal inspiration
- Inspiratory capacity (IC)—maximum volume of gas that can be inspired from resting expiratory level
- Expiratory reserve volume (ERV)—additional volume of gas that can be expired from resting expiratory level
- Vital capacity (VC)—maximal volume of gas that can be expired after a maximal inspiration

Lung volumes not directly measured by a spirometer:
- Residual volume (RV)—volume of gas left in the lungs at the end of a maximal expiration
- Functional residual capacity (FRC)—volume of gas left after a passive expiration
- Total lung capacity (TLC)—volume of gas in the lungs after a maximal inspiration

Given these definitions, the FRC, V_T, and VC can be calculated using the following relationships:

FRC	= TLC − IC
	= 7.0 − 4.0
	= 3.0 L
V_T	= IC − IRV
	= 4.00 − 3.50
	= 500 ml
VC	= IC + ERV
	= 4.0 + 1.5
	= 5.5 L

72-74. The answers are: 72-C, 73-C, 74-E. *(Physiology; Pulmonary/respiratory)*
The forced vital capacity (FVC) is the volume of gas capable of being exhaled after a maximal inspiration. From B to C, this woman exhaled 4 L of gas. The forced expiratory volume in 1 second (FEV_1) is the volume of gas exhaled in 1 second after a maximal inspiration. In 1 second, the patient exhaled 3 L of gas (B to D). Normal individuals are capable of exhaling between 75% and 80% of their FVC in 1 second:

$$FEV_1/FVC = 75\%-80\%$$

The data obtained from this patient are in the normal range:

$$FEV_1/FVC = 3L/4L = 75\%$$

The constancy of the FEV_1/FVC ratio among individuals implies that for most of an FVC maneuver, the maximal rate at which air can flow is limited by something other than effort. This information is extremely useful for assessing alterations in pulmonary function.

Airflow through the respiratory tree is dependent on the pressure gradient developed between the alveoli and the atmosphere. During a forced expiration, such as the FVC maneuver, the intrapleural pressure becomes positive. This positive pressure is transmitted to the alveoli, so that intra-alveolar pressure becomes positive relative to the atmosphere and rate at which air flows increases. This increase in airflow is limited, however, because the positive intrapleural pressure also compresses the airways, narrowing them and increasing their resistance. Increasing the resistance of an airway decreases the rate at which air flows through it. The factor most responsible for limiting expiratory flow during a forced expiration is the compression of the airways that results from the positive intrapleural pressure.

The residual volume (RV) cannot be determined directly, because a spirometer can only measure the volume of gas either entering or leaving it.

75-77. The answers are: 75-C, 76-E, 77-B. *(Physiology; Pulmonary/respiratory)*
Each gram of hemoglobin is capable of binding with 1.34 ml of O_2. Because the hemoglobin concentration in the anemic individual is only 10 gm/100 ml (10 gm%), the hemoglobin would be 100% saturated when each 100 ml of blood contained 13.4 ml of O_2.

$$100\% \text{ saturation} = 1.34 \text{ (hemoglobin concentration)}$$

Because of coincidental respiratory problems, the Pa_{O_2} in the anemic individual is only 50 mm Hg. Using Curve B, at a P_{O_2} of 50 mm Hg, the hemoglobin is 70% saturated. Thus, the O_2 content of hemoglobin in arterial blood is:

$$\text{Arterial } O_2 \text{ content} = .70 \ (13.4)$$

$$= 9.38 \text{ vol\%}$$

The O_2 content of hemoglobin in venous blood (P_{O_2} = 25 mm Hg; percent saturation = 20%) is:

$$\text{Venous } O_2 \text{ content} = .20 \ (13.4)$$

$$= 2.68 \text{ vol\%}$$

The arterial-venous difference = arterial O_2 content – venous O_2 content

$$= 9.38 - 2.68$$

$$= 6.70 \text{ vol\%}$$

The inadvertent administration of outdated blood (low 2,3-diphosphoglycerate [DPG]) results in a leftward shift of the oxyhemoglobin dissociation curve (Curve A). A shift to the left indicates an increase in the affinity of hemoglobin for O_2. At a venous P_{O_2} of 25 mm Hg, the hemoglobin in venous blood is 50% saturated (i.e., less O_2 was delivered to the tissues).

78. The answer is D. *(Pathology; Pulmonary/respiratory)*
The terms "pink puffer" and "blue bloater" describe the two types of chronic obstructive pulmonary disease (COPD): chronic bronchitis (CB) and emphysema. Approximately 10% to 15% of patients who smoke will develop COPD. CB is primarily a clinical diagnosis defined as a productive cough for 3 months out of a year for 2 consecutive years. Emphysema refers to permanent damage in the respiratory unit (respiratory bronchioles, alveolar ducts, alveoli) with destruction of their elastic tissue support resulting in permanent enlargement of these airspaces. Patients with pure CB

are often referred to as "blue bloaters," because they commonly retain CO_2 (respiratory acidosis) and are cyanotic. Patients with pure emphysema are called "pink puffers," because they frequently have normal to low arterial P_{CO_2} from hyperventilation. Most patients have a combination of the two.

	Emphysema	Chronic bronchitis
Surnames	Type A	Type B
Onset of symptoms	After age 50	After age 35
Appearance	"Pink puffer"	"Blue bloater"
	Thin, weight loss	Obese, cyanotic
Dyspnea	Progressive, constant, and severe	Intermittent, mild to moderate
Productive cough	None to scant	Increased
AP diameter	Increased	Normal to slight increase
Breath sounds	Diminished	Wheezes and rhonchi
Bullae and blebs	Present	Absent
Polycythemia	None	Present
Hypoxemia	Mild to absent	Moderate to severe
Acid-base disorder	Normal to mild respiratory alkalosis	Respiratory acidosis
Lung compliance	Increased	Normal
D_LCO	Decreased	Normal (preserved alveolar capillary bed)
	Matched loss of alveoli and capillary bed	
Ventilation/perfusion	Not as severe as CB, because of loss of alveoli and capillary beds	Severe mismatches, because obstruction is more proximal and affects a greater area of distal lung
Primary location of obstruction	Respiratory unit, respiratory bronchioles, alveolar ducts, alveoli	Nonrespiratory, small caliber airways, segmental bronchi

79. The answer is D. *(Pathology; Pulmonary/respiratory)*
Cor pulmonale refers to right-ventricular hypertrophy that deveops as a result of pulmonary hypertension (PH). PH must develop from either primary vessel disease or as a result of lung disease (e.g., chronic obstructive pulmonary disease, recurrent pulmonary emboli). The definition does not encompass any of the causes of PH owing to heart disease, such as mitral or pulmonic stenosis or left-to-right shunts (ventricular or atrial septal defects).

80-81. The answers are: 80-B, 81-D. *(Pathology; Pulmonary/respiratory)*
In order to solve this problem, the calculation of alveolar P_{O_2} (P_{AO_2}) must be reviewed. The amount of oxygen that is reaching the alveoli (P_{AO_2}) is the amount of oxygen inspired, or P_{IO_2}, subtracted from the amount that is exchanged in the lung with CO_2, which represents the measured $P_{aCO_2}/0.8$. The P_{IO_2} is calculated by multiplying the percent oxygen the patient is breathing by 760 mm Hg, minus the water vapor pressure, which is 47 mm Hg:

$$P_{IO_2} = \% \text{ oxygen } (713 \text{ mm Hg})$$

The amount of oxygen used up in the exchange process is calculated by dividing the measured P_{aCO_2} by 0.8, which represents the average value for the respiratory quotient (the volume of CO_2 produced [200 ml/minute] divided by the volume of O_2 consumed [250 ml/minute] per unit time). Therefore, the formula for calculating the P_{AO_2} is as follows:

$$P_{AO_2} = \% \text{ oxygen } (713) - P_{aCO_2}/0.8$$

At sea level, the percent oxygen is 0.21 and the normal P_{aO_2} is 40 mm Hg; therefore, the normal P_{AO_2} is:

$$P_{AO_2} = 0.21 (713) - 40/0.8 = 100 \text{ mm Hg}$$

The oxygen concentration at high altitude is still 21%, but the barometric pressure is decreased. On Mt. Everest, the barometric pressure, with the water vapor already subtracted out, is ~200 mm Hg. Therefore, the P_{IO_2} is 0.21 × 200 = 42 mm Hg. Because the mountain climber has a P_{aCO_2} of 40 mm Hg and a respiratory quotient of 1, instead of 0.8, the P_{AO_2} is:

$$P_{AO_2} = 0.21 (200) - 40/1 = 2 \text{ mm Hg}$$

which is the amount of oxygen available for exchange with the pulmonary capillaries.

Assuming the patient is hyperventilating (respiratory alkalosis) on top of Mt. Everest and the arterial P_{CO_2} is 20 mm Hg:

$$P_{AO_2} = 0.21 (200) - 20/1 = 22 \text{ mm Hg}$$

therefore, more oxygen is available for exchange with the pulmonary capillaries when the patient hyperventilates. Hyperventilation is the most useful response at high altitude, because lowering alveolar CO_2 automatically increases alveolar O_2, which increases the arterial P_{O_2}. The hyperventilation is caused by the hypoxemic stimulus of the peripheral chemoreceptors (carotid and aortic bodies).

82. The answer is D. *(Pathology; Pulmonary/respiratory)*

In both bronchial asthma and chronic bronchitis from cigarette smoking, small airway inflammation is the primary site of obstruction to airflow.

Bronchial asthma is more commonly associated with immunoglobulin E (IgE)-mediated inflammation. It is characterized by episodic flare-ups, whereas chronic bronchitis is usually present most of the year. Bronchial asthma is more likely to present initially with respiratory alkalosis. Chronic bronchitis is characterized by respiratory acidosis. Chronic bronchitis is associated with a productive cough throughout most of the year, but bronchial asthma characteristically has a nonproductive cough, unless an acute bronchitis or bronchopneumonia develops.

83. The answer is E. *(Pathology; Pulmonary/respiratory)*

Adenocarcinomas are the most common primary lung cancer to develop within an area of peripheral scar. Scar tissue formation may be secondary to previous granulomatous disease (tuberculosis), infarction, or interstitial diseases of the lung associated with fibrosis (e.g., progressive systemic sclerosis). The pathogenesis of scar carcinoma is not completely understood. One theory is that scar tissue blocks lymphatic drainage, causing an increased concentration of anthracotic pigment–containing carcinogens. Persistent hyperplasia of tissue and the potential for a mutation leading to cancer are other mechanisms that may be similar to those of cancers that develop in fistulous tracts draining pus (e.g., chronic osteomyelitis).

84. The answer is C. *(Pathology; Pulmonary/respiratory)*

The patient most likely has a primary adenocarcinoma of the bronchioalveolar type. Adenocarcinomas of the lung either originate from the glandular cells in the bronchi (more centrally located) or from the peripherally located terminal bronchioles/alveoli. This latter site produces a variant of adenocarcinoma, called bronchioalveolar carcinoma.

Bronchioalveolar carcinomas are either present as peripherally located solitary masses or as multiple lesions, which often coalesce to produce a lobar consolidation (often misinterpreted as pneumonia). Unlike squamous-cell carcinomas and small-cell carcinomas, adenocarcinomas and bronchioalveolar carcinomas have no relationship with smoking. The cell of origin is either the Clara cell (most common) or the type II granular pneumocyte (surfactant-secreting cell). The malignant tumor cells line the alveoli like tombstones. The prognosis is dependent on the type of bronchioalveolar cancer. If the cancer is solitary, the prognosis is good, but if multiple, as in this case, the prognosis is poor. Overall, however, the prognosis is better than for the other types of primary lung cancer.

Primary small-cell carcinoma is a centrally located cancer that does not consolidate. Metastatic squamous-cell carcinoma produces nodular lesions in the lung without consolidation. Bronchial carcinoids are centrally located neuroendocrine tumors that do not consolidate. Scar carcinomas are located at the periphery of the lung. Adenocarcinomas commonly develop in these areas of scar formation.

85. The answer is E. *(Pathology; Pulmonary/respiratory)*

Inspiratory stridor usually characterizes extrathoracic obstruction, most commonly in the larynx; however, fixed obstruction anywhere in the tracheobronchial tree also produces this finding. Inhalation is accompanied by a high-pitched sound that extends throughout inspiration. If the larynx is involved, the patient is unable to speak. A Heimlich maneuver is frequently life saving in this situation. Causes of upper airway obstruction include foreign body obstruction, laryngeal cancer, acute epiglottis (*Haemophilus influenzae*), and tracheobronchitis (croup) caused by the parainfluenza virus. Lower airway disease, like bronchial asthma and bronchiolitis, produces inspiratory and expiratory wheezing, rather than inspiratory stridor. The adult or childhood forms of respiratory distress syndrome are not associated with inspiratory stridor.

86. The answer is A. *(Pathology; Pulmonary/respiratory)*
The diffusion capacity evaluates gas exchange through the alveolar–capillary interface using carbon monoxide (CO). The transfer of CO from the alveoli to the red blood cells in the pulmonary capillaries is primarily dependent on the diffusion of CO through the alveolar–capillary interface. The diffusion capacity of the alveolar membrane depends on its thickness and surface area; therefore, if the thickness increases (e.g., interstitial fibrosis) or the surface area is reduced (e.g., pneumonectomy), the D_LCO is reduced. Other diseases with a decrease in D_LCO include emphysema (destruction of the alveolar–capillary interface), *Pneumocystis carinii* pneumonia (extensive alveolar and interstitial infiltrates), and pulmonary embolus (decreased perfusion, causing less CO to be taken up by the pulmonary capillaries).

Unlike emphysema, chronic bronchitis does not directly damage the alveolar–capillary interface; therefore, the D_LCO is normal. A mesothelioma is a malignancy of the pleura, which does not impact on the alveolar–capillary interface. Asthma often increases the D_LCO, because the increase in lung volume increases the cross-sectional area that CO diffuses. A small-cell carcinoma is a centrally located cancer and does not affect the respiratory unit.

87. The answer is B. *(Pathology; Pulmonary/respiratory)*
The patient most likely has a pulmonary embolus involving the left lower lobe. Dullness to percussion in the absence of tactile fremitus indicates the presence of a pleural effusion. Pleuritic chest pain (pain on inspiration) strongly suggests that pulmonary infarction has occurred. Swelling of the calf and a positive Homan sign (dorsiflexion of the foot) and Mose sign (compression of the calf) strongly suggest a deep saphenous vein thrombophlebitis as the most likely source for the embolus. The arterial blood gas exhibits a partially compensated respiratory alkalosis, and the alveolar-arterial gradient is 50 mm Hg, which is medically significant (> 30 mm Hg).

pH	7.48 (alkalemia)
PaCO₂	24 mm Hg (respiratory alkalosis, because PCO₂ is < 33 mm Hg)
PaO₂	70 mm Hg (mild hypoxemia)
HCO₃	19 mEq/L (compensatory metabolic acidosis, because the HCO₃ is < 22 mEq/L)

The alveolar-arterial (A-a) gradient is calculated as follows:

$$P_{AO_2} = \% \text{ oxygen } (713 \text{ mm Hg}) - P_{aCO_2}/0.8$$
$$P_{AO_2} = 0.21 (713) - 24/0.8 = 120 \text{ mm Hg}$$
$$A\text{-}a = 120 - 70 = 50 \text{ mm Hg}$$

A perfusion is normally the first step in the workup of a patient with suspected pulmonary embolism and would likely demonstrate a perfusion defect in the left lower lobe in this case. The ventilation scan should be normal, because the patient has no previous history of pulmonary problems and does not smoke.

Risk factors for venous thrombosis and the potential for a pulmonary embolus (PE) include increased venous stasis and increased coagulability. High risk situations predisposing patients to venous stasis include the postpartum state, bed rest after surgery (most frequent setting), orthopedic surgery, pelvic or abdominal surgery, postprostatectomy, reduced cardiac output (congestive heart failure, myocardial infarction), and increasing age. Increased coagulability is associated with oral contraceptives, the hereditary coagulopathies (antithrombin III deficiency, protein C and S deficiencies), malignancy, presence of the lupus anticoagulant, and tissue injury.

88. The answer is D. *(Pathology; Pulmonary/respiratory)*
The patient has endotoxic shock (warm skin, fever), most likely from gram-negative sepsis related to his urinary retention. He has developed massive intrapulmonary shunting, evidenced by hypoxemia that is not

relieved by 100% oxygen. Cyanosis of the mucous membranes and skin (central cyanosis) also confirms intrapulmonary shunting. All of the pulmonary findings indicate adult respiratory distress syndrome (ARDS).

ARDS is a diffuse pulmonary disease characterized by leaky pulmonary capillaries, diffuse alveolar damage, proteinaceous deposits in alveoli (hyaline membranes), and widespread atelectasis, leading to massive intrapulmonary shunting of blood. A partial list of factors that predispose patients to ARDS includes gram-negative sepsis (very common), hypovolemic shock, massive trauma, drug overdose, burns, aspiration of gastric contents, and disseminated intravascular coagulation. The mechanisms of injury include:

- Neutrophil injury of pulmonary capillaries, because of the release of proteases and free radicals. The generation and release of arachidonic acid metabolites also predispose patients to pulmonary vasoconstriction and increased vessel permeability, leading to leaky capillaries. Proteinaceous fluid exudes into the alveoli and, along with debris from damaged type I and II pneumocytes, congeal to form hyaline membranes. Damage to type II pneumocytes also results in surfactant deficiency and subsequent widespread atelectasis. Perfusion of unventilated lung is manifested by massive intrapulmonary shunting and severe hypoxemia. The hypoxemia does not respond to the administration of 100% oxygen. Although diffusion and perfusion abnormalities are also present, intrapulmonary shunting is the key pathophysiologic event in ARDS.
- Lymphocytes and macrophages also participate by releasing cytokines (tumor necrosis factor, interleukin 1).
- Lung compliance is decreased (still lungs on inspiration) from fluid within the interstitium and alveoli.

Clinical findings include a rapid onset of dyspnea and tachypnea, and cyanosis. Laboratory studies reveal severe hypoxemia and prolonged A-a gradients that are unchanged after administering 100% oxygen. Mortality exceeds 50%, with a median survival of 2 weeks.

A massive pulmonary embolus (PE) would most likely produce sudden death. In addition, the main problem in PE is a perfusion defect, leading to an increase in alveolar dead space, not intrapulmonary shunting, as in this case. Lobar pneumonia can produce many of the pulmonary findings present in this patient, but it does not commonly produce diffuse lung disease or the intrapulmonary shunting noted in this case. The patient is in endotoxic shock, not hypovolemic shock. There is no history of cardiac disease. The clinical picture is one of sepsis leading to respiratory failure.

89. The answer is A. *(Pathology; Microbial biology and infection)*
Both reactivation tuberculosis (TB) and primary squamous-cell carcinoma of the lung are associated with cavitation. Other pulmonary diseases associated with cavitation include systemic fungal infections, *Klebsiella* pneumonias, and silicosis. In silicosis, the large fibrotic nodules undergo central necrosis from ischemia. Furthermore, silicosis predisposes patients to the development of TB within the fibrotic nodules, thus producing cavitation.

Ectopic secretion of a parathormone-like peptide-producing hypercalcemia is seen only in squamous-cell carcinoma. Scar carcinomas are a possible sequela of old scar tissue in TB. Adenocarcinomas are more frequently associated with scar carcinomas than squamous carcinoma. The most common extrapulmonary site for TB is the kidneys, and the adrenal gland is the most common site for metastatic squamous cancer. Silicosis predisposes patients to TB, but not to primary lung cancer.

90-91. The answers are: 90-D, 91-E. *(Pathology; Microbial biology and infection)*
The patient has a tension pneumothorax in the left lung secondary to rupture of a tension pneumatocyst. The tension pneumatocyst is secondary to *Staphylococcus aureus* pneumonia, because the sputum has gram-positive cocci in clusters.

Respiratory complications are the most common problem and cause of death in patients with cystic fibrosis (CF). Patients with CF have a defect in exocrine secretions. In the lung, this defect is characterized by an increased reabsorption of sodium out of the bronchial secretions into the blood; therefore, the secretions become thickened and obstruct the small caliber airways. These inspissated mucous plugs serve as a nidus for infection and atelectasis of the lung distal to the obstruction.

Staphylococcus aureus, Haemophilus influenza, and *Pseudomonas aeruginosa* are the most common pulmonary pathogens in CF. *Staphylococcus aureus* is most commonly acquired in the hospital (nosocomial infection) because of colonization of the upper airways. Other colonizers are gram-negative organisms, such as *Escherichia coli* and *Pseudomonas.* Staphylococcal pneumonia is characterized by a hemorrhagic pulmonary edema, a tendency for abscess formation, and the development of tension pneumatocysts (subpleural blebs), which may rupture and produce a tension pneumothorax, as in this patient.

A pneumothorax refers to air in the pleural cavity, and is most commonly caused by a rupture of a bleb located in the visceral pleura covering the lung. Other causes include trauma or iatrogenesis, such as that caused by the insertion of a subclavian vein catheter. Pain and dyspnea are the most common symptoms. Physical examination reveals hyperresonance and decreased breath sounds on the affected side. In a spontaneous pneumothorax, the trachea is deviated to the side of the lesion as the lung collapses; however, in a tension pneumothorax, the flap of the ruptured bleb is lifted up on inspiration and air enters the pleural space. On expiration, the flap closes and air remains behind, producing a shift in the mediastinal structures to the opposite side. This process leads to severe cardiorespiratory embarrassment. Insertion of a chest tube under water seal relieves the pressure.

Empyema, a lung abscess secondary to bronchopneumonia, or a pulmonary infarction in the left lung are unlikely because of the hyperresonance to percussion and the absence of breath sounds.

92. The answer is A. *(Pathology; Microbial biology and infection)*

The patient has bacterial pneumonia. The sputum Gram stain reveals lancet-shaped diplococci, most likely representing *Streptococcus pneumoniae* (a gram-positive diplococcus; alias, pneumococcus). The radiograph confirms the presence of a lobar consolidation in the right upper lobe.

Streptococcus pneumoniae is the most common cause of community-acquired bacterial pneumonia. Lobar consolidations, if left untreated, progress through the stages of congestion, red hepatization, gray hepatization, and resolution. The patient is probably in the stage of congestion. Penicillin is still the treatment of choice for pneumococcal pneumonia. Usually, the fever, dyspnea, and cough clear within 12 to 36 hours of therapy.

93. The answer is C. *(Pathology; Pulmonary/respiratory)*

Multiple nodular lesions in any organ in an elderly patient most likely represent metastasis. Metastatic disease of the lung is the most common lung cancer (breast cancer is the leading cause). The most common primary lung cancer is adenocarcinoma. Other cancers that commonly metastasize to the lungs are renal adenocarcinoma, testicular cancer (particularly choriocarcinoma), malignant melanoma, gastrointestinal tract malignancies (esophagus, stomach, pancreas, colon), and sarcomas (e.g., osteogenic sarcoma). Pulmonary metastases are present in 20% to 30% of all cancer-related deaths. The pulmonary capillary bed is the only capillary bed that is exposed to the entire circulating blood volume in each cardiac cycle, which explains why it is a common site for trapping tumor emboli. Severe dyspnea associated with lung metastasis is usually caused by diffuse infiltration of the pulmonary lymphatics, which blocks the outflow of fluid from the interstitial tissue and stimulates the J receptors, producing dyspnea. This condition is called lymphangitis carcinomatosa. Hemoptysis and cough are uncommon findings in pulmonary metastasis. Sometimes removal of tumor nodules is associated with a good prognosis (e.g., osteogenic sarcoma).

Granulomatous disease, like tuberculosis or systemic fungal infections, is not characterized by nodular lesions as large as those in the photograph. The term "miliary spread" connotes the presence of small, millet-sized lesions.

Primary cancer is always an unlikely cause when multiple lesions are present, especially involving both lungs. Multiple areas of liquefactive necrosis, as in a bronchopneumonia, are unlikely because the lesions in the patient's lungs do not appear to be part of the pulmonary parenchyma or to have a soft consistency. Multiple areas of coagulation necrosis would be highly unlikely, because pulmonary infarctions are hemorrhagic and wedge shaped.

94. The answer is B. *(Pharmacology; Pharmacology and pharmacokinetic processes)*
Cromolyn and nedrocromil are drugs that prevent calcium-mediated release of inflammatory mediators from mast cells and leukocytes. These drugs are effective prophylactic treatments for allergic reactions and asthma. Albuterol and ipratropium are bronchodilators, whereas phenylephrine is a vasoconstrictor and nasal decongestant. Diphenhydramine antagonizes histamine H_1 receptors.

95. The answer is B. *(Pharmacology; Pharmacodynamic and pharmacokinetic processes)*
Isoproterenol is a potent agonist at adrenergic β-1 and β-2 receptors, and may cause more tachycardia and arrhythmias than selective β-2 agonists, such as albuterol. The selective bronchodilating effect of β-2 agonists is enhanced when these drugs are administered as aerosols, rather than by systemic routes of administration. Because these drugs are not absolutely selective for β-2 receptors, systemic absorption may result in tachycardia or skeletal muscle tremor.

96. The answer is D. *(Gross anatomy; Nervous/special senses)*
The right recurrent laryngeal nerve branches from the right vagus nerve in the neck and passes under the right subclavian artery in the neck to ascend to the larynx. At no time is this nerve in the thorax, and therefore it is not in the superior mediastinum. In contrast, the left recurrent laryngeal nerve branches from the left vagus nerve in the superior mediastinum, passes under the aortic arch, and then ascends from the thorax into the neck. Both recurrent laryngeal nerves lie within the tracheoesophageal groove in the neck. The right recurrent laryngeal nerve is not affected by intrathoracic pathologies.

97. The answer is D. *(Gross anatomy; Musculoskeletal)*
The muscles that abduct the vocal folds, or cords, are the posterior cricoarytenoids. The adductors are the lateral cricoarytenoids, the oblique arytenoids, and the interarytenoids. The cricothyroid stretches the folds and the thyroarytenoids slacken the folds. The sensory innervation of the mucosa by the vagus nerve serves as the afferent limb of the cough reflex. The vocal folds are widely abducted during respiration, tightly adducted during deglutition, and are situated close to one another during phonation. Paralysis or paresis of laryngeal muscles results in hoarseness.

98. The answer is A. *(Gross anatomy; Cardiovascular)*
By definition, occlusion of any end artery leads to cell death. The difference between an anatomic end artery and a functional end artery is that an anatomic end artery has no collateral vessels, whereas a functional end artery has collateral vessels, but they are inadequate to maintain cell life in the event of occlusion.

99. The answer is A. *(Gross anatomy; Pulmonary/respiratory)*
The respiratory diverticulum is an endodermal evagination of the foregut that gives rise to the parenchymal epithelial surfaces of the respiratory system. The stroma of the entire respiratory system, including the cartilaginous, fibrous, and muscular elements of the respiratory system, are derived from mesoderm. Thyroid epithelium is derived from the thyroglossal duct, another endodermal evagination of the foregut that is rostral to the respiratory diverticulum.

100. The answer is A. *(Pathology; Cardiovascular)*
Mönckeberg's medial calcification has the least clinical significance. It represents dystrophic calcification (often visible on x-ray) in the media of muscular arteries. Luminal occlusion does not occur.

Arteriolosclerosis is a subclass of arteriosclerosis that primarily involves the arterioles. Hyperplastic arteriolosclerosis involves smooth muscle hyperplasia in the muscle wall with an "onion skin" appearance. This change is particularly prominent in the kidneys of patients with malignant hypertension or progressive systemic sclerosis. Hyaline arteriolosclerosis is a small vessel disease associated with diabetes mellitus and essential hypertension.

A port wine nevus flammeus in the trigeminal distribution is often associated with Sturge-Weber syndrome. Glomus tumors (glomangiomas) are painful, benign tumors of modified smooth muscle that are usually located directly beneath the nail bed.

101. The answer is A. *(Pathology; Cardiovascular)*
Lipoprotein (a) [Lp(a)] is low-density lipoprotein (LDL) combined with a plasminogen inhibitor. Atherogenesis is enhanced by Lp(a) because plasmin is not available to break up platelet-containing clots that contribute to the formation of atheromatous plaques. Its presence indicates a high risk for myocardial infarction, cerebrovascular accidents, and saphenous vein graft stenosis in coronary artery bypass operations. It is an independent risk factor for coronary artery disease in patients with familial hypercholesterolemia.

102. The answer is E. *(Pathology; Gastrointestinal)*
Alcoholics more commonly have increased triglyceride rather than cholesterol. In the metabolism of alcohol, production of reduced nicotinamide adenine dinucleotide (NADH) is increased, which causes an increase in dihydroxyacetone phosphate (DHAP). DHAP is converted into glycerol 3-phosphate, which is the carbohydrate backbone of triglyceride.

103. The answer is A. *(Pathology; Cardiovascular)*
Saturated fat should comprise no more than 10% of the total calories consumed each day. In a 2500 kcal diet, 10% saturated fat represents no more than 28 gms/day. Saturated fat intake has a greater effect than does dietary cholesterol on levels of serum cholesterol and low-density lipoprotein (LDL). Saturated fats raise LDL levels by interfering with the body's ability to remove cholesterol from cells.

104. The answer is A. *(Histology; Pulmonary/respiratory)*
Alveolar pores connect neighboring alveoli and function in equalizing pressure in the alveoli and enable collateral circulation of air. They do not constitute part of the blood–air barrier.

105. The answer is B. *(Histology; Cardiovascular)*
Endothelial cells are derived from mesenchyme and thus have vimentin intermediate filaments, which provide structural stability to the cell. Keratin filaments are present in cells derived from epithelia (e.g., ducts, epidermis).

Gap junctions and tight junctions are present between endothelial cells. Weibel-Palade granules are found in endothelial cells of arterial vessels larger than capillaries. These are small, rod-shaped structures about the width of centrioles. These granules contain factor VIII (von Willebrand factor), a blood coagulation protein. Angiotensin I is converted to angiotensin II by an enzyme found in lung endothelial cells. Angiotensin II is the active agent in blood pressure regulation.

106. The answer is B. *(Pathology; Cardiovascular)*
This patient exhibits signs of chronic deep saphenous vein insufficiency. Visible on the inner right knee is an area of discoloration that represents acute superficial thrombophlebitis. This inflammation of the vein is characterized by aching pain, bleeding, and ulceration. Signs of chronic venous insufficiency are apparent in the lower leg. These signs include discoloration secondary to hemosiderin deposition in the subcutaneous tissue, ulceration, and lower leg varicosities. These skin changes are called stasis dermatitis, and indicate the presence of underlying chronic deep saphenous vein insufficiency.

This patient's varicose veins are of the secondary type, which result from damage (i.e., thrombophlebitis, thrombosis) or obstruction of the deep venous system, with backup of pressure into the superficial venous system via incompetent valves in the penetrating branches around the ankles. The pressure is not relieved by calf muscle contraction because the valves are incompetent. Therefore, the pressure is transmitted into the capillaries in the subcutaneous tissue around the ankle, causing edema, hemorrhage with hemosiderin deposition into the subcutaneous tissue, ulceration from ischemia, and secondary varicose veins.

107. The answer is E. *(Pathology; Cardiovascular)*
The patient's hands exhibit Raynaud phenomenon. There is a symmetric cyanosis of the second to fifth fingers on both sides.

Raynaud disease and phenomenon refer to arterial insufficiency of the hands and feet. In Raynaud disease, this results from primary vasospasm of the digital vessels. Raynaud phenomenon occurs secondary to some other process. Cold temperatures and stress may trigger the color changes of the fingers, which progress from white to blue to red. Pain and tingling frequently accompany these changes. Placing the hands in warm water often reverses the condition.

Raynaud phenomenon is associated with collagen vascular diseases, such as progressive systemic sclerosis (PSS) and CREST syndrome; Takayasu disease; and cryoglobulinemia. It is also associated with thromboangiitis obliterans (Buerger disease). However, thromboangiitis obliterans occurs most commonly in males who smoke.

108. The answer is D. *(Pathology; Biochemistry and molecular biology)*
The diagram with the proper substitutions is shown.

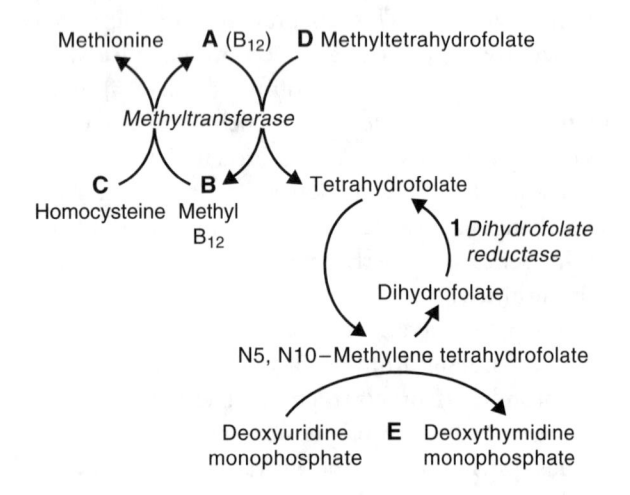

Methylcobalamin (B_{12}; B) is a cofactor for a methyltransferase enzyme that yields its methyl group to homocysteine (C), which transfers it to methionine. Cobalamin (A) retrieves a methyl group from N5-methyltetrahydrofolate (N^5-methyl-FH_4; D), which is the primary circulating form of folic acid in plasma. N^5-methyl-FH_4 is converted into tetrahydrofolate (FH_4), which is critical (via one of its intermediates $N^{5,10}$-methylene-FH_4) in converting deoxyuridine monophosphate (dUMP) into deoxythymidine monophosphate (dTMP; E), which is an immediate precursor of DNA. In the synthesis of dTMP, an oxidized dihydrofolate is produced, which must be reduced back to FH_4 by dihydrofolate reductase (enzyme 1) to replenish the FH_4. Dihydrofolate reductase can be blocked by methotrexate, trimethoprim, 6-mercaptopurine, and cyclophosphamide, resulting in a macrocytic anemia that cannot be corrected with folic acid therapy because of the enzyme block. FH_4 can also be replenished from 1 carbon derivatives derived from serine and foraminoglutamic acid (FIGLu). In addition to forming dTMP, FH_4 also serves as a go-between for 1 carbon transfers for purine synthesis and the production of methionine (requires B_{12} as a cofactor).

The defect in DNA synthesis results in a nuclear maturation delay in all actively dividing cells in the body, including red blood cell precursors, granulocytes, megakaryocytes, and gastrointestinal cells, without altering cytoplasmic maturation of the cells; therefore, the cells remain large. In the bone marrow, the enlarged cells are called megaloblasts, thus the term megaloblastic anemia. The megaloblastic cells in the bone marrow are defective and are destroyed by macrophages in the bone marrow sinusoids, resulting in a massive ineffective erythropoiesis, granulopoiesis and thrombopoiesis, and the pancytopenia noted in the peripheral blood of patients with B_{12} and/or folate deficiency.

B_{12} (A not D) is also involved in propionate metabolism. Propionate is the end-product of β-oxidation of uneven fatty acids.

$$\text{Propionyl CoA} \xrightarrow{B_{12}} \text{methylmalonyl CoA} \rightarrow \text{succinyl CoA (tricarboxylic acid cycle)}$$

In B_{12} deficiency, the build-up of methylmalonate and propionate causes abnormal myelin formation, which produces irreversible changes in the central nervous system and spinal cord (subacute combined degeneration).

In addition, a deficiency of either methyl-FH_4 or B_{12} leads to an accumulation of homocysteine, which damages endothelial cells, thus predisposing to vessel thrombosis.

109. The answer is E. *(Pathology; Biochemistry and molecular biology)*
The peripheral smear exhibits microcytic cells with considerable shape and size variation. Central pallor is increased, indicating a decrease in hemoglobin concentration in the cells. The serum ferritin is reduced; therefore, the patient must have iron deficiency anemia. Overall, the most common cause of iron deficiency is bleeding (see classification following). A 15-year-old boy who eats junk food would not likely develop iron deficiency but could develop multiple vitamin deficiencies (scurvy).

A classification of causes of iron deficiency is as follows:

1. Nutritional or increased demand

- Pregnancy/lactation. Approximately 480 mg of iron are required for the normal increase in red blood cell mass in pregnancy, with an additional 400 mg needed for the fetus and placenta and 85 mg for the loss of blood at delivery minus 250 to 500 mg saved by secondary amenorrhea for 9 months. This leaves an approximate net loss of 500 mg of iron per pregnancy, if iron supplements are not provided in prenatal vitamins. Lactation also decreases iron stores (2.5–3.0 mg/day). Human milk is low in iron (0.3 mg/L), but it has a greater bioavailability than cow's milk.
- "Milk babies." A 14-month-old child who primarily drinks cow's milk, which is low in iron, will very likely develop iron deficiency. It is the most common cause of anemia in children between 9 months and 2 years of age.
- Prematurity deprives the fetus of 3 to 4 mg/day. All premature infants must have iron supplementation.

2. Chronic blood loss (overall the most common cause)

- Menorrhagia. The most common cause of iron deficiency in young women, menorrhagia is usually due to dysfunctional uterine bleeding between menarche and 20 years of age, pregnancy-related problems between 20 and 40, and colon cancer after 40 years of age.
- Gastrointestinal disorders (most common overall site). In a man less than 50 years of age, peptic ulcer disease is the most common cause of iron deficiency. In men and women over 50 years of age, colon cancer is a common cause.
- Genitourinary carcinoma. Renal and bladder cancers predominate.
- Pulmonary disease. Goodpasture syndrome and idiopathic pulmonary hemosiderosis predominate.
- Chronic intravascular hemolysis causes the individual to lose hemoglobin and iron in the urine.

3. Malabsorption

- Malabsorption syndromes. Celiac disease and Whipple disease predominate.
- Crohn disease. Iron deficiency is more common than B_{12} deficiency.
- Achlorhydria. Cannot break up food to release heme iron.
- Postgastrectomy. Forty percent of these patients develop iron deficiency over 10 to 15 years.

110. The answer is D. *(Pathology; Hematopoietic/lymphoreticular)*
An adult with fever and a white blood cell count of 20,000 cells/μl with a left shift has an infection, which is not an indication for a bone marrow examination. Pancytopenia; documenting a myeloproliferative disease; ruling out multiple myeloma as the cause of a monoclonal spike; and evaluating a child with fever, bone pain, anemia, thrombocytopenia, and an increased lymphocyte count for possible acute lymphoblastic leukemia are all valid indications for a bone marrow examination. Other indications include the work-up of metastatic disease to bone and staging for Hodgkin and non-Hodgkin lymphomas. As a general rule, bone marrow aspirates/biopsies are not usually necessary in:

- Diagnosing iron deficiency or anemia of chronic disease, because iron studies are adequate
- Thalassemia syndromes, because hemoglobin electrophoresis is the gold standard test
- B_{12} or folate deficiency, because B_{12}, red blood cell folate, and serum folate levels are usually sufficient
- Hemolytic anemias, because the reticulocyte count (marker for all hemolytic anemias), the direct and indirect Coombs test (autoimmune anemias), osmotic fragility (congenital spherocytosis), Heinz body preparations (glucose-6-phosphate dehydrogenase deficiency), and hemoglobin electrophoresis (sickle cell and its variants) are sufficient in most cases

Most bone marrow examinations are performed on the posterior iliac crest, since both a bone marrow aspirate and biopsy may be obtained in this area. A "dry tap" means that nothing was obtained on the bone marrow aspirate. A "dry tap" may be due to faulty positioning of the needle (most common cause), a marrow packed with cells (leukemia), a marrow replaced by fibrous tissue (agnogenic myeloid metaplasia and myelofibrosis secondary to metastatic disease), or an aplastic marrow. Normal components of a bone marrow examination and report include:

- An evaluation of cellularity. Normally, there is an approximately 30% fat to 70% cell ratio (varies with age).
- Calculation of the myeloid to erythroid ratio, which is normally 3:1
- An evaluation of hematopoietic cell morphology and adequacy of numbers of megakaryocytes (platelet production)
- An estimate of the status of marrow iron stores with the Prussian blue stain. Absent iron characterizes iron deficiency; excess iron the anemia of chronic inflammation; and ringed sideroblasts, a sideroblastic anemia.

111-114. The answers are: 111-D, 112-E, 113-E, 114-E. *(Pathology; Pulmonary/respiratory)*

The patient has chronic obstructive disease secondary to smoking, complicated by the development of pulmonary hypertension, right ventricular hypertrophy (cor pulmonale), and right heart failure. He has a mixture of both chronic bronchitis and emphysema.

Chronic obstructive pulmonary disease (COPD) is progressive airway obstruction caused by destruction of lung parenchyma or irreversible damage to conducting pathways. Cigarette smoking is the main cause of COPD. Approximately 10% to 15% of patients who smoke will develop COPD. Emphysema and chronic bronchitis (CB) are the two major types of COPD. An overlap frequently exists between the two entities. CB is primarily a clinical diagnosis defined as a productive cough for 3 months out of a year for 2 consecutive years. Emphysema refers to permanent damage in the respiratory unit (respiratory bronchioles, alveolar ducts, alveoli) with destruction of elastic tissue support, resulting in permanent enlargement of these airspaces. Patients with pure CB are often referred to as "blue bloaters," because they commonly retain CO_2 (respiratory acidosis) and are cyanotic. Patients with pure emphysema are called "pink puffers," because they frequently have normal to low arterial P_{CO_2} from hyperventilation.

The patient qualifies as having CB. He has had a productive cough most of the year for the last 7 years.

The key pathologic findings in CB that produce obstruction are in the nonrespiratory small caliber airways. The larger airways also have changes, but are not primarily responsible for obstruction. Pathologic findings include:

- Mucous gland hyperplasia/hypertrophy with excess mucus production
- Goblet-cell hyperplasia (large airways) and metaplasia (bronchioles), because they do not normally have goblet cells
- Increased Reid index (ratio of the thickness of the submucosal gland layer to the thickness of the bronchial wall)
- Squamous metaplasia and loss of cilia
- Mucous plug impaction with patchy areas of atelectasis. Mucous plugs are viscous because of an increased DNA content (high molecular weight) from lysed cells
- Bronchial smooth muscle hypertrophy
- Inflammation with peribronchiolar fibrosis. When this condition is present, the positive intrathoracic pressure on expiration causes these airways to collapse, producing an obstructive bronchitis. This condition occurs when the patient develops air trapping with an increase in the anteroposterior diameter, but not to the degree that is seen in emphysema, which involves the respiratory unit.

Destruction of elastic tissue and abnormal dilatation of the bronchi would not be expected in this patient, because these findings indicate bronchiectasis. Bronchiectasis is characterized by more copious amounts of sputum than described by the patient.

The reason for this patient's recent acute exacerbation of coughing most closely relates to acute bronchitis secondary to *Moraxella (Branhamella) catarrhalis*, which is a gram-negative diplococcus. *Klebsiella pneumoniae* is a gram-negative, fat rod with a mucous capsule, and *Haemophilus influenzae* is a gram-negative coccobacillus.

Pulmonary function studies in this patient exhibit a classic obstructive pattern. The total lung capacity and residual volume are both increased from air trapping (7.10 liters, 3.90 liters, respectively). The forced vital capacity (FVC), forced expiratory volume in 1 second (FEV$_1$), and FEV$_1$/FVC ratio are decreased (3.0 liters, 1.0 liter, 33%, respectively) because of the trapping of air. The carbon monoxide diffusion is decreased (20 ml/min/mm Hg), indicating a component of emphysema with destruction of the alveolar–capillary bed. In pure CB, the D$_L$CO is normal, because the disease does not extend into the respiratory unit.

The arterial blood gases on 30% oxygen exhibit a primary respiratory acidosis with a partially compensated metabolic alkalosis and an alveolar-arterial gradient of 73 mm Hg, which is medically significant (> 30 mm Hg).

pH	7.24 (acidemia)
Pao_2	46 mm Hg (hypoxemia)
Paco_2	76 mm Hg (respiratory acidosis, because it is > 33 mm Hg)
HCO$_3$	32 (compensatory metabolic alkalosis, because it is > 28 mEq/L)

$$P_{AO_2} = \% \text{ oxygen } (713) - P_{aCO_2}/0.8$$
$$P_{AO_2} = 0.30 (713) - 76/0.8 = 119 \text{ mm Hg}$$
$$A\text{-}a = 119 - 46 = 73 \text{ mm Hg}$$

There is a secondary polycythemia (hemoglobin 19 g/dl), primarily caused by hypoxemia and the stimulus for erythropoietin release by the kidney. Erythropoietin enhances red blood cell production in the bone marrow.

The findings the patient presented with 3 years later are consistent with cor pulmonale, which refers to pulmonary hypertension (PH) with subsequent development of right ventricular hypertrophy. PH must be related either to primary disease of the vessel or disease resulting from pulmonary causes. This excludes cardiac causes for PH in the definition of cor pulmonale (e.g., mitral stenosis, volume overflow from a left-to-right shunt).

PH is characterized by intimal fibrosis and medial smooth muscle hypertrophy/hyperplasia arising as a primary or secondary process. Normally, the pulmonary circulation is a high flow, low pressure (25/8 mm Hg), and low resistance circuit. The pathophysiology of PH involves the following:

- A reduction in cross-sectional area of the pulmonary artery bed secondary to loss of pulmonary vasculature (emphysema), vasoconstriction (hypoxemia, acidosis), or obstruction (pulmonary embolization)
- Increased pulmonary venous pressure (constrictive pericarditis, mitral stenosis)
- Increased pulmonary blood flow (left-to-right shunts; a ventricular septal defect)
- Increased blood viscosity (polycythemia)

Grossly, the pulmonary arteries exhibit atherosclerosis caused by increased pressure, and thickening owing to an increase in muscle in the media and fibrous tissue in the intima. Clinically, patients have right heart failure with:

- Right ventricular hypertrophy
- An accentuated pulmonic component of the second heart sound (P$_2$), because of closure of the valve under pressure
- Cannon a-waves in the jugular venous pulse caused by the right atrium contracting against increased resistance in the right ventricle

A radiograph shows enlargement of the right and left main and lobar pulmonary arteries with tapering off of the vessels before they reach the periphery (pruning). RVH and right atrial enlargement are present, which increases the contour of the right heart border. The prognosis is poor (some survive 5–6 years).

115. The answer is E. *(Pathology; Microbial biology and infection)*
The low-power section of lung reveals a foamy alveolar infiltrate and the silver stain demonstrates the cysts of *Pneumocystis carinii*.

Pneumocystis carinii pneumonia is an opportunistic infection. It is the most common initial presentation in AIDS. The organisms are not visible on Gram stain, but are readily identified with the methenamine silver and Giemsa stains, or by direct immunofluorescence of clinical material. Recent studies have shown that *Pneumocystis* is a fungus and not a protozoan. Clinically, patients present with low-grade fever and severe dyspnea. The most cost effective way of securing the diagnosis is with an induced sputum (sensitivity 80%). Bronchoalveolar lavage and lung biopsy are excellent methods to identify the organisms when induced sputum samples are negative. Trimethoprim-sulfamethoxazole is considered the drug of choice. Pentamidine is also used, but is more toxic.

His arterial blood gases reveal a respiratory alkalosis with a partially compensated metabolic acidosis. The alveolar-arterial gradient is markedly prolonged (139 mm Hg), primarily because of intrapulmonary shunting from widespread atelectasis.

pH	7.52 acidemia
P_{O_2}	50 hypoxemia
P_{CO_2}	20 respiratory alkalosis (< 33 mm Hg)
HCO_3	16 metabolic acidosis (< 22 mEq/L)

The alveolar-arterial gradient is calculated as follows:

- P_{AO_2} = % oxygen (713) – P_{aCO_2}/0.8
- P_{AO_2} = 0.30 (713) – 20/0.8 = 189 mm Hg
- A-a = 189 – 50 = 139 mm Hg (> 30 mm Hg is medically significant)

The special test is an ELISA screen for human immunodeficiency antibody. The anemia is consistent with the anemia of inflammation (chronic disease). This anemia is characterized by an iron blockade. Iron is stored in the macrophages and is made unavailable to the bone marrow for erythropoiesis. The increased iron stores are reflected in increased serum ferritin levels.

He most likely contracted the disease from his blood transfusions 11 years ago, rather than from anal intercourse (1 year ago), which would not fit the usual time frame for developing overt AIDS (4–10 years after exposure to the virus).

116. The answer is E. *(Pathology; Pulmonary/respiratory)*
Wheezing may result from localized or diffuse narrowing of the airways from the level of the larynx to the bronchioles. The narrowing may be caused by bronchoconstriction (e.g., bronchial asthma), partial obstruction (e.g., foreign body, tumor, extrinsic compression, mucous plugs), or mucosal edema. Wheezes are more commonly heard during expiration, because airways have a tendency to narrow as a result of positive intrathoracic pressure. Inspiratory and expiratory wheezing signifies obstruction.

Wheezing would be expected in cases of:

- Left heart failure, because of peribronchiolar edema (cardiac asthma)
- Pulmonary embolus, because of the release of chemical mediators from platelets in the clot (thromboxane A_2, platelet activating factor)
- Bronchiolitis owing to the respiratory syncytial virus
- Serotonin, a bronchoconstrictor that is released into the systemic circulation in the carcinoid syndrome

Additional causes of wheezing include bronchial asthma and hypersensitivity pneumonitis (Farmer's lung).

Newborns with respiratory distress syndrome do not exhibit wheezing, because the primary problem is a deficiency of surfactant and widespread atelectasis, rather than small airway disease.

117. The answer is C. *(Pathology; Pulmonary/respiratory)*
The patient has sarcoidosis, which is a granulomatous multisystem disorder of unknown etiology. It is more common in blacks than whites and in females than males between 20 and 35 years of age. The lungs, eyes, and skin are most frequently involved. Immunologic abnormalities are present, most notably a defect in cellular immunity. T-helper cell counts in the peripheral blood are reduced. This change may reflect a redistribution of these cells to areas of noncaseating granuloma formation (type IV hypersensitivity), rather than an actual decrease in number. Because skin tests that are used to evaluate cellular immunity require T-helper cells to produce a positive response (similar to a positive PPD skin test in tuberculosis), these patients do not host an immune reaction and are considered anergic.

Sarcoidosis is the second most common cause of restrictive lung disease. The lungs are involved in 90% of cases. It is a diagnosis of exclusion, because no single test confirms the disease. The diagnosis requires the demonstration of noncaseating granulomas, however, and the appropriate clinical picture. Clinical findings include:

- Fever
- Dyspnea, caused by stimulation of the J receptors in the interstitium by fibrosis, which initiates rapid,

shallow breathing, minimizing the work of breathing in restrictive lung diseases
- Uveitis (inflammation of the uveal tract) with visual problems
- Enlargement of the lacrimal and salivary glands (uveoparotid fever; Mikulicz syndrome)
- Erythema nodosum (subcutaneous inflammation of adipose tissue)
- Facial lesions
- Cranial nerve palsies
- Myocarditis
- Diabetes insipidus (hypothalamic involvement)
- Presence of noncaseating granulomas in many organs, such as the lymph nodes (scalene 80%; mediastinal 90%), lungs, liver (60%; most common noninfectious cause of granulomatous hepatitis), bone marrow, and spleen. In the lymph nodes, the noncaseating granulomas contain asteroid bodies and calcified, shell-like structures called Schaumann bodies. Neither of these structures are pathognomonic for sarcoidosis.

Laboratory findings include:

- Abnormal pulmonary function tests with a restrictive pattern. All lung volumes and capacities are decreased. Although both the forced expiratory volume 1 second (FEV_1) and forced vital capacity (FVC) are decreased, the increased lung recoil is able to express most of the air out in 1 second; therefore, the FEV_1 and the FVC are often the same and the ratio of FEV_1/FVC is frequently normal to increased. The D_LCO is reduced because of diffusion abnormalities related to interstitial fibrosis.
- Hypercalcemia (10% of cases), owing to the granulomas synthesizing the 1,25-α–hydroxylating enzyme for vitamin D synthesis
- Hypercalciuria (40%), which predisposes to renal stones
- Increased angiotensin-converting enzyme (approximately 90% with active disease), which is a good monitor of disease activity and response to steroid therapy
- Polyclonal gammopathy, because of an increase in B cells
- Anergy (previously discussed)
- A positive Kveim skin test (positive in 75%). This is an impractical test, because it involves injecting sarcoid antigens subcutaneously and biopsying the area in a few weeks to see if an inflammatory reaction has occurred.
- Radiographic findings consist of bilateral hilar adenopathy alone ("potato nodes") and/or interstitial lung disease with nodular densities.

Corticosteroids are the mainstay for treatment. Death from pulmonary insufficiency is uncommon (5%).

118. The answer is E. *(Pathology; Pulmonary/respiratory)*
Metastasis is a rare cause of hemoptysis, which is defined as the expectoration of blood. Hemoptysis originating from the upper respiratory tract (e.g., nasal passages) must be distinguished from blood originating in the lower tract. Cough associated with hemoptysis is most suggestive of a lower respiratory tract origin; chronic bronchitis is the most common overall cause from this location. Pneumonia, primary lung cancer, and bronchiectasis are also common causes. Massive hemoptysis is more likely to be secondary to primary lung cancer, tuberculosis, bronchiectasis, or an aspergilloma (aspergillus living in an old tuberculous cavity). In mitral stenosis, pulmonary venous congestion commonly results in rupture of the pulmonary capillaries with expectoration of rusty-colored sputum (alveolar macrophages with hemosiderin) or gross blood. Goodpasture syndrome and idiopathic pulmonary hemosiderosis also present with hemoptysis; the former is caused by immunologic destruction of pulmonary capillaries by antibasement-membrane antibodies, and the latter is of unknown etiology.

119. The answer is B. *(Pathology; Pulmonary/respiratory)*
Obstruction is least likely to play a primary role in the pneumoconioses, which are the most common type of restrictive (rather than obstructive) lung disease. Pneumoconioses involve the inhalation of pollutants (e.g., asbestos, silica, anthracotic pigment) that may predispose a patient to pulmonary damage. Restrictive lung diseases are usually associated with interstitial fibrosis and its effect on the reduction of lung compliance (ability to distend on inspiration).

Bronchiectasis is an acquired type of obstructive lung disease in which a cystic dilatation of the bronchus is secondary to a combination of obstruction (mucous plugs, cancer, scar tissue) and infection, which weakens the wall of the bronchus. The saccular dilatations are filled with pus. Cystic fibrosis is the most common cause in the United States, while tuberculosis leads the list in other countries.

Primary lung cancer commonly produces an obstructive lesion in the airways that predisposes patients to atelectasis (air resorbs out of the obstructed lung), inspiratory stridor, and endogenous lipoid pneumonia. The latter disease is characterized by golden-colored areas of consolidation in the lung caused by the accumulation of cholesterol-laden alveolar macrophages. These areas are sometimes referred to as golden pneumonia.

120. The answer is B. *(Pathology; Pulmonary/respiratory)*
Progressive massive fibrosis is the crippling lung disease associated with coal worker pneumoconiosis (CWP). There is an increased incidence of tuberculosis in CWP, but not in primary lung cancer. The same is true for silicosis.

A 70-year-old patient who smokes, with a past history of tuberculosis, could develop a scar carcinoma (usually an adenocarcinoma) in the area of scar tissue formation, or one of the more common centrally located bronchogenic carcinomas (e.g., squamous-cell carcinoma, small-cell carcinoma).

A 56-year-old patient who smokes, with a history of asbestos exposure 20 years ago, is predisposed to both primary lung cancer and a mesothelioma; however, primary lung cancer would be the most common cancer, because smoking acts as a carcinogen in combination with asbestos. The same relationship does not hold true for mesotheliomas.

A 48-year-old woman who works around beryllium has an increased risk of lung cancer. A 51-year-old man who works in a uranium mine is exposed to radon, which is the gas of uranium. Inhalation of this gas is associated with an increased incidence of primary lung cancer.

Other predisposing agents to primary lung cancer include polycyclic hydrocarbons in cigarette smoke, arsenic, gold, chromium, nickel, and cadmium.

121. The answer is E. *(Pathology; Microbial biology and infection)*

Asbestos exposure does not confer an increased risk for contracting tuberculosis (TB), but does increase the risk for primary lung cancer and mesothelioma.

Overcrowding in urban areas is one of the most important risk factors responsible for the resurgence of TB in the United States. Other risk factors include:

- Hispanic and Native American heritage
- Migrant-worker status and homelessness
- Infection with AIDS (both *Mycobacterium tuberculosis* and *M. avium-intracellulare*)
- Treatment with coritcosteroid therapy (reactivation of TB)
- Low socioeconomic status
- Residence in a long-term care facilitiy

122. The answer is B. *(Pathology; Pulmonary/respiratory)*
The risk for lung cancer following cessation of smoking is halved in 10 years when compared with the risk for patients who continue smoking. The following list summarizes the benefits of smoking cessation:

- Coronary heart disease risk is halved in 1 year when compared with the risk for patients who continue smoking, and is halved in 15 years when compared with the risk for patients who do not smoke.
- Stroke risk is reduced compared with that of non-smoking patients 5 to 15 years after quitting.
- Pancreatic cancer risk is reduced in 10 years when compared with that of patients who continue smoking.
- The risk of death from chronic obstructive lung disease is reduced after a prolonged time when compared with the risk for patients who continue smoking.
- The risk for peptic ulcers, peripheral arterial disease, and cancer of the larynx are reduced when compared with the risk for patients who continue smoking.
- The risk for bladder cancer is halved in a few years when compared with the risk for patients who continue smoking.
- Cancers of the mouth, throat, and esophagus are halved in 5 years when compared with the incidence in patients who continue smoking.

123. The answer is E. *(Pathology; Microbial biology and infection)*

A higher dose of tubercle bacilli is required to reinfect an individual who has a positive purified protein derivative (PPD) skin test, indicating that there is partial immunity against reinfection. Most cases of secondary tuberculosis (TB), however, are caused by reactivation of dormant tubercle bacilli related to a breakdown in immunity (e.g., patient is placed on corticosteroids).

The area of induration, rather than the area of erythema, is measured after 48 to 72 hours. This is an example of delayed-reaction hypersensitivity. A negative PPD does not exclude active or inactive disease, because some patients with TB may be anergic because of defective cellular immunity. This is often the case in patients with AIDS, who have an area of induration ≥ 5 mm that is considered positive. A positive PPD does not distinguish active from inactive disease. Atypical mycobacteria, like *Mycobacterium avium-intracellulare*, often produce a weakly positive PPD. Patients previously vaccinated with bacille Calmette-Guérin (BCG) may have a false-positive PPD reaction. Current theories state that these patients are considered to have true infection and should receive follow-up care.

124. The answer is E. *(Pharmacology; Pharmacodynamic and pharmacokinetic processes)*
Diphenhydramine has significant antimuscarinic (atropine-like) activity, and may cause dry mouth, blurred vision, urinary retention, and tachycardia. Sedation is commonly observed in most patients, but children may paradoxically experience excitement.

125. The answer is D. *(Pharmacology; Pharmacodynamic and pharmacokinetic processes)*
Pseudoephedrine, a mixed direct- and indirect-acting sympathomimetic, may raise blood pressure and cardiac oxygen requirement, potentiate the cardiac effects of thyroid hormone, and increase urinary retention in patients with prostatic hypertrophy. It is not contraindicated in cases of urinary incontinence, and may actually be useful in select patients with the condition because it contracts the bladder sphincter.

126. The answer is D. *(Pharmacology; Pharmacodynamic and pharmacokinetic processes)*
Guaifenesin and other expectorants, such as organic iodides (iodinated glycerol), act to liquefy tenacious sputum and facilitate its removal. Large doses of guaifenesin are required to effectively decrease mucous viscosity and increase mucous membrane hydration. These agents do not suppress the cough reflex and are therefore pharmacologically distinct from the antitussives. Fluid intake should be increased to optimize the effectiveness of expectorants.

127-130. The answers are: 127-C, 128-B, 129-D, 130-A. *(Pathology; Cardiovascular)*
Heart murmurs have a longer duration than heart sounds. They may occur during systole, diastole, or both. They are classified as pathologic (secondary to structural abnormalities), physiologic (secondary to a physiologic alteration in the body), or innocent (unrelated to structural or physiologic alterations).

Stenosis refers to a dysfunction in the opening of a valve. Stenotic murmurs occur when valves are normally opening. The aortic and pulmonary valves open during systole; the mitral and tricuspid valves open during diastole. Stenotic murmurs of the aortic and pulmonary valves, in which a stream of blood moves rapidly through a narrow opening, are of the ejection, crescendo–decrescendo, or diamond-shaped type. Stenotic murmurs of the mitral and tricuspid valves produce a diastolic rumble immediately after an opening snap, which results from the sudden opening of the nonpliable valve under increased atrial pressure. The murmur depicted in (*B*) indicates mitral stenosis, and the murmur depicted in (*C*) indicates aortic stenosis.

Regurgitation, or insufficiency, refers to a dysfunction in the closing of a valve. Regurgitant murmurs occur when the valves are normally closing. The mitral and tricuspid valves close during systole; the aortic and pulmonary valves close during diastole. Regurgitant murmurs of the mitral and tricuspid valves are pansystolic (holosystolic) because blood rushes at an even intensity into the atria from the ventricles. S_1 and S_2 heart sounds are often obscured by the murmurs. Regurgitant murmurs involving the aortic and pulmonary valves occur during diastole, and they produce a high-pitched blowing murmur after the S_2 heart sound. The murmur depicted in (*A*) indicates aortic regurgitation, and the murmur depicted in (*D*) indicates mitral regurgitation.

The 65-year-old man with diminished pulses, a history of angina and syncope with exercise, and a murmur radiating into the carotid arteries most likely has aortic stenosis (C). The most frequent cause of aortic stenosis is dystrophic calcification of a congenital bicuspid aortic valve. Less frequent causes include rheumatic fever and degenerative disease. Aortic stenosis is the most common valvular lesion associated with angina and/or syncope with exertion. Angina occurs because of the increased demand of the left ventricle for more oxygen. In addition, filling of the coronary artery is decreased. Reduced cardiac output couples with increased blood flow to the muscles decreases available blood flow to the brain, resulting in syncope.

The 46-year-old woman with a malar flush, dysphagia for solids, hoarseness, an accentuated P_2 heart sound, and rust-colored sputum has mitral stenosis (B). The murmur of mitral stenosis is characterized by an opening snap as the non-pliable valve eventually gives way under increased left atrial pressure. This is quickly followed by a mid-diastolic rumbling murmur. The closer the opening snap is to the second heart sound, the worse the stenosis, because it indicates the great amount of force necessary to open the stenotic valve. Obstruction to blood flow from the left atrium into the left ventricle decreases diastolic filling of the left ventricle and diminishes cardiac output. Left atrial hypertrophy and dilatation result from volume overload secondary to incomplete emptying of the atrium during diastole. Enlargement of the left atrium frequently results in dysphagia for solids, because it compresses the esophagus, and hoarseness, because it irritates the recurrent laryngeal nerve (Ortner syndrome). Pressure from the left atrium is transmitted back into the lungs, producing pulmonary edema and congestion, which leads to hemoptysis, cough, and rust-colored sputum. Pulmonary venous hypertension, manifested by an accentuated P_2, predisposes to right ventricular hypertrophy and right-sided heart failure.

The 62-year-old woman with chronic ischemic heart disease and signs of left-sided heart failure most likely has developed a congestive cardiomyopathy with mitral regurgitation from dilation of the mitral valve ring (D). Other causes of mitral regurgitation include rheumatic fever (the most common cause), mitral valve prolapse, infective endocarditis, rupture of the papillary muscle, and left ventricular dilation from heart failure. In mitral regurgitation, left ventricular end-diastolic pressure and volume are increased. Volume overload results in left ventricle hypertrophy and dilatation. Left atrial pressure also increases, which often progresses into left-sided heart failure.

The 73-year-old man with bounding pulses, bobbing of the head, and a prominent pulsating mass on his anterior chest and who tested positive for syphilis, most likely has a syphilitic aortic aneurysm, which is the most common manifestation of tertiary syphilis. Aortic arch aneurysms stretch the aortic valve ring, resulting in aortic insufficiency (A). Other diseases that result in stretching of the ring include Marfan syndrome, ankylosing spondylitis, and coarctation of the aorta. Rheumatic fever, the most common cause of aortic regurgitation, results in fusion of the commissures, leading to valvular incompetence. In aortic regurgitation, volume overload occurs in the left ventricle, resulting in left ventricular dilatation and hypertrophy. Stroke volume increases due to Frank-Starling mechanisms. Loss of blood into the left ventricle during diastole drops the diastolic blood pressure as well, thus increasing the pulse pressure, which is the difference between systolic and diastolic pressure. The increased pulse pressure is responsible for many of the signs of a hyperdynamic circulation, including a bounding arterial pulse and bobbing of the head during systole.

131-133. The answers are: 131-A, 132-C, 133-A. *(Pathology; Cardiovascular)*
The normal jugular venous pulse consists of three positive waves (a, c, and v) and two negative waves (x and y). The a wave (A) represents atrial contraction. It occurs just after the P wave on an electrocardiograph and just before the first heart sound. The a wave is absent in atrial fibrillation. A cannon a wave, which is abnormally tall, is the result of restriction in filling the right side of the heart with blood. Such restrictions may occur in conjunction with tricuspid stenosis, pulmonary hypertension, constrictive pericarditis, pulmonic stenosis, or pericardial effusions.

The c wave (B) represents ventricular contraction and the bulge of the tricuspid valve into the right atrium. It occurs just after S_1. The x wave (D) is a large negative wave that occupies most of systole. It is the result of downward displacement of the tricuspid valve as blood is ejected out of the right ventricle into the pulmonary artery when there is right atrial relaxation during ventricular systole. A large x wave occurs in the same setting as a large a wave. The x wave is absent in tricuspid regurgitation, because blood is regurgitated back into the right atrium. The v wave (*C*) represents right atrial filling during systole while the tricuspid valve is closed. Large v waves occur when the tricuspid valve is incompetent. Tricuspid regurgitation allows blood reflux into the right atrium, and atrial septal defect shunts blood into the right atrium. The y wave (*E*) is a negative wave that occupies most of diastole. It represents opening of the tricuspid valve with rapid flow of blood from the right atrium into the right ventricle. A slow y-wave descent occurs in tricuspid stenosis, because there is difficulty in getting blood through the stenotic valve.

134-140. The answers are: 134-B, 135-A, 136-B, 137-D, 138-F, 139-E, 140-E. *(Physiology; Cardiovascular)* The figure shows the relationship between the pressures in the aorta, left ventricle, and left atrium during a cardiac cycle.

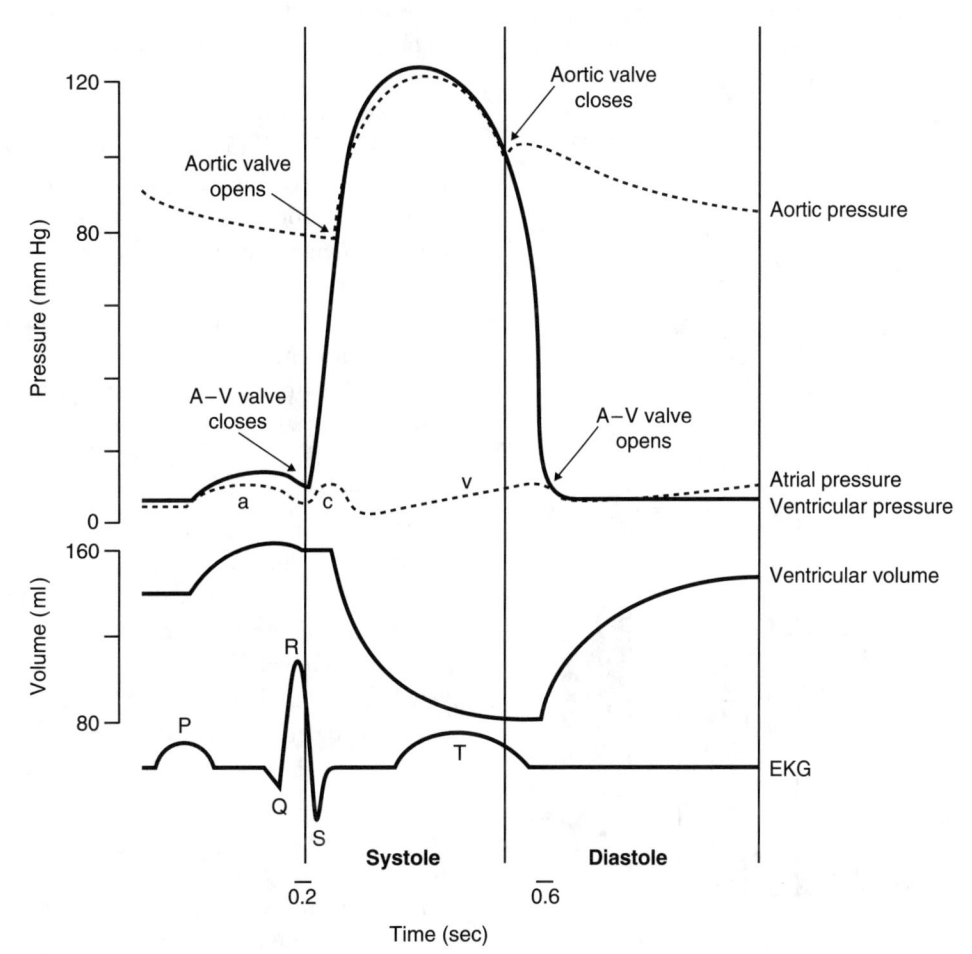

Toward the end of diastole, when the ventricle is approximately 80% filled with blood, the atria contract (between A and B) and complete the filling of the ventricle. At this time (B), the volume of the ventricle is the greatest that it will be in this cardiac cycle and is the end-diastolic volume, an index of preload. Remember that the P wave in an electrocardiogram (ECG) [atrial depolarization] precedes atrial contraction (the "a" wave) and occurs at A. In the left ventricle, the arrival of an action potential initiates ventricular systole. The increasing ventricular pressure results in the closure of the mitral valve and the first heart sound (B). Ventricular contraction continues while both the aortic and mitral valves are closed. This is the phase known as isovolumic contraction (C). The pressure in the ventricle increases until it exceeds the pressure in the aorta, causing the aortic valve to open. Ejection begins at D. As ejection proceeds, the pressure in the ventricle and aorta increase. Eventually, the pressure in the ventricle decreases to below that in the aorta; the aortic valve closes; and the second heart sound is heard (E). The volume of blood in the ventricle at the end of ejection is the end-systolic volume (E). The ventricle relaxes while both the aortic and mitral valves are closed. This is called the period of isovolumic relaxation (F). The pressure in the ventricle continues to decrease as it relaxes, until ventricular pressure decreases to below that in the atrium and the mitral valve opens (G) and filling begins again.

141-143. The answers are: 141-D, 142-B, 143-C. *(Pathology; Hematopoietic/lymphoreticular)*

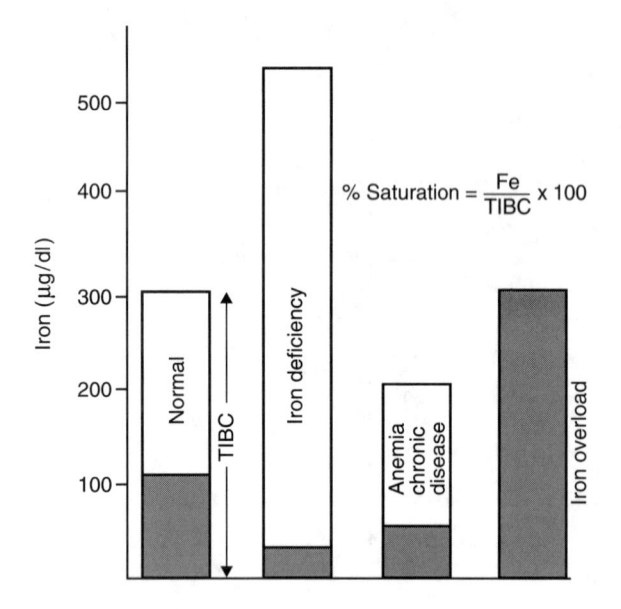

The serum iron in patient plasma is measured after stripping iron off the transferrin molecules (see figure). Note in the normal bar of the graph, that the normal serum iron is 100 μg/dl. In the iron deficiency bar of the graph, the serum iron is decreased (< 35 μg/dl). In the anemia of chronic disease (ACD) bar of the graph, the serum iron is also decreased. Patients with iron overload (hemosiderosis or hemochromatosis) have an increased serum iron (300 μg/dl in the figure).

Transferrin is the same as the total iron binding capacity (TIBC) and is measured by adding iron to the test tube to saturate all available binding sites. The iron is stripped off and measured, but, in this case, it represents the TIBC rather than the serum iron. In the normal bar of the graph, note that the normal TIBC is 300 μg/dl. Patients with iron deficiency have an increased TIBC (> 500 μg/dl in the figure). In ACD, the TIBC is decreased. Patients with iron overload have a normal TIBC (300 ug/dl in the illustration) with values often the same as the serum iron.

The percent saturation (serum iron/TIBC × 100) is decreased in iron deficiency, decreased in ACD, and increased in iron overload.

All storage iron is in the form of ferritin or hemosiderin, the latter representing packets of ferritin, which is a protein–iron complex. Ferritin is found in all tissues, particularly within the hepatocytes and in macrophages in the spleen and bone marrow. Ferritin in hepatocytes derives primarily from transferrin delivery of iron from the intestine via the portal vein, but ferritin stores in the macrophages primarily derives from the normal breakdown of senescent red blood cells. Normally, there is a very small amount of ferritin in the circulation, which is an excellent indicator of iron stores. However, it is not a very good indicator of storage iron in ACD, because it is released as an acute phase reactant from the hepatocytes.

A 28-year-old Asian woman with a mild microcytic anemia and a normal hemoglobin electrophoresis most likely has mild α-thalassemia. Mild α- and β-thalassemias have defects in globin chain synthesis and do not have any problems with iron; therefore, all the iron studies should be normal.

A patient with long-standing rheumatoid arthritis with a guaiac-negative stool most likely has ACD. In ACD, the serum iron, TIBC, and percent saturation are decreased, and the serum ferritin is normal to high.

A patient who has been transfused over 100 times most likely has transfusion-induced hemosiderosis. Iron overload produces an increase in serum iron, a normal TIBC, an increased percent saturation, and increased serum ferritin. The TIBC is not increased in iron overload, so choice C rather than E is the correct answer.

144-150. The answers are: 144-A, 145-B, 146-A, 147-C, 148-D, 149-C, 150-A. *(Physiology; Cardiovascular)*

In panel A, the stenotic aortic valve limits the rate of ejection so that the aortic pressure increases slowly. The ability of the ventricle to develop a high systolic pressure is the result of the concentric hypertrophy, induced by the pressure overload. The unusually large difference between ventricular and aortic pressures during ejection leads to a crescendo–decrescendo murmur.

In panel B, the incompetent aortic valve allows reentry of blood into the left ventricle during diastole, resulting in a rapid decline in blood pressure and, consequently, a low diastolic blood pressure. The ventricle ejects a large stroke volume, which results in a high systolic pressure. The volume overload experienced by the ventricle, as a result of the regurgitant blood, causes an eccentric form of hypertrophy. Note the wide pulse pressure.

In panel C, the stenotic mitral valve limits entry of blood into the ventricle. Atrial pressures are elevated throughout the cardiac cycle. The turbulence that results as blood moves through the stenotic mitral valve during atrial contraction causes a presystolic murmur.

In panel D, the incompetent mitral valve allows blood to reenter the atrium during ventricular systole. The atrium is filling ("v" wave) during ventricular systole. An exaggerated "v" wave denotes mitral regurgitation.

151-157. The answers are: 151-F, 152-B, 153-D, 154-G, 155-B, 156-A, 157-B. *(Physiology; Cardiovascular)*

A pressure–volume loop describes the instantaneous relationship between ventricular volume and ventricular pressure during a single cardiac cycle. A systolic pressure of 120 mm Hg indicates that this is the left ventricle. The loop must be read in a counterclockwise direction.

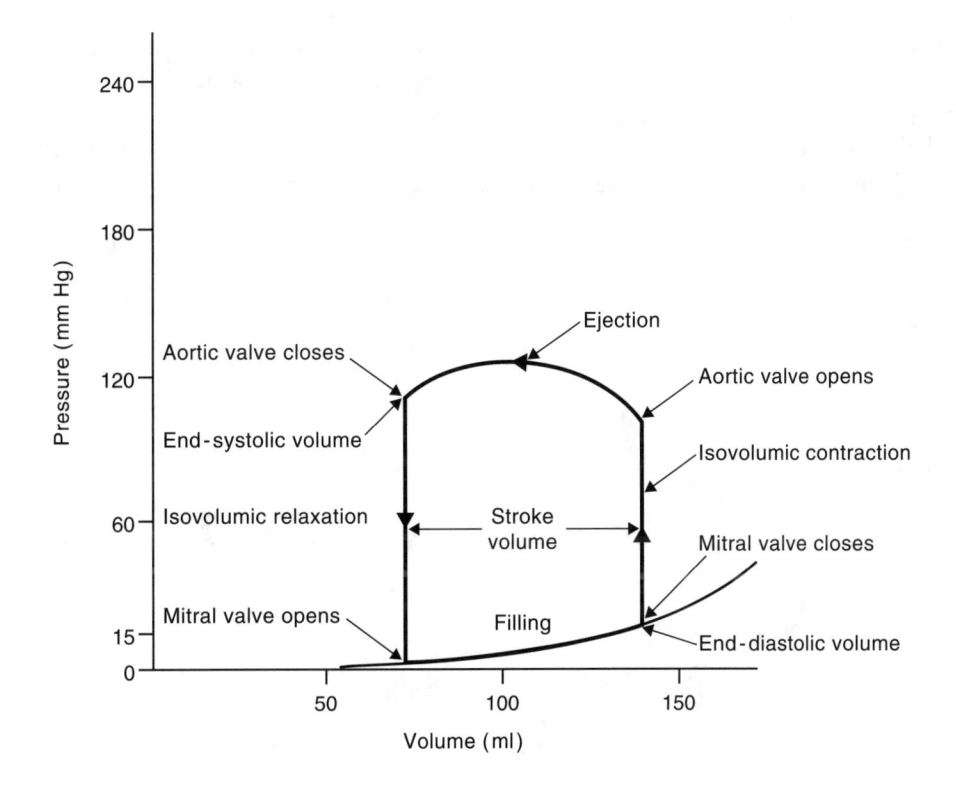

At point A, the mitral valve opens and ventricular filling begins. At point B, the volume of the ventricle has reached its maximum volume, the end-diastolic volume, and systole begins. The mitral valve closes, and the period of isovolumic contraction ensues. Ventricular pressure rises until it exceeds that in the aorta (aortic diastolic blood pressure at D), at which time the aortic valve opens and ejection begins (E). As ventricular pressure declines below that in the aorta, the aortic valve closes (F; second heart sound), and the period of isovolumic relaxation starts (G). The volume in the ventricle at this time is the end-systolic volume. The width of the loop indicates the stroke volume.

158-160. The answers are: 158-C, 159-E, 160-B. *(Pharmacology; Pharmacodynamic and pharmacokinetic processes)*
The changes in theophylline clearance in this patient represent an age-related decrease in childhood, an induction of clearance of drug use in adolescence, and an inhibition of clearance by cimetidine during ulcer therapy.

Theophylline is eliminated almost entirely by hepatic drug metabolism, which can be induced by alcohol, tobacco, marijuana, and therapeutic drugs, such as phenobarbital and rifampin. Theophylline metabolism can be significantly inhibited by cimetidine, erythromycin, and allopurinol, thereby requiring a reduction in dosage. Theophylline clearance is noticeably higher during childhood until approximately age 16, and children require larger doses per kilogram of body weight. Other factors that reduce theophylline clearance include heart failure, liver disease, infections, and old age. Neither cromolyn nor ipratropium has any significant effect on theophylline clearance.

161-163. The answers are: 161-C, 162-B, 163-E. *(Pharmacology; Pharmacodynamic and pharmacokinetic processes)*
The bronchodilators, including adrenergic β-2 agonists (albuterol), muscarinic blockers (ipratropium), and theophylline, block the early response (bronchoconstriction) to mediators released from mast cells. Corticosteroids, such as beclomethasone, block the late response of inflammation. Both steroids and cromolyn act to inhibit the release of mediators from mast cells and are used prophylactically in asthma and related conditions, including allergic rhinitis and conjunctivitis. Antihistamines are also useful in treating allergic rhinitis and conjunctivitis, but have little use in treating asthma, most likely because mediators other than histamine have a larger role in this disorder.

Test 3

QUESTIONS

DIRECTIONS:

Each of the numbered items or incomplete statements in this section is followed by answers or by completions of the statement. Select the ONE lettered answer or completion that is BEST in each case.

1. Which of the following statements concerning the circulatory system is true?

   (A) Occlusion of a functional end artery does not result in cell death
   (B) Veins generally have thicker walls than arteries
   (C) All veins have valves
   (D) Blood always flows in the same direction through a region of an artery
   (E) Blood in a portal vein passes from one capillary bed to another

2. A patient sustains a knife wound through the right fourth intercostal space, 7 cm to the right of the sternum. The area most likely penetrated by the knife is the

   (A) right upper lobe of the lung
   (B) right middle lobe of the lung
   (C) right lower lobe of the lung
   (D) right atrium of the heart
   (E) right ventricle of the heart

3. To auscultate the tricuspid valve, the best location for the stethoscope is

   (A) the fifth intercostal space, immediately to the right of the sternum
   (B) the fifth intercostal space, immediately to the left of the sternum
   (C) the fifth intercostal space, 7 cm to the left of the sternum
   (D) the second intercostal space, immediately to the right of the sternum
   (E) the second intercostal space, immediately to the left of the sternum

4. The development of the respiratory system begins during the fourth week of development as an evagination of the

   (A) first branchial pouch
   (B) first branchial cleft
   (C) ventral wall of the foregut
   (D) dorsal wall of the midgut
   (E) sixth branchial arch

5. The azygos vein drains into which of the following veins?

   (A) Superior vena cava
   (B) Inferior vena cava
   (C) Right brachiocephalic vein
   (D) Left brachiocephalic vein
   (E) Hemiazygos vein

6. Which of the following measures is most likely to decrease the risk of myocardial infarction?

   (A) Cessation of smoking
   (B) Postmenopausal estrogen replacement
   (C) Mild to moderate alcohol consumption
   (D) Maintainance of an exercise program
   (E) Low dose aspirin administration

7. You would expect an increase in cardiac output in a patient

(A) with a decrease in lung compliance
(B) with hypertrophic cardiomyopathy who is standing
(C) with aortic regurgitation who is not experiencing heart failure
(D) who is on positive end-expiratory pressure therapy
(E) with a left ventricular aneurysm

8. In mitral valve prolapse, you would expect the ejection click to be more accentuated and the murmur to move closer to the first heart sound in a patient

(A) with marked anxiety
(B) who is pregnant
(C) who is passive leg lifting in the supine position
(D) on a beta blocker
(E) who is squatting

9. Which of the following disorders would most likely obstruct blood flow, embolize, and produce constitutional symptoms of fever, weight loss, and anemia?

(A) Acute endocarditis involving the aortic valve
(B) Mural thrombus in the left ventricle
(C) Mural thrombus in the left atrium
(D) Vegetations in acute rheumatic fever
(E) Cardiac myxoma

10. Which of the following laboratory findings would you expect in a patient with severe cyanotic congenital heart disease?

(A) Normal hemoglobin and hematocrit
(B) Same risk for cerebral abscesses as in acyanotic congenital heart disease
(C) Increased alveolar–arterial gradient
(D) Correction of hypoxemia with the patient breathing 100% oxygen
(E) Normal oxygen saturation of arterial blood

11. Two months after an acute anterior myocardial infarction, a 68-year-old man has an abnormal bulge on the precordium during systole. The most common complication associated with this abnormality is

(A) rupture
(B) chronic heart failure
(C) infective endocarditis
(D) valvular dysfunction
(E) paradoxical embolization

12. A 65-year-old man with a history of unstable angina died suddenly while at home. Which of the following abnormalities is an autopsy most likely to reveal?

(A) Atherosclerotic stroke
(B) Massive pulmonary embolus
(C) Severe coronary artery atherosclerosis
(D) Pallor of the left ventricle myocardium
(E) Severe atherosclerosis with a fresh coronary thrombus overlying a plaque

13. Which of the following types of arrhythmia is most likely to predispose a patient with an acute myocardial infarction to a terminal arrhythmia and death?

(A) Atrial premature beats
(B) Ventricular extrasystoles
(C) Paroxysmal atrial tachycardia
(D) Atrial fibrillation
(E) Sinus arrhythmia

14. Which of the following signs or symptoms is associated with left heart failure?

(A) Dependent pitting edema
(B) Jugular venous distention
(C) Positive hepatojugular reflux
(D) Pillow orthopnea
(E) Splenomegaly

15. A newborn baby girl has a raised, red lesion on the right side of her upper lip. You would advise the parents to have the lesion

(A) removed by a plastic surgeon
(B) cauterized by a dermatologist
(C) frozen with liquid nitrogen
(D) injected with alpha interferon
(E) left alone and followed on a periodic basis

16. With which of the following adult diseases is childhood obesity most closely associated?

(A) Heart disease
(B) Renal disease
(C) Hypothyroidism
(D) Psychosis
(E) Restrictive lung disease

17. A 35-year-old man has a 7 cm pulsatile mass on the upper portion of his right thigh. He has a wide pulse pressure. The mixed venous oxygen content in the right heart is increased. Compression of the mass results in a sinus bradycardia and an increase in the diastolic blood pressure. The most likely etiology for this lesion is

(A) a congenital malformation
(B) metastasis from a primary lesion in the kidney
(C) previous penetrating injury
(D) a primary vascular tumor
(E) previous sepsis

18. A 72-year-old man presents with a sudden onset of left flank pain. In the emergency room, he has hypotension and a pulsatile mass in the abdomen. The mechanism for this patient's disease most closely relates to

(A) a neoplastic process
(B) elastic tissue fragmentation
(C) atherosclerosis
(D) infection
(E) an immunologic reaction

19. The following image is an autopsy finding depicting cross-sections of the carotid arteries from a 48-year-old man who died suddenly at home. The most likely predisposing factor for this patient's catastrophic event is

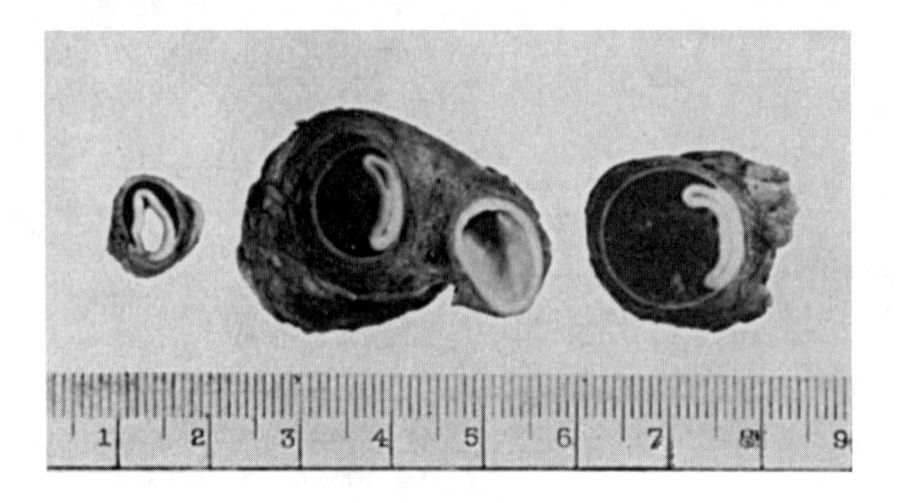

(A) hypertension
(B) atherosclerosis
(C) immune complex vasculitis
(D) granulomatous vasculitis
(E) trauma

20. Which clinical picture is most likely in a patient with a ruptured free wall of the left ventricle and cardiac tamponade?

(A) 0 to 12 hours postinfarction, normal total creatine kinase (CK) level, absent MB isoenzyme fraction, and $\frac{1}{2}$ flip of lactate dehydrogenase (LDH) isoenzyme
(B) 3 to 7 days postinfarction, increased total CK level, absent MB isoenzyme fraction, and $\frac{1}{2}$ flip of LDH isoenzyme
(C) 12 to 24 hours postinfarction, increased total CK level, absent MB isoenzyme fraction, $\frac{1}{2}$ flip of LDH isoenzyme absent
(D) 7 to 10 days postinfarction, normal total CK level, MB isoenzyme fraction present, $\frac{1}{2}$ flip of LDH isoenzyme absent

21. When are cardiac arrhythmias associated with acute myocardial infarction most likely to occur?

(A) 0 to 12 hours postinfarction
(B) 12 to 24 hours postinfarction
(C) 1 to 3 days postinfarction
(D) 3 to 7 days postinfarction
(E) 7 to 10 days postinfarction

22. A 45-year-old male patient complains that he is often tired and has a headache almost every morning. His wife says that her sleep is disturbed because of the patient's loud snoring. Physical examination reveals leg edema, hypertension, and cardiac arrhythmia. From which disorder is this patient most likely suffering?

(A) Sleep-wake schedule disorder
(B) Obstructive sleep apnea
(C) Narcolepsy
(D) Delayed sleep phase syndrome
(E) Nocturnal myoclonus

23. What percentage of premature deaths in the United States can be related directly to smoking?

(A) 2%
(B) 5%
(C) 10%
(D) 15%
(E) 25%

24. In the United States, which substance is associated with the most deaths annually?

(A) Heroin
(B) Crack cocaine
(C) Alcohol
(D) Marijuana
(E) Phencyclidine (PCP)

25. The red blood cell abnormalities represented in this figure are associated with which of the following disorders?

(A) Hereditary spherocytosis
(B) Disseminated intravascular coagulation
(C) Autoimmune thrombocytopenic purpura
(D) Autoimmune hemolytic anemia
(E) Iron deficiency anemia

26. A patient with these cells would be expected to have which of the following disorders?

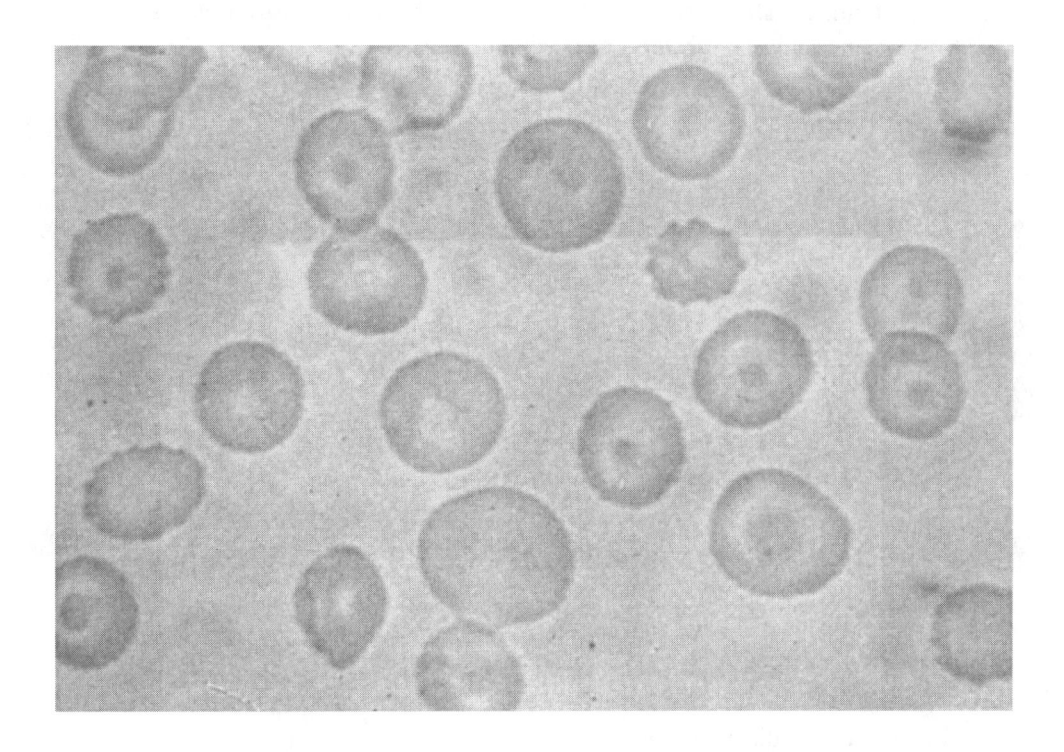

(A) Alcohol-related liver disease
(B) Autoimmune hemolytic anemia
(C) Glucose-6-phosphate dehydrogenase deficiency
(D) B_{12} deficiency
(E) Congenital spherocytosis

27. This figure represents a bone marrow aspirate from a 69-year-old woman who presents with sternal tenderness and a fracture of the left ninth rib after coughing. A radiograph of the rib reveals a lytic area in the center of the fracture site. This woman would be expected to have which disorder?

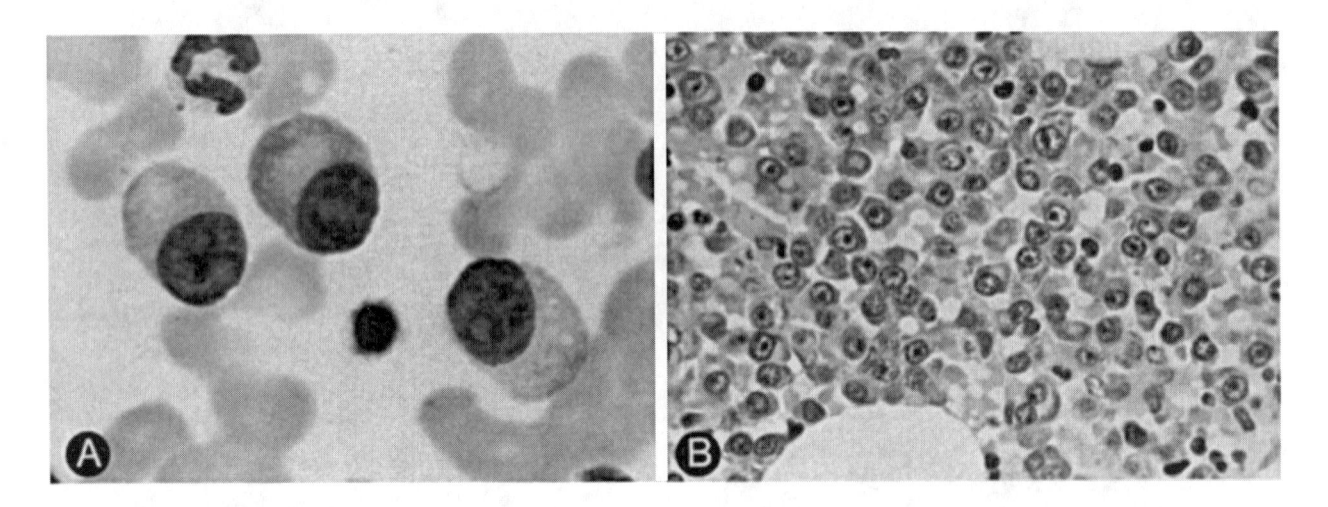

(A) κ-Light or λ-light chain in her urine
(B) Polyclonal gammopathy
(C) Low to normal erythrocyte sedimentation rate
(D) Primary increase in immunoglobulin M

28. This figure represents a lymph node biopsy from a 19-year-old man with a single, nontender lymph node in the lower neck. This patient most likely has which of the following disorders?

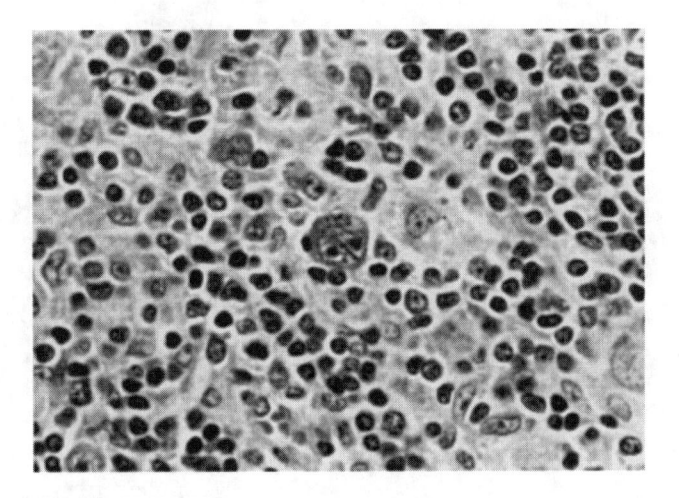

(A) Nodular sclerosis Hodgkin disease
(B) Burkitt lymphoma
(C) Lymphocyte predominant Hodgkin disease
(D) Poorly differentiated lymphocytic lymphoma
(E) Reactive lymph node hyperplasia

184 Body Systems Review I: Hematopoietic/Lymphoreticular, Respiratory, Cardiovascular

29. An increase in the cells represented would be expected in a patient with which of the following disorders?

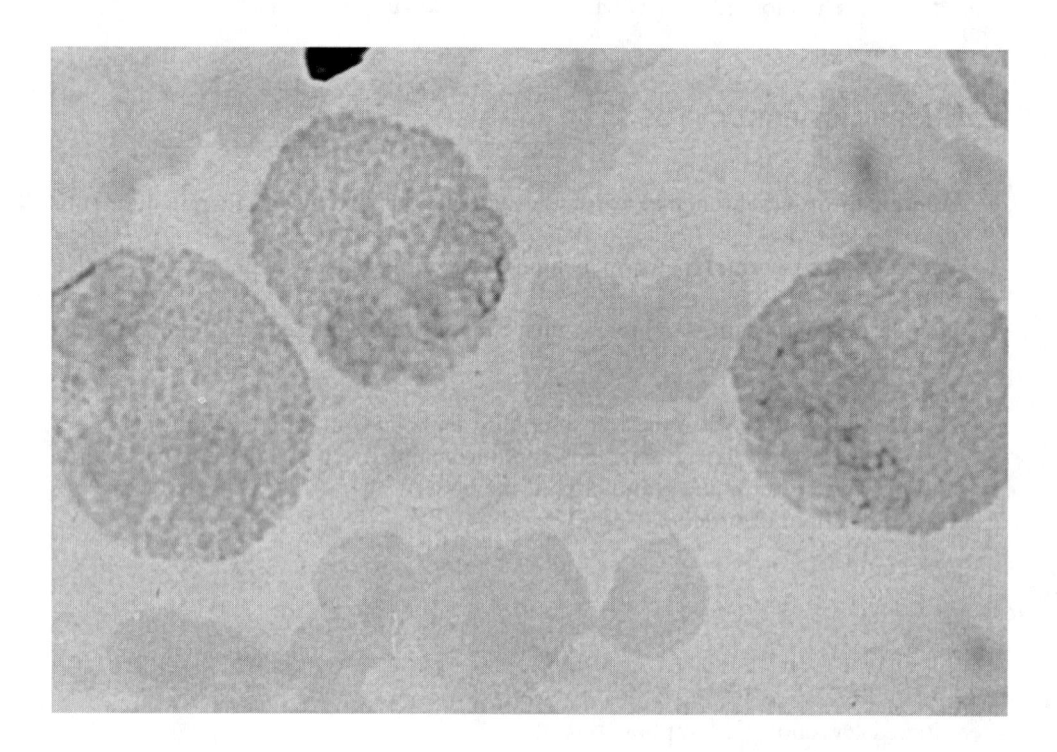

(A) Viral pneumonia
(B) Bacterial pneumonia
(C) Chronic inflammation
(D) Infectious mononucleosis
(E) Bronchial asthma

30. The representative findings in this peripheral smear would most likely be seen in a patient with which of the following disorders?

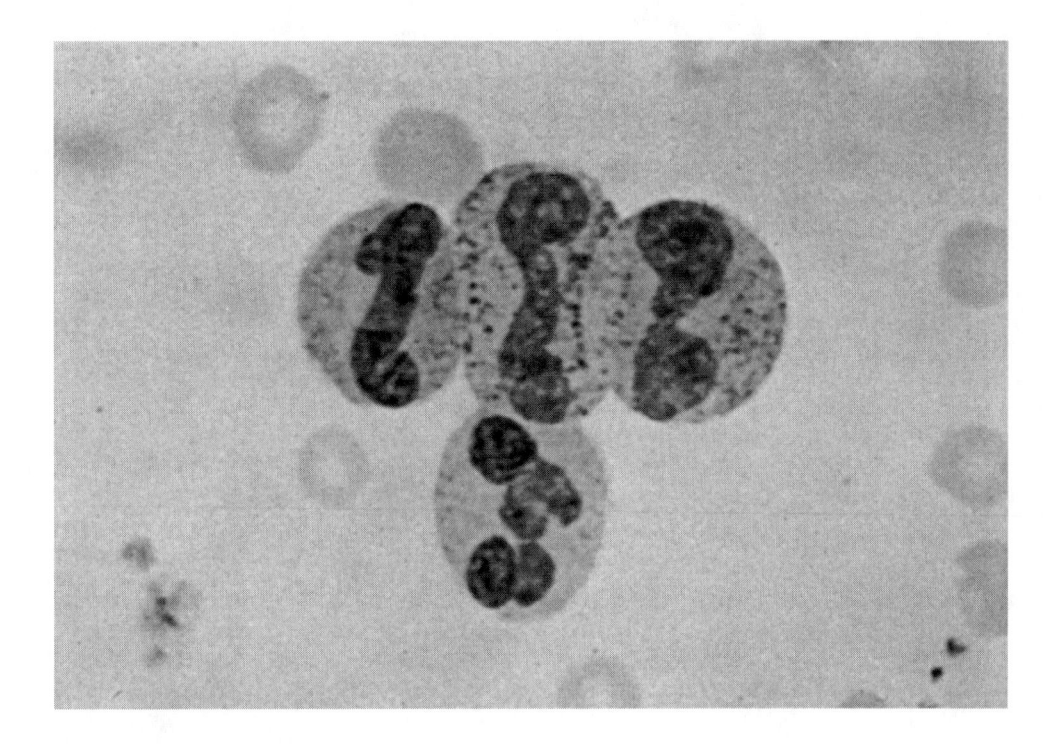

(A) Acute appendicitis
(B) Whooping cough
(C) Disseminated strongyloidiasis
(D) Tuberculosis
(E) Viral gastroenteritis

31. This figure represents a lymph node biopsy from a 17-year-old patient with tender cervical lymphadenopathy. The patient has a brother with Hodgkin disease. The most likely diagnosis is

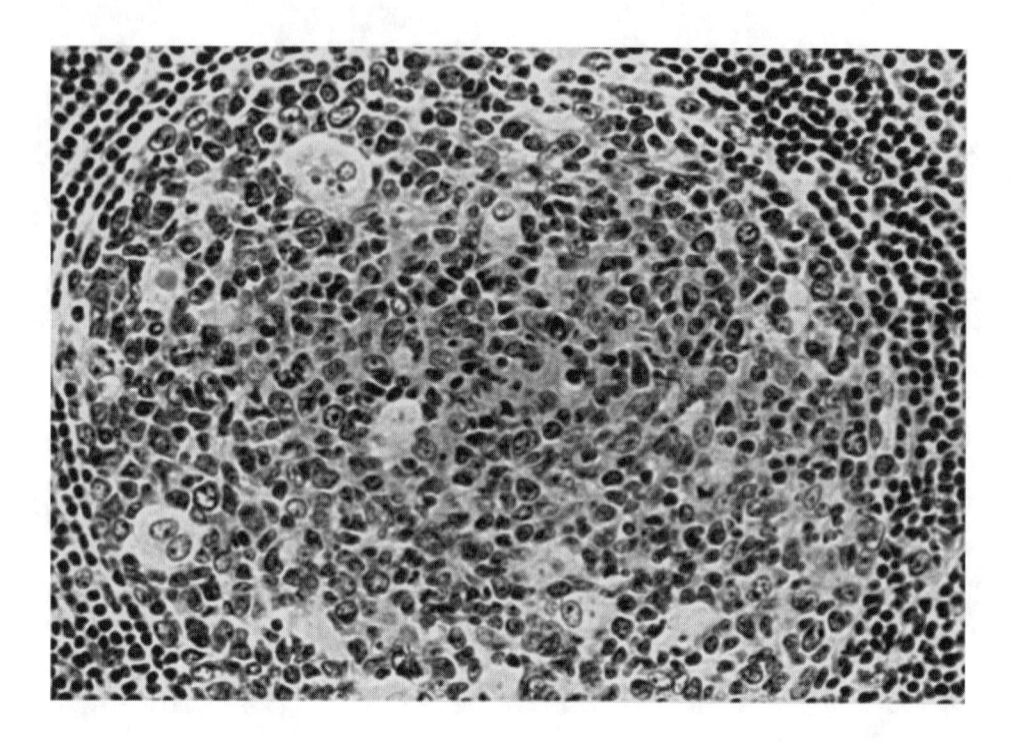

(A) nodular, poorly differentiated malignant lymphoma (follicular lymphoma)
(B) lymphocyte-predominant Hodgkin disease
(C) reactive hyperplasia
(D) metastatic carcinoma
(E) granulomatous inflammation

32. Which of the following disorders is most likely associated with erythroid hyperplasia in the bone marrow?

(A) Anemia of chronic disease
(B) β-Thalassemia minor
(C) Bone marrow 7 to 10 days after a gastrointestinal bleed
(D) Iron deficiency
(E) Chronic renal disease

33. A 45-year-old afebrile woman, who had a radical mastectomy 2 years ago, presents with point tenderness over her vertebra in the lower lumbar spine. Her complete blood cell count reveals a mild normocytic anemia and a 5% corrected reticulocyte count. The peripheral smear exhibits occasional progranulocytes, myelocytes, and nucleated red blood cells. A biochemical profile is unremarkable with the exception of an isolated elevation of alkaline phosphatase unassociated with an increase in γ-glutamyltransferase. The mechanism for this patient's clinical and laboratory findings is most closely related to

(A) a hemolytic anemia
(B) chronic inflammation
(C) a myelophthisic process
(D) a leukemic process
(E) a myeloproliferative disease

34. Using the Fisher-Race Rh nomenclature, the Rh phenotype studies of a woman with an Rh-positive husband and an Rh-negative child are as follows ('+' indicates the presence of the antigen)

C	c	D	E	e
+	+	+	+	+

The woman's Rh genotype would most likely be

(A) Cde / cde
(B) CDE / Cde
(C) CDE / cde
(D) CDE / cDE
(E) cDe / cdE

35. In a paternity case, which of the following children's blood groups is compatible with someone other than the putative father?

	Mother	Putative Father	Child
(A)	A	B	O
(B)	O	A	A
(C)	A	AB	O
(D)	A	B	AB
(E)	AB	B	A

36. A 60-year-old afebrile man has group O, Rh-negative blood. The man has a massive lower-gastrointestinal bleed and must be given group O, Rh-positive blood, because no group O, Rh-negative blood is available. In the pretransfusion workup, the patient has a negative antibody screen and a compatible major crossmatch with 8 units of group O, Rh-positive blood. Midway through the second unit of blood, the patient develops hives. The transfusion is stopped and a transfusion workup reveals the following on a post-transfusion specimen of patient blood.

Patient temperature:	100.0°F
Patient blood pressure:	130/86 mm Hg
Patient pulse:	100 beats/minute
Patient plasma:	no evidence of hemoglobinemia
Patient antibody screen:	negative
Patient direct Coombs':	negative
Patient urine:	negative dipstick for blood

The patient's transfusion reaction was most likely caused by which mechanism?

(A) Type I, immunoglobulin E (IgE)-mediated hypersensitivity reaction
(B) Type II, cytotoxic antibody-mediated hypersensitivity reaction
(C) Type III, immune complex-mediated hypersensitivity reaction
(D) Type IV, cellular immunity-mediated hypersensitivity reaction
(E) Anaphylactic reaction

37. A group O, Rh-negative woman with a negative antibody screen and no previous administration of Rh immunoglobulin delivers a group A, Rh-positive infant. The infant develops indirect hyperbilirubinemia 8 hours after birth. Which of the following statements is correct concerning the mother and baby?

(A) The baby most likely has physiologic jaundice of the newborn
(B) The mother is not a candidate for Rh immunoglobulin
(C) The mechanism of jaundice in the newborn is intravascular hemolysis
(D) The mother's blood group offers her protection against Rh sensitization
(E) The newborn's cord blood would most likely have a negative direct Coombs'

38. In a child with severe anemia, the expansion of the marrow cavity with alterations in the facial features is most likely associated with which of the following disorders?

(A) Lead poisoning
(B) Acute lymphocytic leukemia
(C) β-Thalassemia major
(D) Severe iron deficiency
(E) Hemoglobin Bart disease

39. Extramedullary hematopoiesis would most likely be associated with which of the following disorders?

(A) Severe iron-deficiency anemia
(B) Congenital spherocytic anemia
(C) Agnogenic myeloid metaplasia
(D) β-Thalassemia minor
(E) Acute blood loss

40. Which of the following signs is a marker of an erythropoietin-stimulated bone marrow?

(A) Low reticulocyte count
(B) Increased myeloid:erythroid ratio
(C) Decreased serum ferritin
(D) Polychromasia (shift cells) in the peripheral blood
(E) Decreased radioactive plasma iron turnover

41. A patient with low serum haptoglobin concentration and increased urine hemosiderin most likely has which of the following disorders?

(A) Severe calcific aortic stenosis
(B) Congenital spherocytosis
(C) α-Thalassemia minor
(D) Aplastic anemia
(E) Liver disease

42. Anemia associated with a corrected reticulocyte count less than 2% would most likely be seen in which of the following patients? A patient

(A) with anemia of chronic inflammation who is taking iron
(B) with rheumatoid arthritis who has a normocytic anemia and a positive direct Coombs test
(C) who received an intramuscular injection of B_{12} for treatment of pernicious anemia one week ago
(D) with a gastrointestinal bleed 6 days ago
(E) with agnogenic myeloid metaplasia

43. Neutrophilic leukocytosis, lymphopenia, and eosinopenia are most likely associated with which of the following disorders?

(A) Endotoxic shock
(B) Typhoid fever
(C) Whooping cough
(D) Paroxysmal nocturnal hemoglobinuria
(E) Cushing syndrome

44. A routine Wright-Giemsa stain of peripheral blood detects

(A) peripheral blood reticulocytes
(B) Heinz bodies
(C) coarse basophilic stippling
(D) globin chain inclusions in severe β-thalassemia
(E) residual RNA filaments

45. In a patient undergoing warfarian therapy, which of the following factors is measured in the partial thromboplastin time (PTT) rather than the prothrombin time (PT)?

(A) II
(B) VII
(C) IX
(D) X
(E) XI

46. Which of the following proteins is unrelated to the production of amyloid?

(A) Albumin
(B) Prealbumin
(C) Calcitonin
(D) Light chains
(E) β-Protein

47. Arrange the following events in hemostasis associated with a bleeding time into the proper sequence.

1. Platelet aggregation
2. Platelet adhesion
3. Synthesis of thromboxane A_2
4. Temporary hemostatic plug

(A) 3-2-1-4
(B) 2-3-1-4
(C) 1-2-3-4
(D) 2-1-3-4
(E) 3-1-2-4

48. Which of the following disorders is more likely associated with thrombocytopenia than thrombocytosis?

(A) Chronic iron deficiency
(B) Polycythemia rubra vera
(C) Nonhematologic malignancy
(D) Tuberculosis
(E) Congestive splenomegaly

49. A patient having a routine electrocardiogram (EKG) moves her left hand. In which of the following leads would an electrical disturbance take place?

(A) Leads I and II
(B) Leads I and III
(C) Leads II and III
(D) Lead I only
(E) Lead II only

Questions 50-52

The tracing depicted below was obtained from a jugular vein during a single cardiac cycle.

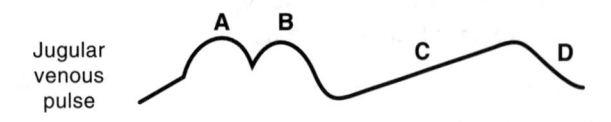

Jugular venous pulse — A, B, C, D

50. The fall in pressure associated with the portion of the trace marked D is caused by which of the following actions?

(A) Opening of the mitral valve
(B) Opening of the aortic valve
(C) Closing of the tricuspid valve
(D) Opening of the tricuspid valve
(E) First heart sound

51. Which of the following statements is true regarding the rise in pressure associated with the portion of the trace marked A? The rise

(A) is caused by the bulging of the tricuspid valve into the atrium during right ventricular contraction
(B) is caused by atrial contraction
(C) is caused by the build-up of blood in the atrium as the ventricle contracts
(D) occurs early in diastole
(E) occurs below the P wave of the electrocardiogram (EKG)

52. Which of the following statements is true concerning the rise in pressure associated with the portion of the trace marked C? The rise

(A) is caused by atrial contraction
(B) occurs prior to the QRS complex of the EKG
(C) occurs during ventricular ejection
(D) occurs during isovolumic relaxation of the ventricle
(E) occurs while the tricuspid valve is open

53. In which of the following cases would the tracing depicted below be expected?

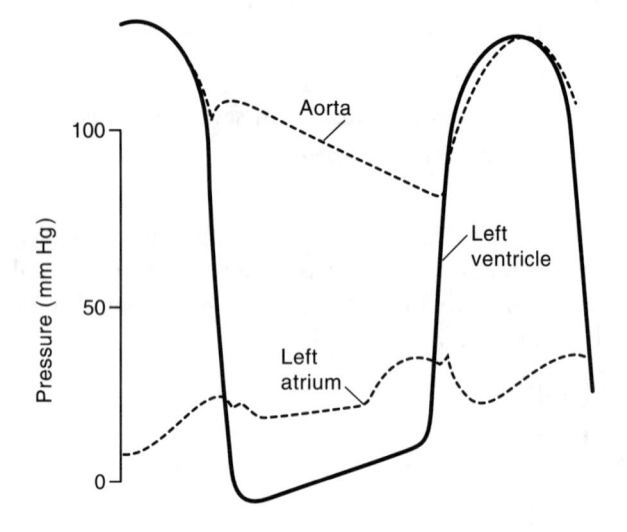

(A) Aortic stenosis
(B) Mitral stenosis
(C) Aortic insufficiency
(D) Mitral insufficiency
(E) Heart failure

54. The duration of a ventricular myocyte action potential is

(A) approximately twice as long as the relative refractory period
(B) as long in duration as the QRS complex
(C) nearly as long as the QT interval
(D) twice as long as in skeletal muscle

55. This tracing would be expected in which of the following disorders?

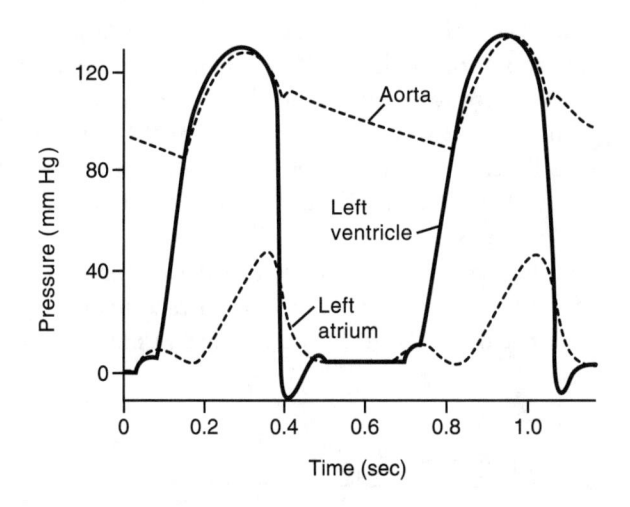

(A) Mitral insufficiency
(B) Aortic stenosis
(C) Aortic insufficiency
(D) Mitral stenosis
(E) Essential hypertension

56. In an electrocardiogram (EKG) recording, the electrical potential measured during the QRS complex was +5 mV in Lead I and + 9 mV in Lead II. What was the electrical potential in Lead III?

(A) 4 mV
(B) 14 mV
(C) 5 mV
(D) 0 mV
(E) 7 mV

57. Norepinephrine, which is released by sympathetic nerves, increases the heart rate by

(A) hyperpolarizing the sinoatrial node
(B) depolarizing the atrioventricular node
(C) increasing the rate of diastolic depolarization (phase 4)
(D) decreasing the conductance of the sinoatrial node to Ca^{++}
(E) increasing the conductance of the sinoatrial node to K^+

58. The plateau (phase 2) of the ventricular myocyte action potential

(A) describes when Ca^{++} is the only ion moving through the membrane
(B) describes when Ca^{++} influx is balanced by Na^+ efflux
(C) describes when Ca^{++} and Na^+ influx is balanced by K^+ efflux
(D) can be prolonged by sympathetic nerve stimulation
(E) can be prolonged by a drug that increases the amount of cytosolic Ca^{++}

59. The action potential of the the sinoatrial node

(A) results from the activation of rapid Na^+ channels
(B) results from opening of the slow Ca^{++} channels
(C) is equivalent in height and conduction velocity to that of the atrial muscle
(D) results from an increase in conductance to K^+
(E) results from a rhythmic stimulation of the sinoatrial node by the sympathetic nervous system

60. Which interval best approximates the duration of ventricular systole?

(A) PR interval
(B) Duration of the QRS complex
(C) Interval between the first and second heart sounds
(D) Interval between the first heart sound and the opening of the aortic valve
(E) Interval between the opening and closing of the aortic valve

61. Initiation of the action potential (phase 0) in ventricular myocytes is

(A) due to an increase in the Ca++
(B) results from the closing of the K⁺ channels
(C) due to activation of the rapid Na+ channels
(D) due to the intrinsic automaticity of myocytes
(E) dependent on the presence of a sympathetic innervation

Questions 62-64

This diagram depicts the relationship between the electrocardiogram (EKG) recorded on the surface of the body and the duration of a single ventricular myocyte action potential. The duration of the ventricular myocyte refractory period is nearly as long as the action potential, approximately 300 msec. The QRS complex of an EKG represents the electrical potential on the surface of the body produced by the wave of depolarization as it spreads through the ventricle and each myocyte depolarizes (phase 0). The T wave represents ventricular repolarization (phase 3 in each myocyte). Note that the ST segment occurs during the plateau phase.

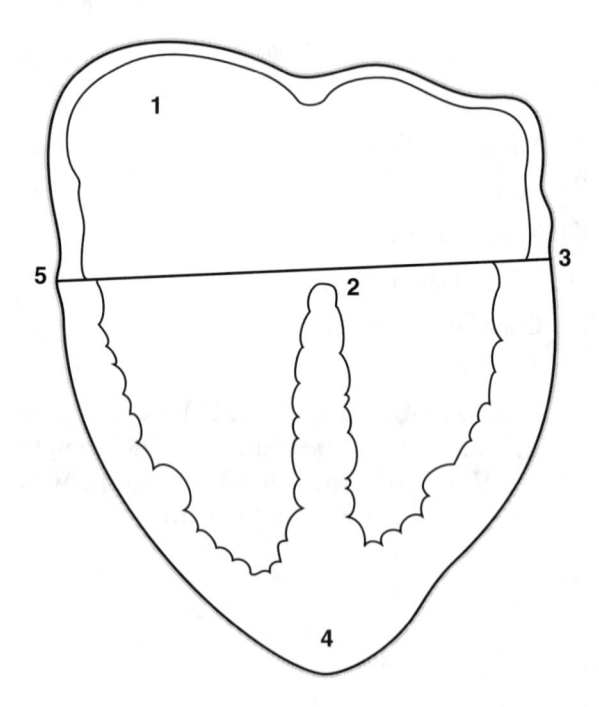

1. Sinoatrial node
2. Left bundle branch
3. Posterior basal subepicardium of the left ventricle
4. Purkinje fiber in subepicardium of the right ventricle
5. Anterior basal subepicardium of the right ventricle

62. The proper sequence in which the heart depolarizes is

(A) 1-2-3-4-5
(B) 1-2-5-4-3
(C) 1-2-4-5-3
(D) 2-1-3-4-5
(E) 5-4-3-2-1

63. The time it takes for a wave of depolarization to travel from 1 to 2 is

(A) 10–20 msec
(B) 120–210 msec
(C) 300–400 msec
(D) 60–80 msec
(E) 500 msec

64. The time it takes for a wave of depolarization to travel from 2 to 3 is

(A) 10–20 msec
(B) 60–80 msec
(C) 100–200 msec
(D) 300–400 msec
(E) 500 msec

65. The photograph represents a high-power view of a lung biopsy from a 58-year-old man. The patient's chest radiograph shows diffuse interstitial disease in both lungs and the presence of calcified pleural plaques. This patient most likely was exposed to

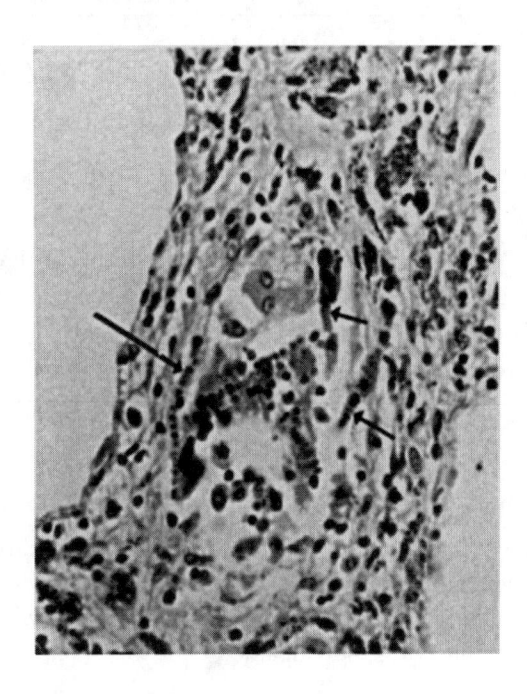

(A) thermophilic actinomycetes
(B) silica
(C) anthracotic pigment
(D) asbestos
(E) cotton mill dust

66. Based on the information below, this patient would be expected to have which of the following disorders?

	Predicted	Observed
Forced Vital Capacity (Liters)	5.0	3.0
Compliance (Liters/cmH$_2$O)	0.2	0.4
Maximal Expiratory Flow (Liters/sec) at equivalent lung volumes	6.0	3.0

(A) A forced expiratory volume in 1 second (FEV$_1$) higher than predicted
(B) A residual volume lower than predicted
(C) Chronic bronchitis
(D) Pulmonary interstitial fibrosis
(E) Emphysema

67. Of the three oxyhemoglobin dissociation curves shown, only curve A has a normal hemoglobin concentration of 15 gm%. Curve B was obtained from which of the following individuals?

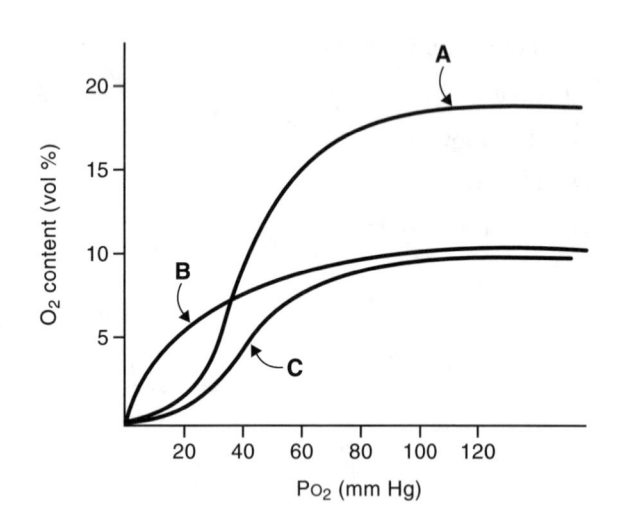

(A) An individual with anemia (low hematocrit)
(B) A resident at a high altitude
(C) An individual with chronic pulmonary disease
(D) An individual with carbon monoxide poisoning
(E) An individual's exercising muscle

Questions 68-70

This spirogram of the forced vital capacity (FVC) was obtained from a male patient weighing 70 kg.

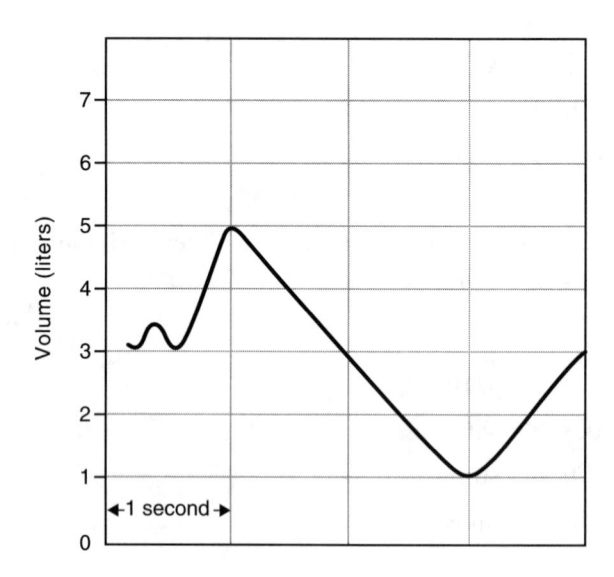

68. The forced expiratory volume in 1 second (FEV$_1$) to FVC ratio is

(A) 50%
(B) 75%
(C) 80%
(D) 100%
(E) not able to be determined

69. The FEV$_1$ to FVC ratio is typical of

(A) healthy individuals
(B) patients with obstructive disease
(C) a restrictive pattern of pulmonary disease
(D) a diffusion abnormality
(E) patients with diffuse interstitial fibrosis

70. Which of the following statements about this patient is correct?

(A) Had the functional residual capacity been measured, it would have been less than predicted
(B) At comparable lung volumes, the observed flow would be higher than predicted
(C) Had the residual volume been measured, it would have been greater than predicted
(D) The expiratory flow rate (liters/sec) is greater than predicted
(E) Had the total lung capacity been measured, it would have decreased

Questions 71-72

A patient with severe obstructive disease is placed on a respirator. The tidal volume and peak inspiratory pressure are set correctly, but the rate is mistakenly set too high (35/min).

71. The initial repercussion of this high rate is

(A) underinflation of the lung
(B) hyperinflation of the lung
(C) no change in lung volume
(D) a further constriction of the airways that worsens the condition
(E) a sympathetic discharge leading to bronchoconstriction

72. The probable outcome in this case is not disastrous because

(A) at high lung volumes, airway resistance falls
(B) at low lung volumes, airway resistance falls
(C) sympathetic discharge leads to bronchoconstriction
(D) at high lung volumes, lung compliance falls
(E) the functional residual capacity was too low when ventilation began

73. Which of the following statements is true regarding solitary coin lesions in the lung?

(A) The majority of these lesions are malignant, regardless of age or smoking history
(B) Metastatic cancers are more likely to be solitary than primary lung cancers
(C) No increase in size in two years suggests malignancy because of the prolonged doubling time in most cancers
(D) Calcification is highly predictive of malignancy regardless of the configuration of calcifications
(E) A previous infection with Histoplasma capsulatum is the likely cause if the patient lives in the Midwest

74. The estimated occurrence of pulmonary metastases by primary malignancy is greatest for which type of cancer?

(A) Breast cancer
(B) Prostate cancer
(C) Choriocarcinoma
(D) Malignant melanoma
(E) Gastrointestinal carcinoma

75. This photograph shows a hematoxylin- and eosin-stained section of lung tissue at 1000 oil magnification from a 26-year-old man with severe dyspnea. The patient had a bone marrow transplantation three weeks ago for aplastic anemia. The best treatment for this patient is

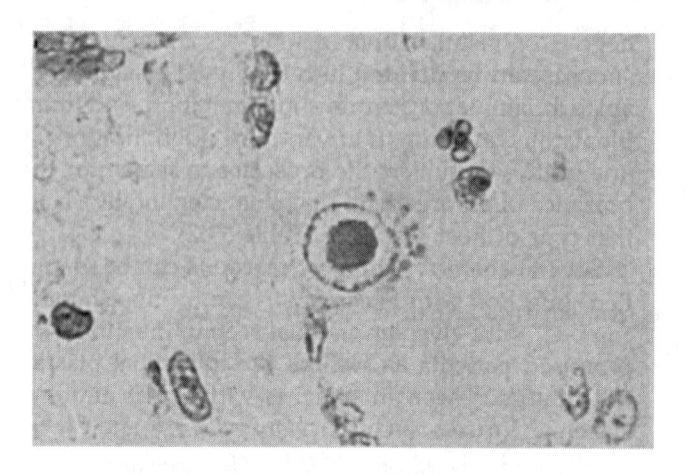

(A) ganciclovir
(B) amphotericin B
(C) amantadine
(D) ribavirin
(E) erythromycin

76. Drug-resistant *Streptococcus pneumoniae* cause serious respiratory tract infections and are most often sensitive to

(A) penicillin G
(B) ampicillin
(C) erythromycin
(D) cefotaxime
(E) vancomycin

77. A patient presents with dry cough, dyspnea, diffuse reticular infiltrates on radiologic exam, and a CD4 cell count of 185/mm^3. A methenamine silver stain of sputum reveals large extracellular organisms. Because the patient is allergic to sulfonamides, suitable therapy may consist of

(A) trimetrexate
(B) pentamidine
(C) rifampin
(D) trimetrexate or pentamidine
(E) all of the above agents

78. What is the appropriate treatment for an individual with a recent positive tuberculin (Mantoux) test but who shows no signs of tuberculous disease? This patient should receive

(A) no drug therapy unless the patient becomes ill
(B) a combination of isoniazid and rifampin for 3 months
(C) isoniazid alone for 6 to 12 months
(D) another Mantoux test in 3 to 6 months
(E) only drug prophylaxis if the patient has other risk factors

79. Which statement regarding drug-resistant strains of *Mycobacterium tuberculosis* is true?

(A) Such strains have not yet appeared in the United States
(B) Resistance is usually limited to one or two drugs
(C) The highest prevalence of isoniazid resistance occurs in Southeast Asia
(D) Patients infected with drug-resistant strains do not need to be quarantined because of the infection's low transmissibility

80. Rimantadine may be preferred over amantadine for influenza prophylaxis because of its

(A) reduced tendency to cause central nervous system side effects
(B) greater in vitro activity against influenza A
(C) longer half-life and reduced frequency of administration
(D) higher efficacy in elderly patients
(E) greater effectiveness in symptomatic patients

DIRECTIONS:

Each of the numbered items or incomplete statements in this section is negatively phrased, as indicated by a capitalized word such as NOT, LEAST, or EXCEPT. Select the ONE lettered answer or completion that is BEST in each case.

81. All of the following statements concerning the heart are true EXCEPT

(A) the sternocostal surface is comprised mostly of the right ventricle
(B) the right border is comprised mostly of the right ventricle
(C) the diaphragmatic surface is comprised mostly of the left ventricle
(D) the left border is comprised mostly of the left ventricle
(E) the posterior surface is comprised mostly of the left atrium

82. The posterior mediastinum contains all of the following EXCEPT

(A) the hemiazygos vein
(B) the azygos vein
(C) the thoracic duct
(D) the brachiocephalic veins
(E) the greater splanchnic nerves

83. All of the following are true of pneumothorax EXCEPT

(A) the elastic fibers in the lung cause the lung to collapse
(B) the parietal pleura and visceral pleura are separated from one another
(C) with a tension pneumothorax, the mediastinum may deviate toward the side with the pneumothorax
(D) a pneumothorax may occur without any injury to the chest wall
(E) a stab wound in the right fourth intercostal space, 3 inches to the right of the sternum, may cause a pneumothorax

84. All of the following statements concerning the thorax and heart are true EXCEPT

(A) the parietal layer of the serous pericardium lines the inner surface of the fibrous pericardium
(B) the phrenic nerves are lateral to the fibrous pericardium
(C) accumulation of fluid in the pericardial cavity causes the heart chambers to fill excessively during diastole
(D) the fibrous pericardium blends with the adventitia of the roots of the great vessels
(E) the left atrium forms the anterior wall of the oblique pericardial sinus

85. Valvular vegetations are LEAST likely to be associated with

(A) malignancy
(B) ischemic heart disease
(C) systemic lupus erythematosus
(D) congenital heart disease
(E) intravenous drug abuse

86. Right ventricular hypertrophy (RVH) is LEAST likely to be associated with

(A) ventricular septal defect
(B) atrial septal defect
(C) long-term chronic obstructive pulmonary disease
(D) tricuspid stenosis
(E) tetralogy of Fallot

87. Which of the following disorders is LEAST likely to cause high output failure?

(A) Thiamine deficiency
(B) Addison disease
(C) Hyperthyroidism
(D) Paget disease of bone
(E) Fluid overload

88. Which of the following disorders is LEAST likely to be associated with left heart failure?

(A) Severe anemia
(B) Right ventricular infarct
(C) Myocarditis
(D) Aortic regurgitation
(E) Chronic ischemic heart disease

89. Atherosclerosis is primarily operative in all of the following disorders EXCEPT

(A) Leriche syndrome
(B) laminar necrosis in the cerebral cortex
(C) renovascular hypertension
(D) Prinzmetal angina
(E) transient ischemic attacks

90. Which of the following disorders is LEAST likely to be associated with lymphedema?

(A) Turner syndrome
(B) Lymphogranuloma venereum
(C) Postradical mastectomy
(D) Filariasis
(E) Erysipelas

91. Vagal stimulation of the heart will do all of the following EXCEPT

(A) decrease the rate of diastolic depolarization (phase 4)
(B) increase the $^+K+$ conductance of the sinoatrial node
(C) hyperpolarize the sinoatrial node
(D) increase the P–P interval of the electrocardiogram (EKG)
(E) increase the threshold of the sinoatrial node

92. The tracing below depicts the forced vital capacity (FVC) and forced expiratory volume in 1 second (FEV_1) in a patient with asthma before (tracing A) and after (tracing B) the administration of a bronchodilator. Analysis of the response suggests that the bronchodilator accomplished all of the following results EXCEPT

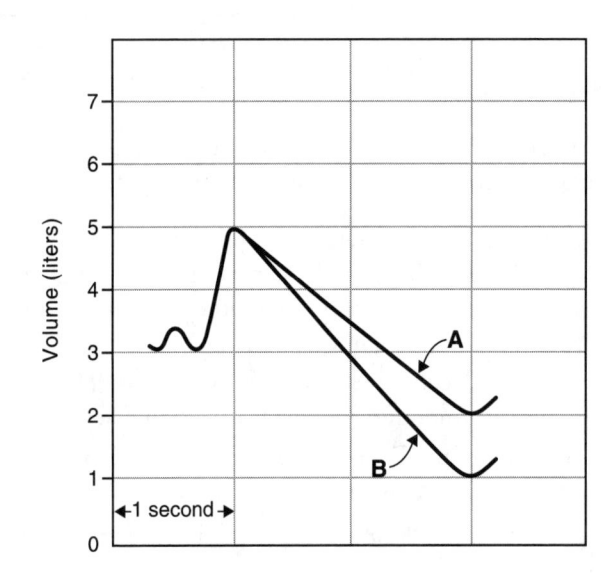

(A) decreased the residual volume
(B) increased the vital capacity
(C) improved the FEV_1 to FVC ratio
(D) increased the forced expiratory flow 25% to 75%
(E) increased the FVC

93. An individual with a hemoglobin concentration of 7.5 gm% would exhibit all of the following characteristics EXCEPT

(A) a normal arterial oxygen tension (PO$_2$)
(B) a lower than normal venous PO$_2$
(C) an O$_2$ % saturation of hemoglobin in the normal range
(D) a normal alveolar ventilation
(E) a normal O$_2$ content (vol%)

94. As blood passes through the tissues, all of the following actions occur EXCEPT that

(A) bicarbonate concentration of blood decreases
(B) the affinity of hemoglobin for O$_2$ decreases
(C) the ability of hemoglobin to bind CO$_2$ increases
(D) the ability of hemoglobin to bind H+ increases
(E) chloride enters the red blood cells (RBC)

95. This figure represents gas exchange in the lung. The letters indicate key components in the biochemical reactions.

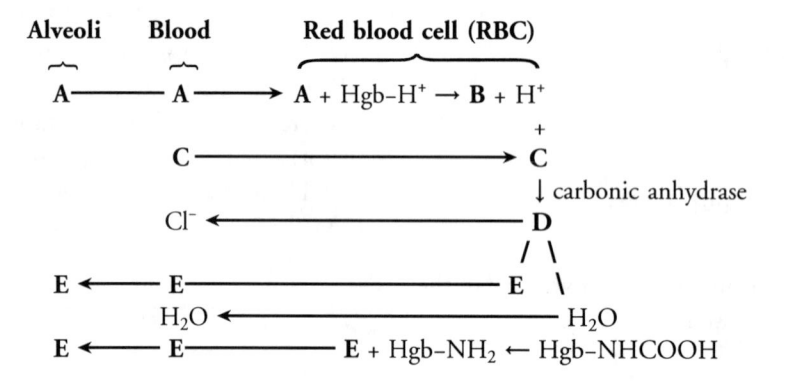

```
        Alveoli    Blood       Red blood cell (RBC)
        A────────── A ────→ A + Hgb-H⁺ → B + H⁺
                                         +
                   C ──────────────────→ C
                                         ↓ carbonic anhydrase
                   Cl⁻ ←───────────────── D
                                         / \
        E ←─────── E ───────────────── E   \
                   H₂O ←─────────────────── H₂O
        E ←─────── E ───────── E + Hgb-NH₂ ← Hgb-NHCOOH
```

Which of the following statements is NOT correct concerning the process depicted in the figure?

(A) Compound A enters the blood and the red blood cell (RBC) because the partial pressure is higher in the alveoli than in these other compartments
(B) Compound C is the main carrier for carbon dioxide and combines with hydrogen ions (H$_+$) to form compound D, which is carbonic acid
(C) Compound E enters the pulmonary capillary, where it dissolves in the blood and establishes a concentration gradient favoring diffusion into the alveoli
(D) Compound E is more soluble than compound A and has less difficulty passing through the alveolar capillary interface
(E) An increase of compound E in the alveoli does not alter the concentration of compound A

96. All of the following antitubercular drugs are correctly paired with a characteristic adverse effect EXCEPT

(A) isoniazid—peripheral neuritis
(B) ethambutol—optic neuritis
(C) pyrazinamide—hepatitis
(D) rifampin—inhibits P450 drug metabolism
(E) kanamycin—nephrotoxicity

97. Pneumonia caused by *Chlamydia pneumoniae* infection may be treated effectively with any of the following antimicrobial agents EXCEPT

(A) trimethoprim-sulfamethoxazole
(B) azithromycin
(C) erythromycin
(D) doxycycline
(E) clarithromycin

98. When used to treat respiratory tract infections in pregnant patients, which of the following antimicrobials is LEAST likely to produce adverse effects on the fetus or mother?

(A) Doxycycline
(B) Clarithromycin
(C) Ciprofloxacin
(D) Trimethoprim-sulfamethoxazole
(E) Cefuroxime

99. Pulmonary fibrosis is a potentially serious adverse effect associated with the administration of all of the following drugs EXCEPT

(A) amiodarone
(B) clindamycin
(C) methysergide
(D) nitrofurantoin
(E) bleomycin

100. An acute exacerbation of a chronic pulmonary infection in a patient with cystic fibrosis may be effectively treated with any of the following antimicrobial agents EXCEPT

(A) ciprofloxacin
(B) tobramycin
(C) ceftriaxone
(D) piperacillin
(E) ceftazidime

Questions 101-102

The following schematic represents gas exchange at the tissue level. The letters indicate key components in the biochemical reactions, and the number represents an enzyme.

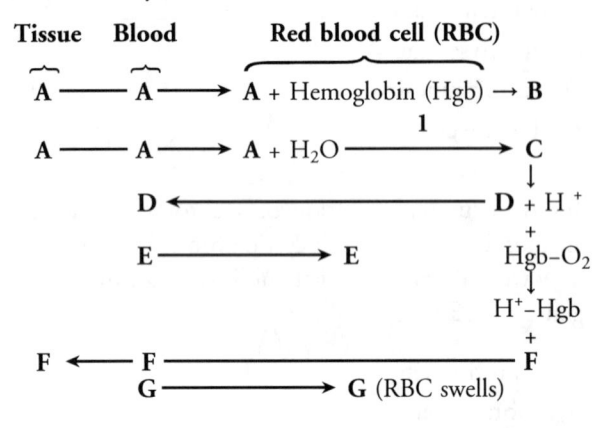

```
Tissue   Blood       Red blood cell (RBC)
  A ──── A ────→ A + Hemoglobin (Hgb) ──→ B
                           1
  A ──── A ────→ A + H₂O ──────────────→ C
                                          ↓
           D ←─────────────────────── D + H⁺
                                          +
           E ────────→ E              Hgb-O₂
                                          ↓
                                       H⁺-Hgb
                                          +
  F ←─── F ──────────────────────────── F
         G ────────→ G (RBC swells)
```

101. Which of the following statements is NOT correct concerning the biochemical and physiologic events that are occurring at the tissue level?

(A) Compound A dissolves in the blood and has a higher partial pressure in the pulmonary capillaries than the partial pressure in the alveoli
(B) Compounds A (in the blood), B, and D are the main carriers of carbon dioxide, with compound B carrying the greatest amount in the blood
(C) Enzyme 1 is also found in the brush border of the proximal tubules and in the cytosol of the distal tubules
(D) Compound D is the major extracellular fluid buffer of hydrogen ions
(E) Compound E, entering the red blood cell, maintains electroneutrality when compound D leaves the cell

102. Which of the following statements is NOT correct concerning the biochemical and physiologic events that are occurring?

(A) The reactions from D to F are responsible for the Bohr effect
(B) The partial pressure of compound F in the capillary determines the total amount of compound F that is delivered to tissue
(C) The partial pressure of compound F in capillary blood is increased in the presence of acidosis and fever
(D) The amount of compound F that is released into tissue is decreased in cases of carbon monoxide poisoning, methemoglobinemia, and respiratory acidosis
(E) When the partial pressure of compound F in the capillary is the same as the partial pressure of compound F in the tissue, gas exchange ceases

DIRECTIONS:

Each set of matching questions in this section consists of a list of four to twenty-six lettered options (some of which may be in figures) followed by several numbered items. For each numbered item, select the ONE lettered option that is most closely associated with it. To avoid spending too much time on matching sets with a large number of options, it is generally advisable to begin each set by reading the list of options. Then for each item in the set, try to generate the correct answer and locate it in the option list, rather than evaluating each option individually. Each lettered option may be selected once, more than once, or not at all.

Questions 103-105

Select the embryonic origin of the adult structure.

(A) First branchial arch
(B) Second branchial arch
(C) Third branchial arch
(D) Fourth branchial arch
(E) First branchial pouch
(F) Second branchial pouch
(G) Third branchial pouch
(H) Fourth branchial pouch
(I) First branchial cleft
(J) Second branchial cleft
(K) Third branchial cleft
(L) Fourth branchial cleft

103. Thyroid cartilage

104. Tympanic cavity

105. Cricothyroid muscle

Questions 106-111

The following are schematics of congenital heart disease (CHD). Match the clinical scenario with the most appropriate CHD.

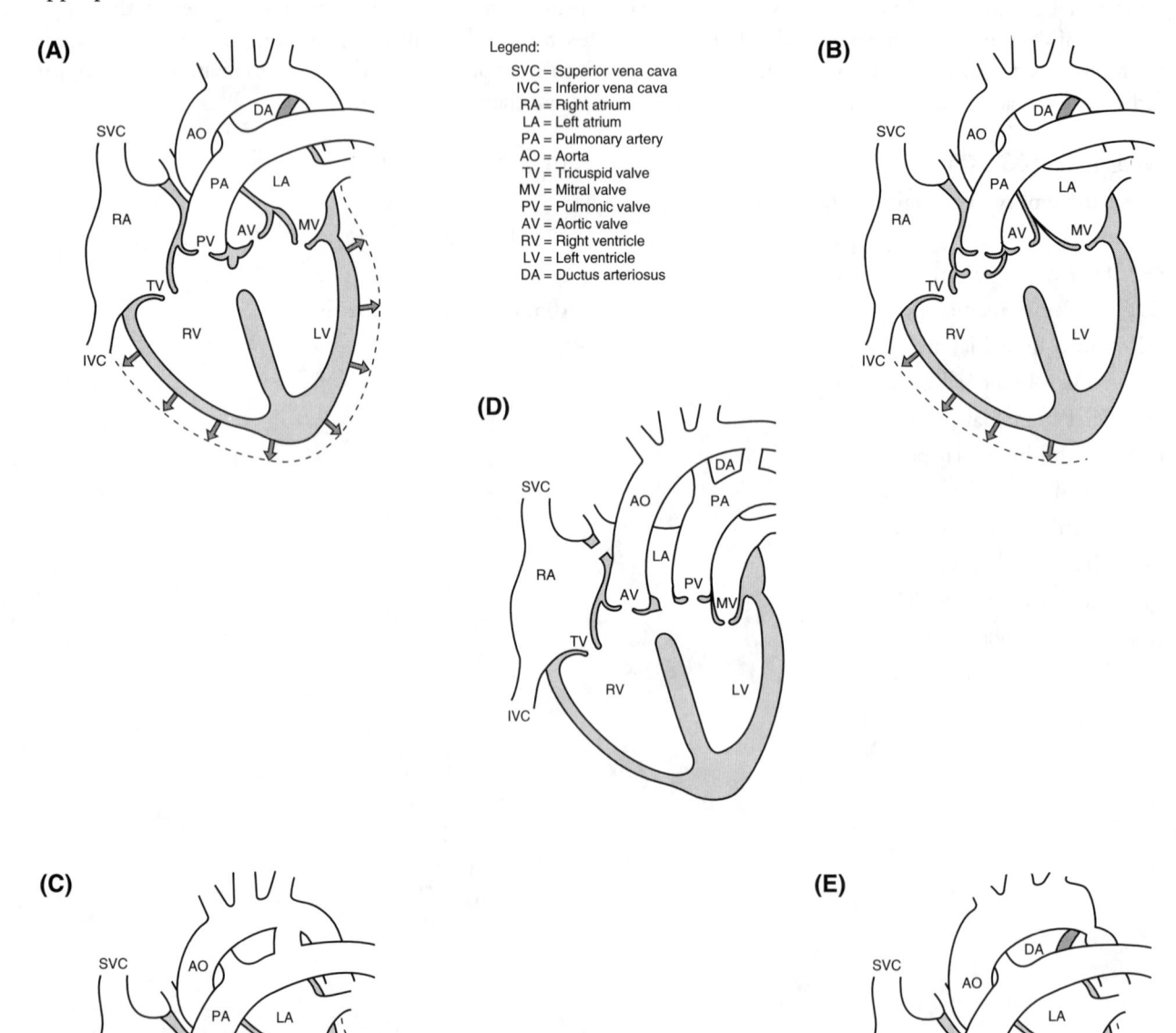

(A)

(B)

Legend:
SVC = Superior vena cava
IVC = Inferior vena cava
RA = Right atrium
LA = Left atrium
PA = Pulmonary artery
AO = Aorta
TV = Tricuspid valve
MV = Mitral valve
PV = Pulmonic valve
AV = Aortic valve
RV = Right ventricle
LV = Left ventricle
DA = Ductus arteriosus

(D)

(C)

(E)

106. The most common CHD that can present without cyanosis if the valvular abnormality is not severe

107. In this CHD, reversal of the initial left-to-right shunt results in a differential cyanosis between the upper and lower extremities

108. This CHD is associated with diastolic hypertension and an increased incidence of a congenital bicuspid aortic valve and berry aneurysms in the central nervous system

109. The most common CHD associated with Eisenmenger syndrome

110. This CHD is fatal without left-to-right and right-to-left shunting of blood

111. This CHD is associated with congenital rubella

Questions 112-122

Match each description with the appropriate disorder.

(A) Dermatopathic lymphadenopathy
(B) Cat-scratch disease
(C) Sinus histiocytosis
(D) Large cell, immunoblastic lymphoma
(E) Small-cleaved, poorly differentiated lymphocytic lymphoma (follicular lymphoma)
(F) Well differentiated lymphocytic lymphoma (small lymphocytic lymphoma)
(G) Burkitt lymphoma (small, noncleaved lymphoma)
(H) Lymphoblastic lymphoma
(I) Mycosis fungoides
(J) Waldenström macroglobulinemia
(K) Heavy chain disease
(L) Hand-Schüller-Christian disease
(M) Letterer-Siwe disease
(N) Eosinophilic granuloma
(O) Urticaria pigmentosa
(P) Plasmacytoma
(Q) Chloroma
(R) Richter syndrome

112. High-grade, malignant B-cell lymphoma that most often develops in the setting of abnormal immune states (e.g., systemic lupus erythematosus, Sjögren syndrome, renal transplants, and AIDS)

113. Common low-grade adult malignant lymphoma that frequently has a leukemic phase and a propensity for metastasis to the bone marrow

114. Produces granulomatous microabscesses in lymph nodes that require silver stains to confirm the specific diagnosis

115. High-grade T-cell malignant lymphoma with mediastinal involvement that is most commonly seen in children

116. Collection of cells in bone or the upper-respiratory tract that frequently evolves into a malignancy associated with a monoclonal gammopathy

117. Monoclonal proliferation of lymphoplasmacytoid cells, generalized lymphadenopathy, and the hyperviscosity syndrome

118. Neoplastic infiltration of the small bowel leading to malabsorption and the presence of a monoclonal spike on a serum protein electrophoresis

119. Skin disease associated with a dermal infiltrate of mast cells

120. Associated with a leukemic phase, which is identified with chronic lymphocytic leukemia

121. Good prognostic sign in ipsilateral axillary lymph nodes draining breast cancer

122. Pigmented macrophages in lymph nodes that are frequently confused with metastatic malignant melanoma

Questions 123-125

Match each of the following descriptions with the appropriate disorder.

(A) Felty syndrome
(B) Gamna-Gandy bodies
(C) Congenital asplenia
(D) Systemic lupus erythematosus
(E) Hairy cell leukemia
(F) Sago and lardaceous spleens

123. Association with malformations of the heart in greater than 80% of cases

124. Hyperplastic arteriolosclerosis of the splenic arterioles

125. Association with a "sugar-coated" spleen in a patient with cirrhosis

Questions 126-127

Match each of the following case descriptions with the appropriate timing sequence.

	Bleeding Time	Platelet	PT	PTT
(A)	Prolonged	Normal	Normal	Normal
(B)	Prolonged	Normal	Normal	Prolonged
(C)	Normal	Normal	Normal	Prolonged
(D)	Normal	Normal	Prolonged	Normal
(E)	Normal	Normal	Prolonged	Prolonged

126. Patient with menorrhagia whose platelets do not aggregate when exposed to ristocetin

127. Patient with uremia with a hemorrhagic diathesis

Questions 128-132

The left ventricular pressure and volume loops depicted below were obtained from normal individuals. The solid line represents the control loop; the dashed line was obtained after a particular event. Match each of the following situations with the appropriate dashed-line loop.

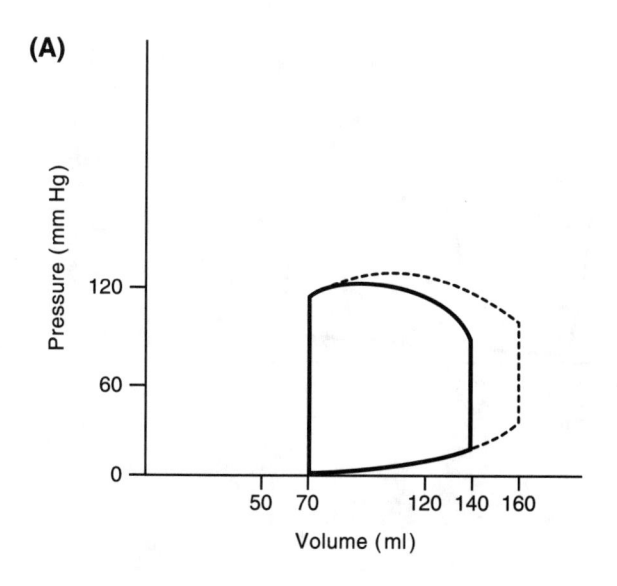

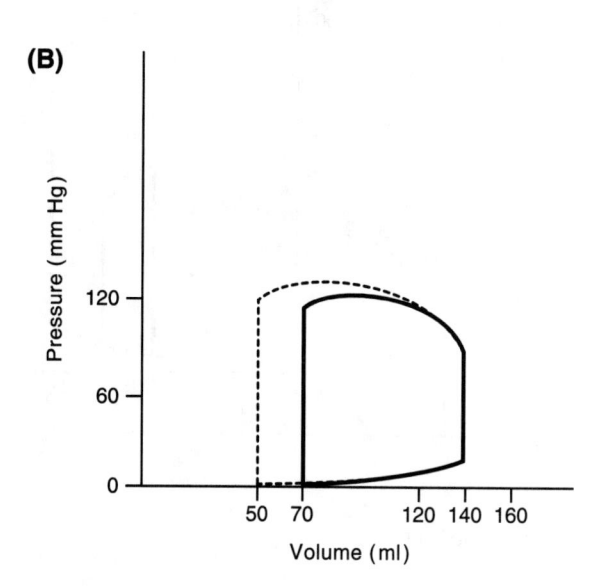

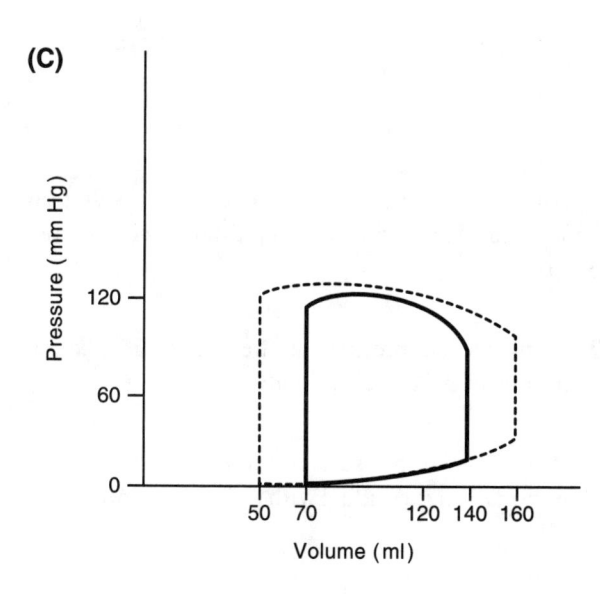

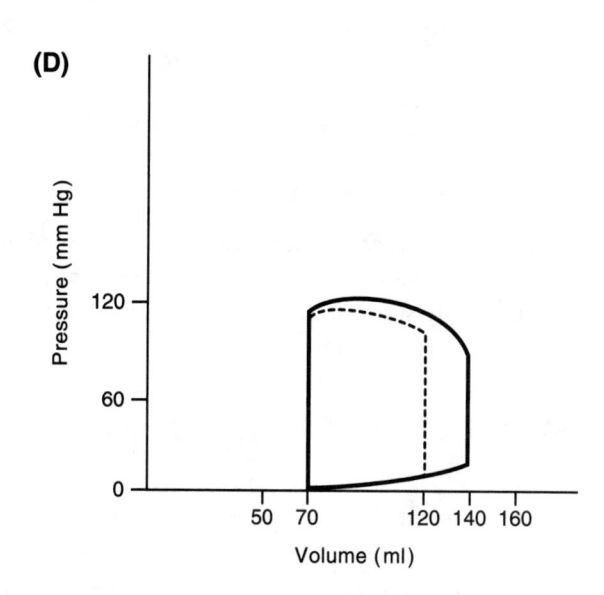

(A)

(B)

(C)

(D)

128. Moving from the erect to the supine position

129. The administration of a positive inotropic agent

130. Moderate exercise

131. The administration of a venodilator

132. Mitral stenosis severe enough to compromise filling

Questions 133-135

Match the following left ventricular pressure–volume loops to the appropriate condition.

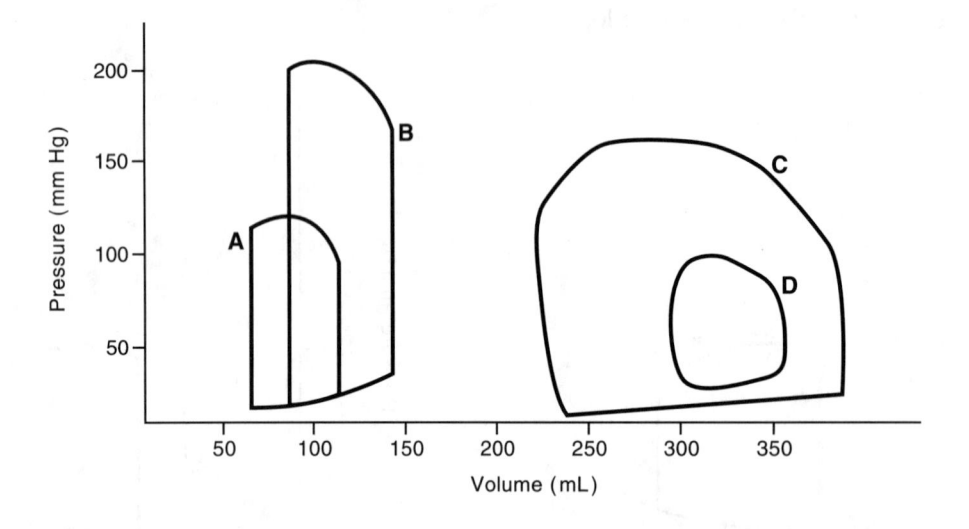

133. Aortic stenosis

134. Aortic regurgitation

135. Heart failure

Questions 136-139

All the pressure measurements given were recorded at the apex of the lung in individuals in the erect position. Match the pressures to the appropriate circumstance.

(A) 0 cm H_2O
(B) –10 cm H_2O
(C) –15 cm H_2O
(D) –30 cm H_2O

136. Intrapleural pressure at the end of an inspiration of 500 ml in a normal individual

137. Intrapleural pressure at the end of an inspiration of 500 ml in an individual with pulmonary interstitial fibrosis

138. Intraalveolar pressure at the end of a maximal inspiration in a normal individual

139. Intraalveolar pressure at the end of a maximal inspiration in an individual with pulmonary interstitial fibrosis

Questions 140-143

These photographs show various lung pathogens. Pathogen A is a touch preparation stained with Diff-Quik × 1000 oil magnification. Pathogen B is stained with lactophenol cotton blue × 400 magnification. Pathogen C is stained with hematoxylin and eosin × 400 magnification. Pathogen D is stained with Gomori methenamine silver × 205 magnification. Match the clinical descriptions with the most appropriate pathogen.

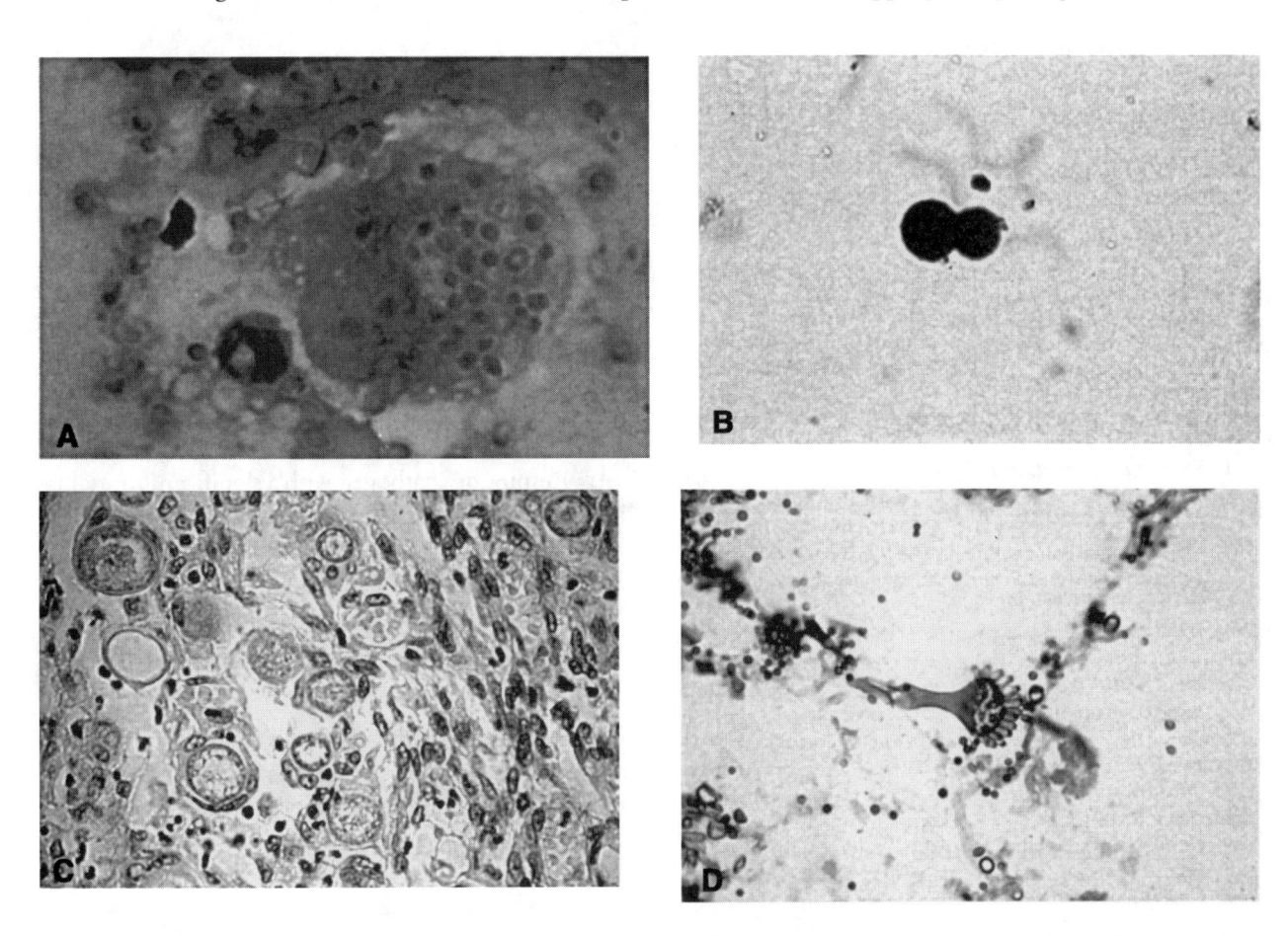

140. Dimorphic fungus with lung disease characterized by multiple granulomas with extensive dystrophic calcification

141. Lives in old tuberculous cavities and is associated with massive hemoptysis, bronchial asthma, and invasive vasculitis

142. Along the Southeast coast, this dimorphic fungus, which produces both skin and lung disease, occurs predominantly in males

143. Dimorphic fungus that was a prevalent pulmonary pathogen after the recent Los Angeles earthquake

Questions 144-157

Match the following microbial pathogens with the best clinical description.

(A) Rhinovirus
(B) Respiratory syncytial virus
(C) Parainfluenza
(D) Influenza
(E) Cytomegalovirus
(F) Rubeola
(G) *Chlamydia trachomatis*
(H) *Chlamydia pneumoniae* (TWAR)
(I) *Chlamydia psittaci*
(J) *Coxiella burnetii*
(K) *Mycoplasma pneumoniae*
(L) *Staphylococcus aureus*
(M) *Streptococcus pneumoniae*
(N) *Hemophilus influenzae*
(O) *Klebsiella pneumoniae*
(P) *Pseudomonas aeruginosa*
(Q) *Escherichia coli*
(R) *Legionella pneumophila*
(S) *Actinomyces israelii*
(T) *Nocardia asteroides*
(U) *Histoplasma capsulatum*
(V) *Coccidioides immitis*
(W) *Blastomyces dermatidis*
(X) *Aspergillus fumigatus*
(Y) *Candida albicans*
(Z) *Cryptococcus neoformans*

144. Pathogen that commonly produces lung abscesses and is a common secondary invader in the lung in patients with rubeola or influenza

145. Pathogen that is contracted when the infant passes through the birth canal and produces a pneumonia characterized by an abrupt onset of tachypnea, wheezing, hyperaeration, eosinophilia, and a conspicuous lack of fever; often associated with a conjunctivitis

146. Systemic pathogen that is often associated with the presence of indwelling venous/arterial catheters and immunodeficiency states; produces a pneumonia characterized by diffuse nodular infiltrates and evidence of vessel invasion

147. Childhood pathogen that may produce a pneumonia associated with Warthin-Finkeldey multinucleated giant cells

148. Water-loving pathogen that is most commonly seen in men more than age 40 years who smoke; produces a confluent bronchopneumonia with fever, nonproductive cough, hemoptysis, and other systemic symptoms; best visualized with direct immunofluorescence or a silver stain

149. Respiratory pathogen with a significant mortality in patients more than age 55 years who have underlying renal, cardiac, or lung problems; produces a severe, exudative pneumonia with a propensity for secondary bacterial invasion; has an association with Reye syndrome in children

150. Respiratory pathogen that is transmitted without a vector, unlike other pathogens in its family group, through inhalation primarily by individuals who work with birthing sheep, cows, goats, or in the milk industry

151. Respiratory pathogen that is transmitted by direct hand-to-hand transfer (usually by school children) of infected material and by respiratory droplet infection; development of a vaccine is unlikely

152. Respiratory pathogen transmitted by droplet infection and accounts for approximately 10% of community-acquired atypical pneumonias and a smaller percentage of bronchitis cases; not associated with cold agglutinins and responds well to tetracycline, as with other members of its family

153. Strict anaerobe that can produce an empyema that drains through a sinus tract out to the skin surface; characteristic morphology demonstrated by Gram-stained yellow flecks of material in the drainage fluid

154. Pathogen that is commonly contracted in military stations and in crowded situations; produces an interstitial pneumonia and is often associated with erythema multiforme and bullous myringitis

155. Respiratory pathogen that is an example of a zoonosis and is associated with an interstitial pneumonia; incidence has declined by putting tetracycline in animal feed

156. Predominantly a respiratory pathogen, a strict aerobe most commonly seen in patients with defects in cellular immunity, particularly in the setting of heart transplantation; produces microabscesses in the lungs, often with granuloma formation; has a characteristic feature, aside from its unusual Gram-stain morphology, of being partially acid fast

157. Common pathogen recovered in the sputum of patients with cystic fibrosis; second to *Streptococcus pneumoniae* in causing septicemia in children in the absence of a localized infection site; causes septicemia in patients with sickle-cell anemia who have splenic dysfunction, and vaccine is available

Questions 158-162
Identify the antihistamine that best matches the descriptions.

(A) Cyproheptadine
(B) Chlorpheniramine
(C) Promethazine
(D) Meclizine
(E) Aztemizole

158. Less sedating drug used in motion sickness and vertigo

159. Marked sedation and antiemetic activity

160. Blocks serotonin receptors

161. Produces little or no sedation

162. Available in nonprescription allergy and cold products

ANSWER KEY

1. E	28. C	55. A	82. D	109. A	136. C
2. B	29. E	56. A	83. C	110. D	137. D
3. B	30. A	57. C	84. C	111. C	138. A
4. D	31. C	58. C	85. B	112. D	139. A
5. A	32. C	59. B	86. D	113. E	140. A
6. A	33. C	60. C	87. B	114. B	141. D
7. C	34. C	61. C	88. B	115. H	142. B
8. A	35. C	62. C	89. D	116. P	143. C
9. E	36. A	63. B	90. E	117. J	144. L
10. C	37. D	64. B	91. E	118. K	145. G
11. B	38. C	65. D	92. E	119. O	146. Y
12. C	39. C	66. E	93. E	120. F	147. F
13. B	40. D	67. D	94. A	121. C	148. R
14. D	41. A	68. A	95. E	122. A	149. D
15. E	42. A	69. B	96. D	123. C	150. J
16. A	43. E	70. C	97. A	124. D	151. A
17. C	44. C	71. B	98. E	125. B	152. H
18. C	45. C	72. A	99. B	126. B	153. S
19. A	46. A	73. E	100. C	127. A	154. K
20. B	47. B	74. C	101. B	128. A	155. I
21. A	48. E	75. A	102. B	129. B	156. T
22. B	49. B	76. E	103. D	130. C	157. N
23. E	50. D	77. D	104. E	131. D	158. D
24. C	51. B	78. C	105. D	132. D	159. C
25. B	52. C	79. C	106. B	133. B	160. A
26. A	53. B	80. A	107. C	134. C	161. E
27. A	54. C	81. B	108. E	135. D	162. B

ANSWERS AND EXPLANATIONS

1. The answer is E. *(Gross anatomy; Cardiovascular)*
A portal vein connects one capillary bed to another. In humans, the hepatic portal vein connects the intestinal capillaries with the sinusoids of the liver. The blood, along with blood from the proper hepatic artery, enters the central vein after passing through the sinusoids, and then enters the hepatic veins and the inferior vena cava.

The direction of blood flow through an artery is determined by the pressure gradient and may change from moment to moment.

Not all veins have valves. For example, the portal vein and all of the veins that drain into it do not have valves. Valves are most prominent in the veins of the limbs.

2. The answer is B. *(Gross anatomy; Cardiovascular)*
The right border of the heart is approximately 1 to 2 cm to the right of the right border of the sternum; therefore, the knife probably did not hit the heart. The middle lobe of the right lung extends from the level of the fourth costochondral junction to the level of the fifth intercostal space. The middle lobe is most likely at the level of the fourth intercostal space.

3. The answer is B. *(Gross anatomy; Cardiovascular)*
The tricuspid valve can be heard best over the right ventricle—the chamber immediately distal to the valve. This chamber underlies the fifth intercostal space immediately to the left of the chamber. The mitral valve can be heard best approximately 7 cm to the left of the sternum in the fifth intercostal space, over the left ventricle. The aortic valve can be heard best over the ascending aorta in the second intercostal space immediately to the right of the sternum. The pulmonic valve can be heard best over the pulmonary trunk in the first or second intercostal space immediately to the left of the sternum.

4. The answer is D. *(Gross anatomy; Pulmonary/respiratory)*
The respiratory diverticulum is an evagination of the endodermal floor of the foregut. All of the epithelial surfaces of the respiratory system are derived from this evagination.

The first branchial pouch is the endodermal evagination that forms the eustachian tube and the middle ear linings.

The first branchial cleft is an ectodermal invagination that forms the external ear canal.

The sixth branchial arch contributes to the cartilage of the larynx.

5. The answer is A. *(Gross anatomy; Cardiovascular)*
The azygos vein receives blood from the right intercostal veins, the hemiazygos vein, and the ascending lumbar veins on the right. The azygos vein arches over the top of the right main bronchus to empty into the superior vena cava. With right-sided heart failure and increased central venous pressure, enlargement of the azygos vein may occur. This may be detected radiographically at the level of the fourth thoracic vertebra, which is the level of the arch of the azygos.

6. The answer is A. *(Pathology; Cardiovascular)*
Cessation of smoking decreases the risk of myocardial infarction by 50% to 75% within 5 years of cessation. The reduced risk for the other therapies are as follows:

- Postmenopausal estrogen replacement has a 44% decreased risk
- Mild to moderate alcohol consumption has a 25% to 45% decreased risk
- Exercise has a 45% decreased risk
- Prophylactic low dose aspirin has a 33% decreased risk, particularly in the incidence of the first acute myocardial infarction in middle aged men and women; however, there is no reduction in overall total cardiovascular mortality. Patients over 50 years of age with risk factors for coronary artery disease are the group most likely to benefit. It is not good for prophylaxis if the patient has poorly controlled hypertension, because there is a danger for a hemorrhagic stroke.

7. The answer is C. *(Pathology; Cardiovascular)*
An increase in cardiac output is expected in a patient with aortic regurgitation who is not experiencing left heart failure. The excess amount of blood dripping back into the left ventricle in diastole, plus the blood

coming from the right heart, increases the left ventricular end-diastolic volume (increases preload). This causes an increase in stroke volume by the Frank-Starling relationship, whereby the increase in fiber length increases muscle tension, resulting in an increased force of contraction.

A decrease in lung compliance, or reduced stretching of the lung on inspiration, decreases venous return to the heart because there is less of a drop in intrathoracic pressure with a lung that is not fully expanded.

A standing patient with hypertrophic cardiomyopathy has decreased venous return to the heart. This decreases preload in the left ventricle and further exacerbates the degree of obstruction to outflow by the anterior leaflet of the mitral valve compressed against the hypertrophied and asymmetric interventricular septum.

Positive end-expiratory pressure therapy imposes an increased pressure on the thin-walled right heart and vena cava, thus reducing venous return to the heart and cardiac output.

A left ventricular aneurysm commonly results in chronic heart failure, because the scar tissue is noncontractive.

8. The answer is A. *(Pathology; Cardiovascular)*
Anxiety, with an increase in heart rate or decreasing venous return to the heart (standing), causes the click and murmur to occur earlier in systole (closer to S_1). Lying down, squatting, sustained hand grip exercise, or passive leg lifting in the supine position increase venous return and cause the click and murmur to occur later in systole (less gravity and more venous return). The diagnosis is best made by echocardiography. β-adrenergic blocking agents are frequently used for treatment of supraventricular tachycardias. Calcium channel blockers are also used in symptomatic cases.

9. The answer is E. *(Pathology; Cardiovascular)*
Obstruction of blood flow, embolization, and constitutional signs of fever, weight loss, and anemia characterize a cardiac myxoma, the most common primary tumor of the heart. They are derived from mesenchyme and are almost exclusively located in the left atrium. In this location, they produce a ball valve effect by restricting blood flow into the left ventricle, thus decreasing the cardiac output.

Acute endocarditis involving the aortic valve, or a mural thrombus in the left ventricle or left atrium, could embolize, but would not be expected to obstruct blood flow. Fever, weight loss, and anemia could be seen in acute endocarditis.

Vegetations in acute rheumatic fever do not embolize or obstruct.

10. The answer is C. *(Pathology; Cardiovascular)*
In a patient with a cyanotic congenital heart disease with right-to-left shunting of unoxygenated blood into oxygenated blood you would expect:

- Hypoxemia, because unoxygenated blood is mixing with oxygenated blood. Hypoxemia is a stimulus for erythropoietin release, resulting in a second polycythemia with an increase in the hemoglobin, hematocrit, and red blood cell count. In addition, the oxygen saturation (percent of heme groups that are occupied by oxygen) is decreased, because less blood is dissolved in the plasma

- An increase in the alveolar–arterial gradient, because unoxygenated blood is mixing with the oxygenated blood from the lungs, thus increasing the difference between alveolar oxygen and arterial oxygen. Correction of the hypoxemia would not be expected with 100% oxygen, because it would have no effect on reducing the flow of unoxygenated blood into the arterial system

- An increased risk for cerebral abscesses, because the lungs are bypassed and cannot filter out the bacteria

11. The answer is B. *(Pathology; Cardiovascular)*
The patient has a ventricular aneurysm. This is a late complication that develops 4 to 8 weeks after acute myocardial infarction in 10% to 20% of cases. Patients observe an abnormal bulge on the precordium. Aneurysms located on the anterior wall bulge during systole as they fill with blood. The diagnosis is verified by two-dimensional echocardiography or ventriculography. The electrocardiogram shows persistent S-T elevation. Complications include

- Chronic heart failure, if more than 40% of the left ventricle is involved (this is the most common cause of death)

- Thromboembolization
- Ventricular arrhythmias

Rupture is uncommon, because the aneurysm is composed of scar tissue.

A ventriculectomy is the treatment of choice in selected cases.

12. The answer is C. *(Pathology; Cardiovascular)*
The patient had sudden cardiac death syndrome, which refers to death within 1 hour of chest pain. It is most commonly associated with severe, fixed coronary artery disease (80%). A grossly obvious coronary thrombosis is present in only 10% to 35% of cases. Obesity, hypertension, and a recent history of a non-Q wave infarction are known predisposing factors. Patients usually die from a lethal ventricular arrhythmia. Cardiac death syndrome is responsible for more than 50% of deaths caused by ischemic heart disease. The differential for sudden death includes massive pulmonary embolus, conduction system abnormalities, cocaine abuse, aortic stenosis, hypertrophic cardiomyopathy, and mitral valve prolapse.

13. The answer is B. *(Pathology; Cardiovascular)*
An atherosclerotic stroke, although a possible cause of death in elderly patients with atherosclerotic heart disease, does not usually produce sudden death. A massive pulmonary embolus can produce sudden death because of acute right heart strain, but the patient has a history of unstable angina; therefore, a cardiac cause is more likely. Pallor of the left ventricle occurs only after 12 hours. Severe atherosclerosis with a fresh coronary thrombus is unlikely.

Arrhythmias are the most common cause of death in myocardial infarction. Unfortunately, most of these occur within the first few hours of the infarct when patients are not in the hospital. The most common arrhythmias are ventricular extrasystoles, which often progress into ventricular tachycardia and fibrillation.

Although there has been an approximate 40% decline in death owing to myocardial infarction in the United States, 1.5 million people still have heart attacks each year. Males are affected more than females, because atherosclerosis is more common in men. Approximately 25% of patients with myocardial infarction will die, roughly half before they reach the hospital. Most acute myocardial infarctions gaining admission to the hospital are complicated in descending order by arrhythmias, left heart failure, thromboembolism, cardiogenic shock, and cardiac rupture. The purpose of a coronary care unit is to identify these complications and administer therapy to prevent death in the patient.

14. The answer is D. *(Pathology; Cardiovascular)*
Dependent pitting edema, jugular venous distention, splenomegaly, and a positive hepatojugular reflux from liver congestion are all signs of right heart failure with an accumulation of blood behind the failed heart. Pillow orthopnea is a sign of left heart failure that occurs when the patient lies down. Gravitational pressure is reduced, resulting in a greater return of blood to the right heart, as well as to the left. If the left heart is unable to handle the increased load, the blood backs up into the pulmonary system. Fluid in the intersitial space stimulates the J receptors. This results in rapid, shallow breathing and the sensation of dyspnea. Placing pillows under the head reduces venous return, hence the term pillow orthopnea.

15. The answer is E. *(Pathology; Hematopoietic/Lymphoreticular)*
The patient most likely has a "strawberry type" of capillary hemangioma, which occurs in newborns, grows rapidly in a few months, and then begins to regress over the next few years. Because 80% of these lesions regress by 5 years of age, 90% by 9 years of age, and the remainder by adolescence, a hands-off policy is usually recommended.

Capillary hemangiomas are hamartomas, rather than benign neoplasms of mature capillary channels, because they do not have a capsule like most benign tumors. Cavernous hemangiomas are characterized by the presence of large, dilated spaces filled with blood. They are most commonly located in the skin or viscera (liver, spleen, and placenta). The visceral types can spontaneously hemorrhage. In von Hippel-Lindau disease, an autosomal dominant disease, cavernous hemangiomas occur in the cerebellum (cerebellar hemangioblastoma), brain stem, and eyes. They also have an increased incidence of renal adenocarcinoma (secondary polycythemia) and cysts in the liver, kidneys, and pancreas.

16. The answer is A. *(Pathology; Cardiovascular)*
Obesity in children is most closely associated with heart disease, especially in men who were obese as children. In children, obesity is characterized by a hyperplasia of adipose cells; however, obesity beginning in adulthood is characterized by hypertrophy, rather than hyperplasia of adipose cells.

Obesity is frequently a consequence of hypothyroidism because of decreased metabolism of adipose. Renal disease, psychosis, and restrictive lung disease are no more frequent in patients who are obese than in those who are not.

17. The answer is C. *(Pathology; Cardiovascular)*
The patient has an arteriovenous (AV) fistula, which is most frequently the result of a penetrating knife injury. An arteriovenous fistula is an abnormal communication between arteries and veins. It can be congenital or acquired (most common), the latter resulting from penetrating injuries (knife), Paget disease of bone (soft, highly vascular mosaic bone), a rupture of an aneurysm into a vein, or venous communication with an arterial graft. Large AV fistulas bypass the microcirculation and cause increased venous return to the heart. Depending on the size, this leads to high output failure, an increase in heart rate, and a drop in diastolic pressure, because the peripheral resistance decreases. The mixed venous oxygen content in the right heart is increased because of the mixing of oxygenated with unoxygenated blood. Compression of the fistula produces bradycardia (Branham sign), secondary to a sudden increase in peripheral resistance by blocking the run-off into the venous circulation. A thrill is usually felt at the site of the lesion and auscultatory evidence of a bruit extends throughout the entire cardiac cycle. Antiography is the diagnostic test of choice.

A congenital AV fistula can occur in the lungs (shunting can produce cyanosis) and hereditary telangiectasia (Osler-Rendu-Weber disease), but they are not the most common cause of AV fistulas, nor are they commonly found in the thigh. Metastasis from a primary lesion in the kidney is an unlikely cause, even though these tumors tend to be highly vascular. Primary vascular tumors are also an uncommon cause. Previous spesis does not predispose to an AV fistula.

18. The answer is C. *(Pathology; Cardiovascular)*
The patient most likely has a rupture of an abdominal aortic aneurysm, which is most commonly associated with atherosclerosis. An atherosclerotic aneurysm is a sac-like dilatation of the aorta that most often occurs in men after age 55. Recent evidence suggests familial clustering and biochemical alterations in the structural matrix, which predispose the vessel to atherosclerosis as a secondary rather than a primary cause of the vessel weakening. The location in the abdominal aorta below the renal arteries (90% of cases) may relate to the absence of the vasa vasorum in this part of the aorta and insufficient oxygen/nutrients for the outer part of the vessel wall. Aneurysms of the thoracic aorta are also most commonly of atherosclerotic origin. The majority of aneurysms are asymptomatic. Clinical associations include:

- mid-abdominal to lower back pain
- a pulsatile mass
- abdominal bruit (50%)
- a thromboembolic event
- abdominal angina (pain 30 minutes after eating often associated with massive weight loss from fear of eating)
- renovascular hypertension (most common cause of secondary hypertension)
- evidence of peripheral vascular disease
- presence of a popliteal artery aneurysm

The classic rupture triad is:
- abrupt onset of severe back pain (most rupture into the left retroperitoneum)
- hypotension
- a pulsatile mass in the abdomen

Abdominal ultrasound is the gold standard test (sensitivity approaching 100%). The size and risk for rupture influence the treatment selected. Rupture is inevitable over time. Mortality exceeds 90% in patients who rupture outside the confines of a hospital.

A neoplastic process is rarely, if ever, involved in the formation of an aneurysm. Elastic tissue fragmentation is the key finding in dissecting aortic aneurysms in patients with Marfan syndrome. Infection is associated with aortitis (syphilitic) and also in the formation of mycotic aneurysms. An immunologic reaction is the key finding in vasculitis.

19. The answer is A. *(Pathology; Cardiovascular)*
The patient has a dissecting aortic aneurysm, which extended into the carotid arteries and occluded the lumen. Hypertension is the most common primary predisposing factor.

Dissecting aortic aneurysms refer to defects in the aortic wall that predispose to an intimal tear and dissection of blood into the aortic wall through the areas of weakness. The mean age of occurrence is 60 to 65. Men outnumber women in incidence. Weakness in the elastic arteries is most commonly caused by elastic tissue fragmentation (95%) and/or mucoid degeneration (cystic medial necrosis) in the middle and outer part of the media. Predisposing causes for these changes include Marfan syndrome (defect in fibrillin, which provides the scaffolding for elastic fiber deposition), Ehler-Danlos syndrome (defects in collagen synthesis/remodeling), pregnancy (increase in plasma volume), copper deficiency (cofactor in lysyl oxidase, which is necessary for cross-bridging), coarctation of the aorta (wall stress), trauma, and aminoproprionitriles (in sweet peas; lathyrism), which interfere with collagen synthesis. Hypertension applies a shearing force to the intimal surface of the aorta, resulting in an intimal tear, usually within 10 cm of the aortic valve. Blood dissects under arterial pressure through the areas of weakness created by elastic tissue fragmentation and/or cystic necrosis and progresses superiorly and/or inferiorly. The eventual sites of egress of the advancing column of blood include rupture into the pericardial sac (most common cause of death), mediastinum peritoneum, or re-entry through another tear to create a double-barreled aorta. Type A aneurysms (most common type) involve the ascending aorta. Type B aneurysms begin below the subclavian artery. Patients present with an acute onset of severe retrosternal chest pain associated with a searing pain that radiates into the back. Aortic regurgitation, often associated with congestive heart failure, cardiac tamponade, and stroke, are potential complications. The aortic diameter is widened on a radiograph (first step in the work-up) in 80% of cases. Retrograde arteriography is considered the gold standard test. It is the most common cause of death in Marfan syndrome and Ehler-Danlos syndrome. The overall long-term survival rate is 60%.

20. The answer is B. *(Pathology; Cardiovascular)*
Rupture of the free wall of the left ventricle with cardiac tamponade most commonly occurs between 3 and 7 days postinfarction. At this time, an increase in total CK, absence of CK-MB, and an LDH $\frac{1}{2}$ flip would be expected.

The following table depicts the time sequences for the cardiac enzymes in an acute myocardial infarction.

	First onset (hours)	Peak (hours)	Normal (days)
Total creatine kinase (CK)	4–6	12–24	3–4
CK-MB isoenzyme	4–8	24	1.5–3
Total lactate dehydrogenase (LDH)	8–12	72–144	10–14
LDH 1/2 flip	14	48–72	10
Aspartate aminotransferase (AST)	6–12	20–48	5–9

21. The answer is A. *(Pathology; Cardiovascular)*
Most cardiac arrhythmias associated with an acute myocardial infarction occur within the first 2 hours. These arrhythmias are usually ventricular.

22. The answer is B. *(Behavioral science; Pulmonary/respiratory)*
In obstructive sleep apnea, airway obstruction results in snoring as well as failure to breathe during the night. The resulting anoxia causes frequent awakenings during the night so that the patient feels tired in the morning. Decreased oxygen availability may result in leg edema, hypertension, morning headaches, cardiac arrhythmias, and stroke in patients with obstructive sleep apnea.

23. The answer is E. *(Biostatistics; Quantitative methods)*
Twenty-five percent of premature deaths can be related directly to smoking. Smoking is associated with the leading causes of deaths including myocardial infarction, cancer, stroke, emphysema, and bronchitis.

24. The answer is C. *(Behavioral science, Biostatistics; Psychosocial, cultural, and environmental influences)* Approximately 200,000 deaths are directly associated with alcohol abuse annually. Disorders related to alcohol use include heart disease, liver disease, and cancer as well as suicide and homicide.

25. The answer is B. *(Pathology, Hematology; Hematopoietic/lymphoreticular)* The peripheral smear exhibits numerous schistocytes, or fragmented red blood cells (RBCs). These schistocytes occur in microangiopathic hemolytic anemias, where the RBCs are destroyed by mechanical trauma (e.g., prosthetic heart valves, fibrin strands in disseminated intravascular coagulation, thrombotic thrombocytopenic purpura, the hemolytic uremic syndrome, and runner's anemia). These are normocytic anemias with corrected reticulocyte indexes greater than 3%. Because hemolysis is intravascular, there is a drop in serum haptoglobin, hemoglobinuria, and hemosiderinuria. Hereditary spherocytosis, autoimmune thrombocytopenia purpura, autoimmune hemolytic anemia, and iron deficiency anemia are not associated with schistocytes because there is no mechanical damage to the RBCs.

26. The answer is A. *(Pathology, Hematology; Hematopoietic/lymphoreticular)* The peripheral smear demonstrates numerous target cells with a bull's-eye appearance. Target cells are red blood cell (RBC) markers of hemoglobinopathies (e.g., sickle-cell disease, hemoglobin C disease) and disorders associated with the alteration in RBC membrane lipids (e.g., alcoholic liver disease, obstructive jaundice). The bull's-eye effect is caused by excess RBC membrane that bulges in the middle of the cell. Normally, the spleen removes excess RBC membrane; therefore, splenectomy is another cause of target cells in the peripheral blood. Autoimmune hemolytic anemias, glucose-6-phosphate dehydrogenase deficiency, congenital spherocytosis, and B_{12} deficiencies are not associated with target cells.

27. The answer is A. *(Pathology; Microbial biology and infection)* The bone marrow aspirate exhibits malignant plasma cells with eccentrically located nuclei and prominent nucleoli. A perinuclear clearing is also noted. These findings, in addition to the clinical history of sternal tenderness and a lytic pathologic fracture of the rib, are consistent with a diagnosis of multiple myeloma, most likely Bence Jones (BJ) proteinuria (light chains in the urine).

Malignant plasma cell disorders commonly present with an increase in total proteins due to an increase in γ-globulins. One of the key findings in a malignant monoclonal gammopathy is suppression of the other immunoglobulins by CD8 T-suppressor cells. A monoclonal gammopathy is present usually on a serum protein electrophoresis. Monoclonal gammopathies refer to a group of diseases (usually plasma cell or lympho-plasmacytoid) that have an uncontrolled proliferation of plasma cells or closely related cell types, and abnormally high levels in the blood and/or urine of a homogeneous intact immunoglobulin and/or its corresponding light-chain or heavy-chain constituent. In order to determine the exact immunoglobulin or light chain involved, an immunofixation study or immunoelectrophoresis must be performed on the serum and/or urine.

Multiple myeloma (MM) is the most common primary hematologic malignancy of bone. It is more common in males than females, blacks than whites, and in the older population (50 to 70 years old). MM components are identified in 80% to 90% of cases and urine BJ protein is present in 60% to 80% of cases. The distribution of immunoglobulin (Ig) types in MM is IgG (50-70%), IgA (15% to 20%), light chains alone (10% to 5%), and IgD and IgE, which are extremely rare. Any middle-aged to elderly patient with unexplained anemia, bone pain, pathologic fracture, recurrent infection, unexplained hypercalcemia, renal failure without hypertension, or a monoclonal serum/urine protein is suspect for having MM.

28. The answer is C. *(Pathology; Tissue biology and associated response to disease)* The lymph node biopsy reveals a centrally located Reed-Sternberg (RS) cell, the neoplastic cell of Hodgkin disease (HD), with a large binucleate nucleus and nucleoli that are about one-half the diameter of a lymphocyte. The nucleoli are surrounded by a clear halo. RS cells are of B-cell origin in some cases and T-cell origin in others. Given the patient's young age, sex, and the unilateral location of the lymph node in the lower cervical chain, this man most likely has the lymphocyte-predominant variant of HD.

Nodular sclerosis HD, the most common type of HD, is female predominant and usually presents with cervical and mediastinal nodes. Burkitt lymphoma is the most common malignant lymphoma in children and primarily is found in the abdominal cavity rather than the neck. A poorly differentiated lymphocytic lymphoma more commonly occurs in older people and usually does not localize to a single set of lymph node hyperplasia, the lymph nodes. In reactive lymph nodes hyperplasia, the lymph nodes are characteristically painful. Although RS-like cells may be seen in reactive processes, it is not a common occurrence.

29. The answer is E. *(Pathology, Hematology; Immune response)*
The peripheral blood exhibits three eosinophils with prominent granules that do not cover the nucleus. Eosinophilia is defined as an absolute increase greater than 700 cells/ul and is seen in type I hypersensitivity reactions associated with bronchial asthma, hay fever, drug allergies (e.g., penicillin, iodides, aspirin, and sulfonamides), and allergic rhinitis.

30. The answer is A. *(Pathology, Hematology; Immune response)*
The peripheral smear exhibits three band neutrophils and one segmented neutrophil with toxic granulation in the cytoplasm. The neutrophil count would most likely be elevated and a 100-cell differential would demonstrate a left shift, which refers to the presence of immature neutrophils in the peripheral blood (greater than 10% band neutrophils or the presence of any cells less than a band in the smear, such as a metamyelocyte or myelocytes). This type of neutrophil reaction is commonly associated with suppurative bacterial infections (e.g., acute appendicitis and sterile inflammation with necrosis).

Whooping cough, caused by *Bordetella pertussis*, produces an absolute lymphocytosis; disseminated strongyloidiasis is associated with eosinophilia; tuberculosis produces a monocytosis; and viral gastroenteritis produces a lymphocytosis.

31. The answer is C. *(Pathology; Tissue biology and associated response to disease)*
The lymph node biopsy exhibits a nodular-appearing hyperplastic germinal center that is sharply demarcated from the mantle of T cells surrounding the follicle. Both small and large lymphocytes are noted in the center as well as tingible macrophages with phagocytized material in the cytoplasm. These findings are most consistent with reactive lymph node hyperplasia, which often simulates a neoplastic process resulting in a false positive diagnosis of a malignant lymphoma or Hodgkin disease. Key distinguishing factors from a malignant lymphoma are that the reactive hyperplasia nodes are painful and composed of many different cell types. Malignant nodes are hard, nonpainful, and are composed of one cell type. Reactive hyperplasia may involve the germinal follicle (indicating a B-cell response), the paracortex (representing a T-cell response), the sinus histiocytes, or a combination of the above. Painful lymph nodes draining bacterial infections, like a group A streptococcal pharyngitis, is a clinical example of reactive hyperplasia.

A nodular, poorly differentiated malignant lymphoma (follicular lymphoma) would be very uncommon in this young age bracket and it would not be painful. Lymphocyte predominant Hodgkin disease is common in this age bracket, but no Reed-Sternberg cells are noted in the lymph node biopsy. Metastatic carcinoma would be unusual in this age bracket. The nodes would be nontender and hard. Granulomatous inflammation in the neck caused by cat-scratch disease or scrofula (atypical Mycobacteria) produces painful lymphadenopathy, but the biopsy does not exhibit a granulomatous reaction with multinucleated giant cells and epithelioid cells.

32. The answer is C. *(Pathology; Tissue biology and associated response to disease)*
Erythroid hyperplasia in the bone marrow is a response of the erythroid series to erythropoietin stimulation. These stimuli include hypoxemia (e.g., chronic, obstructive pulmonary disease), anemia less than 7 gm/dl, a left-shifted oxygen dissociation curve (e.g., carbon monoxide poisoning), a bleed more than 5 to 7 days, and testosterone. Anemia of chronic disease, β-thalassemia minor, and iron deficiency have problems in hemoglobin synthesis, so these would not be expected

to have a reticulocyte response. In chronic renal disease, there is a decreased concentration of erythropoietin, which results in a normocytic anemia and no reticulocyte response.

33. The answer is C. *(Pathology; Musculoskeletal)*
The patient has metastatic breast cancer to bone. The peripheral blood findings of anemia, the presence of immature white blood cells, nucleated red blood cells, and an increase in reticulocytes indicates a metastatic tumor pushing marrow elements out into the periphery. This is called a leukoerythroblastic smear. Anemia associated with metastasis to the marrow is called myelophthisic anemia. In addition to the hematologic findings, the patient has bone pain and elevation of alkaline phosphatase in the presence of a normal γ-glutamyltransferase (GGT). This indicates that the alkaline phosphatase is of bone origin because the GGT is not present in the bone. Alkaline phosphatase is present in osteoblasts and indicates osteoblastic activity by the metastatic tumor.

The patient does not have a hemolytic anemia. The increase in the reticulocyte count is inappropriate because it is caused by metastasis and not an erythropoietin-stimulated marrow. Chronic inflammation is present in this patient, but the entire clinical picture is related to bone metastasis rather than anemia of inflammation. The immature white blood cells in the peripheral blood is not a leukemic process, which specifically refers to a primary malignancy arising from the bone marrow. Myeloproliferative disorders often have a peripheral blood picture like this patient, but these disorders are primary stem-cell disorders of the bone marrow.

34. The answer is C. *(Pathology, Genetics; Human development and genetics)*
There are five major Rh antigens, mainly DCEce. There is no d antigen, but it is usually indicated in the Fisher-Race nomenclature. Rh antigens are inherited in groups of three on each chromosome (e.g., CDE) and in codominant fashion (i.e., each antigen is capable of expressing itself). Many different combinations may be assembled, such as CDE, CDe, cDE, cde. A person inherits one of the above complexes from each parent. The blood bank tests for each of these antigens and establishes the Rh phenotype (e.g., CcDEE, ccDEe, etc.) and then, from statistical charts, the possible genotypes (e.g., CDE/cDE) and percent probability of the individual are established. Most blood banks only test for the D antigen. A person who is Rh-positive is synonymous with being D-antigen positive, which is present in 85% of the population. A weak variant of the D antigen, called Du, is also tested for by the blood bank, and, if present, the individual is still considered Rh positive.

In the present case, the individual has the following phenotype

C	c	D	E	e
+	+	+	+	+

The most likely genotype for the woman is CDE / cde. Because the husband is Rh positive and the child is Rh negative, she and her husband would have to be heterozygotes for the D antigen, that is Dd rather than DD (homozygotes). Note the possible combinations below.

		Husband	
		D	d
Mother	D	DD	Dd
	d	Dd	dd (child who is Rh negative)

35. The answer is C. *(Pathology, Genetics; Human development and genetics)*
It is not possible for an A mother and an AB father to have a child who is blood group O, even if the mother has an AO rather than an AA phenotype.

		Father	
		A	B
Mother	A	AA	AB
	O	AO	BO

36. The answer is A. *(Pathology, Hematology; Hematopoietic/lymphoreticular)*

The patient has an allergic-transfusion reaction, which is an immunoglobulin E (IgE)-mediated type I hypersensitivity reaction. The patient is reacting to a protein in the donor unit and has developed hives and a fever.

A febrile-transfusion reaction is not associated with hives. These patients have cytotoxic antibodies against antigens on the donor leukocytes, which causes the release of pyrogens. This example is type II, cytotoxic antibody-mediated hypersensitivity reaction. There are no type III (immune complex) or type IV (cellular immunity) transfusion reactions. Anaphylactic reactions present with shock, which is not present in this patient. Although fever is present, this is not a hemolytic transfusion reaction because the antibody screen in the patient's serum is negative for atypical antibodies, and the direct Coombs' test is negative for antibodies on the surface of the patient's red blood cells. There is also no free hemoglobin in the plasma or urine to indicate an intravascular hemolytic transfusion reaction.

37. The answer is D. *(Pathology; Immune response)*
ABO incompatibility would normally protect this mother against Rh immunization (e.g., D antigen), because any group A, Rh-positive fetal cells that entered her circulation during delivery would immediately be destroyed by her anti-A immunoglobulin (Ig) M antibodies. The baby has ABO incompatibility, which is the most common cause of hemolytic disease of the newborn. ABO incompatibility is present in 25% of normal pregnancies and almost always involves a group O mother with a group A or B fetus. The following discussion is based on the mother being group O, Rh negative and the baby group A and Rh positive.

Laboratory findings in this infant would include spherocytes in the infant's peripheral blood (good marker) because splenic macrophages remove a little red blood cell (RBC) membrane when picking off the antibody from the cell surface. Also, there would be a weakly positive direct Coombs' on the babies RBCs due to the presence of maternal IgG anti A,B on their surface. A positive, indirect Coombs' on the newborn's serum would be evident because maternal anti-A,B IgG is present. Recall that newborns normally do not have any isohemagglutinins in their serum at birth because they do not synthesize IgM until after birth.

This group A infant will eventually develop anti-B IgM antibodies over the ensuing months. Therefore, the presence of anti-A,B in the babies serum at birth must be IgG, which came from the mother. This is an indirect sign of ABO incompatibility.

Physiologic jaundice of the newborn, due to immature conjugating enzymes in the liver, develops after 3 days and not within the first 24 hours after birth. The mother is a candidate for Rh immunoglobulin because she is Rh negative and anti-D negative. The mechanism of jaundice in the fetus is extravascular hemolysis by splenic marcrophages in the fetal spleen.

38. The answer is C. *(Pathology, Hematology; Hematopoietic/lymphoreticular)*
In anemias that impose a severe stress on the bone marrow for erythropoiesis (e.g., β-thalassemia major, sickle-cell disease), inactive fatty marrow reverts back to active red marrow. The excess erythropoiesis results in the expansion of the marrow cavity and produces certain physical diagnostic signs, which include a 'hair-on-end' appearance of the skull bones on a radiograph, a chipmunk facies caused by the expansion of the marrow space into the zygomatic bone, and frontal bossing of the skull, resulting in a prominent forehead. Hematopoietic demands beyond the capacity of the bone marrow (e.g., β-thalassemia major) are also associated with extramedullary hematopoiesis (i.e., blood cell formation outside the marrow in the liver, spleen, and other organs).

Lead poisoning produces a microcytic anemia due to a defect in heme synthesis. There is a mild hemolytic component from damage to the red blood cell membrane. One would not expect massive erythropoiesis in this setting. Acute lymphocytic leukemia originates in the bone marrow, most commonly from B lymphocytes. It produces bone pain but is not associated with expansion of the narrow cavity. Severe iron deficiency is not associated with marrow expansion because the proper raw materials for hemoglobin production are absent. Hemoglobin Bart disease is due to a deletion of all four genes involved in α-chain synthesis. This disease most commonly results in intrauterine death with hydrops fetalis.

39. The answer is C. *(Pathology, Hematology; Tissue biology and associated response to disease)*

Hematopoietic demands beyond the capacity of the bone marrow are associated with extramedullary hematopoiesis (i.e., blood cell formation outside the marrow). This formation often occurs in the liver, spleen, and other organs. Agnogenic myeloid metaplasia is an example of this process because the bone marrow undergoes reactive fibrosis, whereas hematopoiesis primarily resides in the spleen. Iron-deficiency anemia and β-thalassemia minor do not stimulate marrow erythropoiesis because both have reduced synthesis of hemoglobin, the former due to an absence of iron and the latter, a reduction in β-chain synthesis. Congenital spherocytosis does not usually produce an anemia severe enough to require extramedullary hematopoiesis. In acute blood loss, the bone marrow would respond to the deficit in red blood cells (RBC) and would not require outside sources for additional RBC synthesis.

40. The answer is D. *(Pathology; Hematopoietic/lymphoreticular)*
Shift cells are not normally found in the peripheral blood unless the marrow is stimulated by erythropoietin or something in the marrow is pushing the cells out, such as metastatic tumor. These cells have a bluish discoloration on the Wright-Giemsa stain because they have more RNA than a peripheral blood reticulocyte. The reticulocyte count is the best index of effective erythropoiesis, or how well the bone marrow is responding to an anemia. It is increased in an erythropoietin-stimulated marrow. Shift cells are basophilic staining red blood cells that are even younger than peripheral blood reticulocytes. These cells are considered marrow reticulocytes as opposed to peripheral blood reticulocytes.

The myeloid:erythroid ratio is calculated by counting 500 myeloid and erythroid cells in a bone marrow aspirate. The normal myeloid:erythroid ratio is 3:1. This ratio is a measure of total granulopoiesis and total erythropoiesis, but it does not evaluate how many of the cells are actually released into the peripheral blood (effective granulopoiesis:erythropoiesis). Decreased serum ferritin indicates a decrease in the marrow iron stores, as in iron deficiency. It does not evaluate effective or ineffective erythropoiesis. The radioactive plasma iron turnover study calculates how much radioactivity labelled iron is removed from the peripheral blood and delivered to developing normoblasts in the bone marrow. Therefore, one would expect an increased plasma iron turnover in an erythropoietin-stimulated marrow.

41. The answer is A. *(Pathology; Pulmonary/respiratory)*
A low serum haptoglobin concentration and increased urine hemosiderin is expected in a patient with severe calcific aortic stenosis. Recall that haptoglobin is synthesized in the liver and forms a complex with free hemoglobin (Hgb). This complex is removed by macrophages, thus decreasing the haptoglobin concentration in severe, acute intravascular hemolysis. When the haptoglobin is used up, free Hgb spills into the urine. Some of it is reabsorbed by renal tubular cells with subsequent formation of hemosiderin. When hemosiderinuria is present in the urine, this indicates chronic intravascular hemolysis. Calcific aortic valves damage red blood cells (RBCs), particularly if they must pass through a narrow opening (stenosis). Schistocytes are formed, indicating mechanical trauma to the cells.

Congenital spherocytosis is a hemolytic anemia with extravascular hemolysis, so low haptoglobin levels and hemosiderinuria are not usually present. α-thalassemia minor is a defect in α-chain synthesis and does not have a hemolytic component. Aplastic anemia is associated with a pancytopenia in the peripheral blood. It is not a hemolytic disease. However, if paroxysmal nocturnal hemoglobinuria is the cause of the aplastic anemia, the complement-induced intravascular hemolysis of RBCs does result in low serum haptoglobin levels and the presence of hemosiderinuria. Because haptoglobin is synthesized in the liver, liver disease reduces haptoglobin levels, but there would not be an associated hemosiderinuria.

42. The answer is A. *(Pathology; Hematopoietic/lymphoreticular)*
Anemia associated with a corrected reticulocyte count less than 2% would most likely be seen in a patient with anemia of chronic inflammation (ACD) who is taking iron. In ACD, iron is blockaded in macrophages and is unavailable for hemoglobin synthesis. Iron therapy does not produce any reticulocyte response.

A patient with rheumatoid arthritis who has a normocytic anemia and a positive direct Coombs test has an autoimmune hemolytic anemia. The corrected reticulocyte count is increased in hemolytic anemias. A patient who received an intramuscular injection of B$_{12}$ for treatment of pernicious anemia one week ago will have an increased reticulocyte response because the missing raw material for DNA synthesis is now available. A patient with a gastrointestinal bleed 6 days ago will have an increased reticulocyte response because the marrow has had enough time for accelerated erythropoiesis. However, if the bleed was less than 5 days ago, the reticulocyte count would not be increased because the marrow has not had time to produce and release RBCs. A patient with agnogenic myeloid metaplasia has an inappropriate increase in the reticulocyte count because it is not related to erythropoietin stimulation of the bone marrow.

43. The answer is E. *(Pathology; Hematopoietic/lymphoreticular)*
Neutrophilic leukocytosis, lymphopenia, and eosinopenia are most likely associated with Cushing syndrome and the hypercortisolism associated with the syndrome. Cortisol decreases neutrophil adhesion in the peripheral blood, thus releasing the marginating pool into the circulation. Cortisol also increases lymphocyte adhesion in the efferent lymphatics and is toxic to eosinophils.

Endotoxic shock results in a neutropenia because endotoxins increase neutrophil adhesion. Absolute neutropenia is a feature of typhoid fever, splenomegaly, and sinus bradycardia. Whooping cough due to *Bordetella pertussis* produces an absolute lymphocytosis. Paroxysmal nocturnal hemoglobinuria, an acquired stem-cell disorder with increased sensitivity of hematopoietic cells to complement destruction, is associated with aplastic anemia and pancytopenia.

44. The answer is C. *(Pathology; Hematopoietic/lymphoreticular)*
A routine Wright-Giemsa stain of peripheral blood detects coarse basophilic stippling, which is pathognomonic for lead poisoning. In general, basophilic stippling refers to clumps of blue-staining ribosomes, most often seen in anemias with defects in hemoglobin synthesis (e.g., iron deficiency, the thalassemias, lead poisoning). The ribosomal clumps in lead poisoning are large because of the inactivation by lead of ribonuclease, which normally breaks down the ribosomes.

In addition, a routine Wright-Giemsa stain specifically identifies Howell-Jolly bodies (nuclear remnants), malarial pigments, marrow reticulocytes (polychromatic cells or shift cells with a bluish-gray discoloration), and intraerythrocytic parasites (malaria, Babesiosis). Special stains are necessary to identify Pappenheimer bodies (hemosiderin requires a Prussian blue stain), reticulocytes (supravital stain detects residual RNA), Heinz bodies (clumps of denatured hemoglobin in glucose-6-phosphate dehydrogenase deficiency), and globin chain inclusions (excess α-chains) in severe β-thalassemia.

45. The answer is C. *(Pathology; Hematopoietic/lymphoreticular)*
Warfarin inhibits epoxide reductase, thus inactivating vitamin K and the vitamin K-dependent factors: prothrombin (II), VII, IX, X, protein C, and protein S. The prothrombin time (PT) measures the extrinsic system down to the formation of a clot (VII, X, V, II, fibrinogen → clot), whereas the partial thromboplastin time (PTT) is a measure of the intrinsic system down to the formation of a clot (XII, XI, IX, VIII, X, V, II, fibrinogen → clot). Therefore, both the PT and PTT are prolonged in a patient on a warfarin derivative, but the deficiency in factor IX is only measured in the PTT, not the PT. Protein C and S are not measured in either the PT or the PTT.

46. The answer is A. *(Biochemistry; Biochemistry and molecular biology)*
Amyloid is a 7.5 nm to 10 nm linear nonbranching structure with hollow cores on electron microscopy and is not derived from albumin. Amyloidosis refers to diseases that deposit a variety of fibrillary proteins into intestinal tissues resulting in organ dysfunction. Amyloid has a twisted β-pleated sheet and a hyaline that appears eosinophilic on hematoxylin and eosin stains of tissue. Amyloid appears red with the Congo red stain and has an apple-green birefringence under polarized light.

Amyloid is derived from many different proteins, including light chains, serum-associated amyloid, prealbumin, β-proteins, and peptide hormones (e.g., calcitonin).

47. The answer is B. *(Pathology, Hematology; Hematopoietic/lymphoreticular)*

The following sequence of events occurs with small vessel injury. Injury first causes the vessel to constrict, which reduces the blood flow and allows circulating platelets to adhere to the von Willebrand factor (VIII$_{VWF}$) via VWF's platelet receptors. VIII$_{VWF}$ is a part of the factor VIII molecule that is synthesized by the vascular endothelium. Platelet adhesion is a stimulus for the generation of thromboxane A$_2$ (TXA$_2$) by the platelet. TXA$_2$ stimulates the platelet-release reaction, which involves the release of numerous chemical mediators from both the dense bodies and α-granules within the platelets. TXA$_2$ is also a potent vasoconstrictor and stimulus for platelet aggregation. Platelet aggregation also occurs when adenosine diphosphate and thrombin are released from the platelet granules after the release reaction. The platelets are loosely held together by fibrinogen-connecting IIb/IIIa receptors on the platelet membranes, thus forming a temporary hemostatic plug. The temporary hemostatic plug is unstable and undergoes further alterations to become a stable clot.

48. The answer is E. *(Pathology, Hematology; Hematopoietic/lymphoreticular)*
Platelet disorders are classified by quantitative problems (thrombocytopenia or thrombocytosis) and qualitative defects (the platelets function improperly). Thrombocytopenia is divided into increased destruction (antibodies, disseminated intravascular coagulation), decreased production (decreased megakaryocytes in the bone marrow; drugs, aplastic anemia), and abnormal distribution (splenomegaly, which sequesters platelets).

Congestive splenomegaly is more likely associated with thrombocytopenia than thrombocytosis because the enlarged spleen traps platelets in the Billroth cords and destroys them. This commonly occurs in congestive splenomegaly secondary to portal hypertension in patients with cirrhosis of the liver.

Chronic iron deficiency, polycythemia rubra vera (a myeloproliferative disease), nonhematologic malignancies, and tuberculosis are more likely to produce thrombocytosis. Other causes of thrombocytosis include stress and essential thrombocythemia, which is a myeloproliferative disease with an increase in granulocytes and platelets. Hematologic malignancies, such as leukemias, almost invariably have thrombocytopenia.

49. The answer is B. *(Physiology; Quantitative methods)*
The arrangement of the three standard limb leads in the Einthoven triangle is shown below.

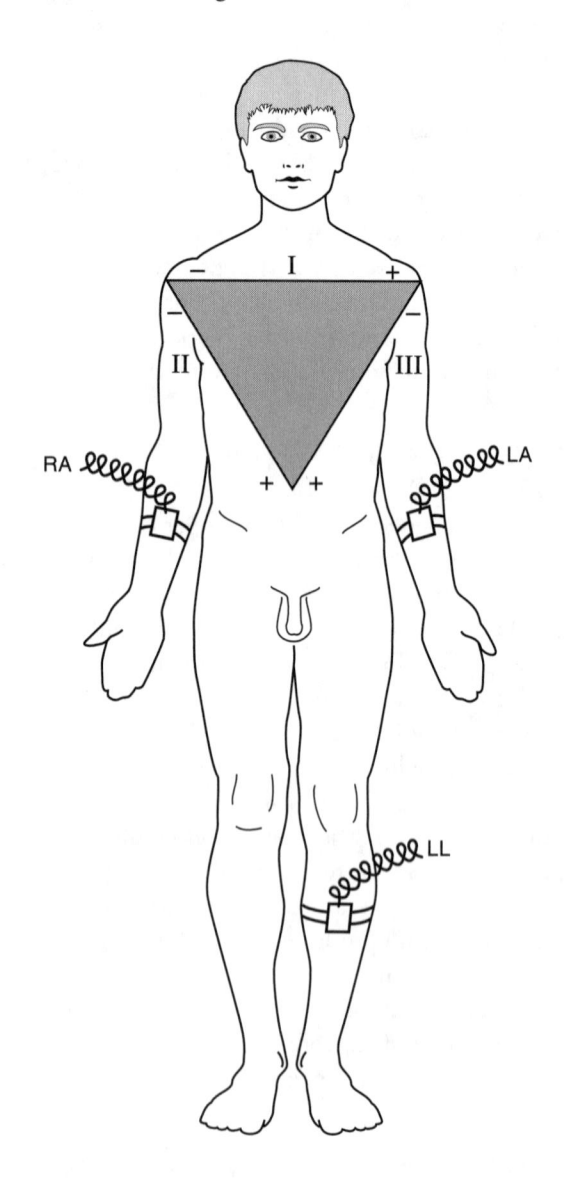

Lead I connects the right arm (- electrode) to the left arm (+ electrode). Lead II connects the right arm (- electrode) to the left leg (+ electrode). Lead III connects the left leg (+ electrode) to the left arm (- electrode). If a patient moves her left arm during the recording of an electrocardiogram (EKG), an electrical disturbance will arise in Leads I and III because both are connected to the left arm.

50-52. The answers are: 50-D, 51-B, 52-C. *(Physiology; Cardiovascular)*

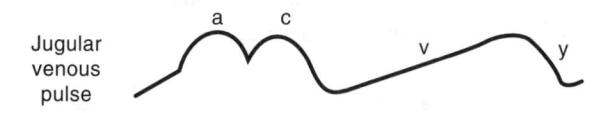

The jugular venous pulse records changes in right atrial pressure during a cardiac cycle. The 'a' wave (A) is associated with atrial contraction. Atrial contraction occurs at the end of diastole and is preceded by the P wave of the electrocardiogram (EKG). The 'c' wave results primarily from the bulging of the tricuspid valve into the atrium during ventricular isovolumic contraction. The 'v' wave represents atrial filling. Note that while the atrium is filling, the tricuspid valve is closed, and the right ventricle is ejecting the stroke volume. The 'y' descent is associated with the opening of the tricuspid valve and rapid emptying of the atrium.

53. The answer is B. *(Physiology; Cardiovascular)*

Narrowing of the mitral valve, as a result of the fusion of the mitral commissures, hinders blood flow from the atrium to the ventricle. The primary effect of mitral stenosis is a rise in left atrial pressure. In mitral stenosis, atrial pressure (a, c, and v waves) is elevated throughout the cardiac cycle. The stenosis also increases the amplitude of the first heart sound. Ventricular filling is usually not impaired until the stenosis becomes very severe; thus, ventricular systolic pressure remains normal.

54. The answer is C. *(Physiology; Cardiovascular)*

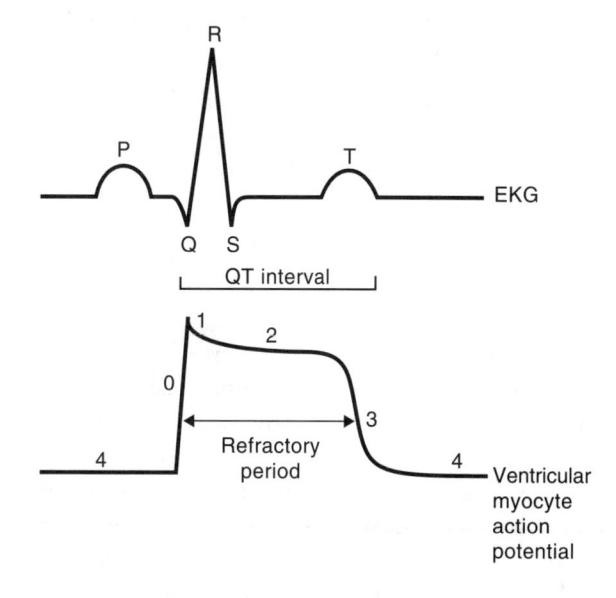

The diagram depicts the relationship between the electrocardiogram (EKG) recorded on the surface of the body and the duration of a single ventricular myocyte action potential. The duration of the ventricular myocyte refractory period is nearly as long as the action potential, approximately 300 msec. The QRS complex of an EKG represents the electrical potential on the surface of the body produced by the wave of depolarization as it spreads through the ventricle and as each myocyte depolarizes (phase O). The T wave represents ventricular repolarization (phase 3 in each myocyte). Note that the ST segment occurs during the plateau phase.

55. The answer is A. *(Physiology; Cardiovascular)*

In mitral insufficiency, the mitral valve, which should be closed during left ventricular contraction, is incompetent. Thus, blood regurgitates into the left atrium when ventricular pressure is higher than atrial pressure (i.e., from the onset of isovolumic ventricular contraction until ventricular isovolumic relaxation is nearly complete). Therefore, atrial pressure rises dramatically throughout ventricular systole.

56. The answer is A. *(Physiology; Cardiovascular)*
Einthoven's Law states that the electrical potentials recorded in each of the standard limb leads exhibit the following relationship:

$$\text{Lead I} + \text{Lead III} = \text{Lead II}$$

$$5 + \text{Lead III} = 9$$

$$\text{Lead III} = 4 \text{ mV}$$

57. The answer is C. *(Physiology; Cardiovascular)*
Norepinephrine, released by sympathetic nerves, increases heart rate by opening the slow Ca^{++} channels and increasing the conductance of the sinoatrial node to Ca^{++}. This action accelerates the rate at which the membrane spontaneously depolarizes (phase 4) and reaches threshold, thus increasing the heart rate. Acetylcholine hyperpolarizes the sinoatrial node and slows the rate at which threshold is reached (decreased rate of phase 4 depolarization), thus slowing the heart. The atrioventricular node is not the pacemaker because its rate of phase 4 depolarization is slower than that of the sinoatrial node (i.e., the sinoatrial node reaches threshold first).

58. The answer is C. *(Physiology; Cardiovascular)*
During the plateau phase (phase 2), the membrane's conductance to K^+ is low. However, the electrochemical gradient for K^+ is so strong that some K^+ does leak out. The K^+ efflux is balanced by the inward movement of Ca^{++} and Na^+ through the slow Ca^{++} channels so that the membrane potential plateaus above 0 mV. Sympathetic stimulation and drugs, which increase the concentration of cytosolic Ca^{++} (digitalis), shorten the duration of the plateau phase by increasing gK^+.

59. The answer is B. *(Physiology; Cardiovascular)*
The action potential of the sinoatrial node and atrioventricular node results from the opening of the slow Ca^{++} channels and the influx of Ca^{++} (some Na^+ may also enter through these slow Ca^{++} channels). An action potential is elicited when the membrane potential reaches threshold. Spontaneous depolarization is an intrinsic property of nodal tissue and can occur in the absence of nerves. The low resting potential results in a small action potential that propagates slowly. The rapid Na^+ channels are inactivated in nodal tissue. An increase in the conductance to K^+ would hyperpolarize the membrane.

60. The answer is C. *(Physiology; Cardiovascular)*
The interval between the closing of the mitral valve (S1) and the closing of the aortic valve (S2) represents ventricular systole. Ventricular systole includes the period of isovolumic contraction (between S1 and the opening of the aortic valve) and ejection (between the opening and closing of the aortic valve).

61. The answer is C. *(Physiology; Cardiovascular)*

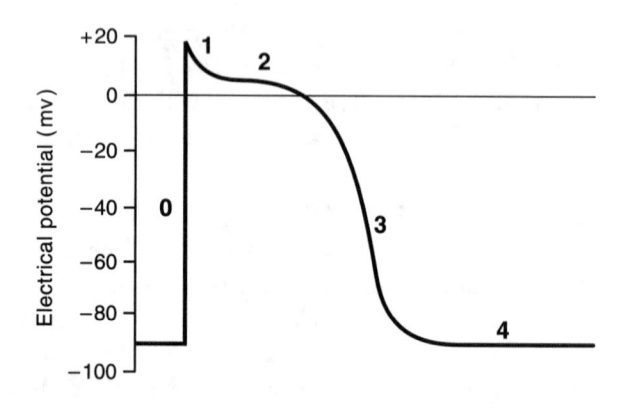

The diagram depicts a typical ventricular myocyte action potential. The stable resting membrane potential (phase 4) is approximately -90 mV. This highly polarized state indicates that the membrane has a high conductance (permeability) to K^+. The rapid reversal of the membrane potential in ventricular myocytes (phase 0) is due to the opening of the rapid Na^+ channels and the approach of the membrane to the equilibrium potential for Na^+ (+60 mV). During phase 1, the membrane potential begins to repolarize. This is caused by the transient opening of K^+ channels, as well as the inactivation of the Na^+ current. Phase 2 (the plateau phase) results from the opening of the slow Ca^{++} channels. The influx of Ca^{++} is balanced by the efflux of K^+, so a stable plateau potential is maintained. During phase 3, the membranes' conductance to K^+ increases and the membrane repolarizes. Myocytes do not possess automaticity and depend on the spread of electrical activity for activation.

62-64. The answers are: 62-C, 63-B, 64-B. *(Physiology; Cardiovascular)*

The cardiac action potential is generated at the sinoatrial node and is conducted through the internodal pathways to the atrioventricular node (approximately 40 msec), the atrioventricular node (an additional 120 msec), the atrioventricular bundle of His, and the left and right Purkinje bundles. The ventricles are entirely depolarized in 60 to 80 msec. The diagram below depicts the time of appearance of the action potential (msec) after its initiation at the sinoatrial node.

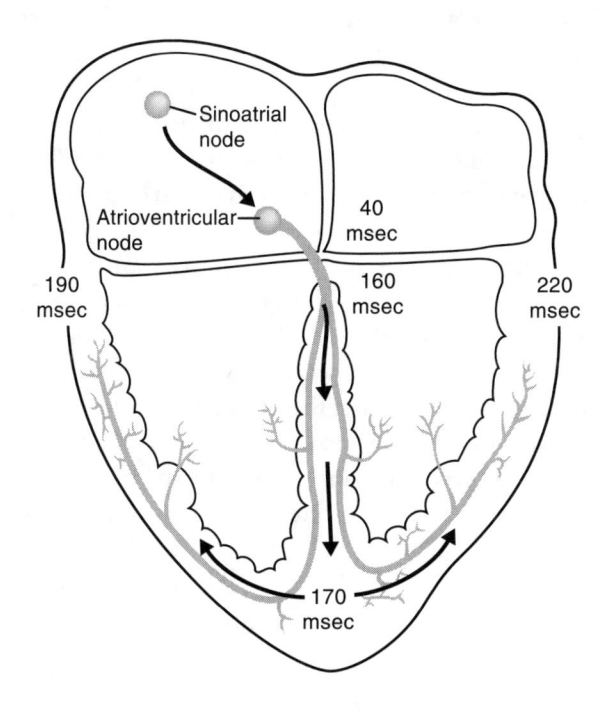

The time it takes for the action potential to travel from the sinoatrial node through atrial muscle and through the atrioventricular node is the P-R interval. It normally varies from 120 to 210 msec. Note that most of the conduction delay is in the atrioventricular node.

The time it takes for the action potential to spread through both ventricles is the duration of the QRS complex and is normally 60 to 80 msec.

65. The answer is D. *(Pathology; Pulmonary/respiratory)*

The lung biopsy reveals asbestos (ferruginous) bodies. Pleural plaques are the most common manifestation of asbestosis caused by the inhalation of asbestos fibers. Asbestos is found in pipe fittings, vehicle brake linings, cement pipes, and insulation material. It is a crystalline silicate with two morphologic forms: serpentine and amphibole. The serpentine chrysotile form is most commonly encountered in industrial exposure. The amphibole group, which includes the mineral crocidolite, is most commonly associated with mesotheliomas. Asbestos bodies lodge lengthwise in small bronchial passages. They are composed of iron (ferruginous bodies) and protein around a core of asbestos. Asbestos, unlike silica, does not damage the phagolysosomes of macrophages, which results in cell death. However, asbestos does activate the alveolar macrophages to produce chemical mediators that stimulate fibrogenesis (e.g., interleukin-1). Asbestos is a complete carcinogen (i.e., it is both an initiator and a promoter) and predisposes to a number of diseases. Some of these diseases include benign pleural plaques, primary lung cancer, mesothelioma, laryngeal and colon cancer, and restrictive lung disease.

Asbestos requires 20 to 30 years from first exposure before cancer first presents itself. The prognosis is extremely poor.

66. The answer is E. *(Physiology; Pulmonary/respiratory)*

The decrease in the maximal expiratory flow suggests increased airway resistance, indicating an obstructive disease. Both chronic bronchitis and emphysema are obstructive diseases, but only emphysema is characterized by a loss of elasticity and an increase in compliance. Compliance is a measure of the distensibility of the lung (i.e., the greater the increase in lung volume for a given fall in intrapleural pressure, the more compliant the lung). As in all obstructive diseases, early closure of the airways during a forced expiration leads to increased residual volume. Another common attribute of obstructive diseases is lower than normal forced expiratory volume in 1 second (FEV_1), which results from the low rate at which expired gas flows.

In restrictive diseases such as interstitial fibrosis, pulmonary compliance is reduced (i.e., the lung is stiffer), and maximal expiratory flow rates at comparable lung volumes are normal or greater.

67. The answer is D. *(Physiology; Pulmonary/respiratory)*

Curve A describes a normal individual who has a hemoglobin concentration of 15 gm%, with an oxyhemoglobin-binding capacity of 1.34 ml O_2 per 1g hemoglobin per 100 ml of blood. Thus, the hemoglobin is fully saturated when 100 ml of blood is carrying 20 ml O_2 (1.34 × 15 = 20 vol%). At an oxygen tension (PO_2) of 100 mm Hg, the hemoglobin is 97.5% saturated [i.e, the hemoglobin carries 19.5 ml of a possible 20 ml of O_2/100 ml (.975 (20) = 19.5 vol%)]. Curves B and C describe individuals whose O_2 content (vol%) at a PO_2 of 100 mm Hg is one-half the normal; thus, these individuals are anemic. Curve C is shifted to the right and is typical of anemias that result from a reduced hematocrit. The rightward shift is caused by an increase in 2, 3-diphosphoglyceric acid (2,3-DPG). Curve B is shifted to the left and is typical of the anemia produced by the presence of carbon monoxide (CO). CO has 210 times the affinity for the same site on hemoglobin as does O_2 and, therefore, reduces the O_2 content of hemoglobin. The leftward shifts indicates that, in the presence of CO, hemoglobin holds on more firmly to O_2, making it unavailable to the tissues.

At a high altitude, increased levels of 2,3-DPG shift the oxyhemoglobin curve to the right, whereas the polycythemia increases the O_2 content above normal at 100% saturation. During exercise or chronic pulmonary disease, the oxyhemoglobin dissociation curve would reflect a normal O_2 content at 100% saturation and a rightward shift.

68-70. The answers are: 68-A, 69-B, 70-C. *(Physiology; Pulmonary/respiratory)*
An individual with normal pulmonary function is capable of exhaling approximately 80% of the forced vital capacity (FVC) in one second. Therefore, the ratio of the gas volume expired in one second (FEV_1) to the FVC is approximately 80%.

A disease process that results in an increased airway resistance will considerably decrease the rate at which air flows during expiration; thus, the FEV_1 will be reduced. Conditions in which airway resistance is increased are categorized as obstructive diseases (e.g., chronic bronchitis, emphysema, asthma). Because these condions lead to either a narrowed airway (chronic bronchitis, asthma) or an untethered airway with a loss of support from neighboring parenchyma (emphysema), a forced expiration causes early closure of the airway that increases the residual volume and reduces the FVC. In obstructve disease, the FEV_1 to FVC ratio is reduced (FEV_1 is reduced to a greater extent than the FVC), and values around 50% are common in moderate disease. The rate at which gas is exhaled in individuals with restrictive disease is not reduced, so that the FEV_1 to FVC ratio remains approximately 80%.

71-72. The answers are: 71-B, 72-A. *(Physiology; Pulmonary/respiratory)*
The airways of a patient with obstructive disease offer a greater than normal resistance to airflow (the rate of air movement through the airways decreases). When a patient with obstructive disease is placed on a respirator, adequate time must be allowed for the patient to complete an exhalation. If the rate is set too high, there is not enough time for exhalation, and the lung will become hyperinflated. This usually does not lead to a rupture of the lung because, as the alveoli expand, the airways widen, the resistance falls, and the rate of airflow on expiration increases. This is a mechanical effect (i.e., nerves are not involved) that results from the increased traction exerted on the airways at high lung volumes by the neighboring tissue.

This diagram describes the relationship between lung volume and airway resistance.

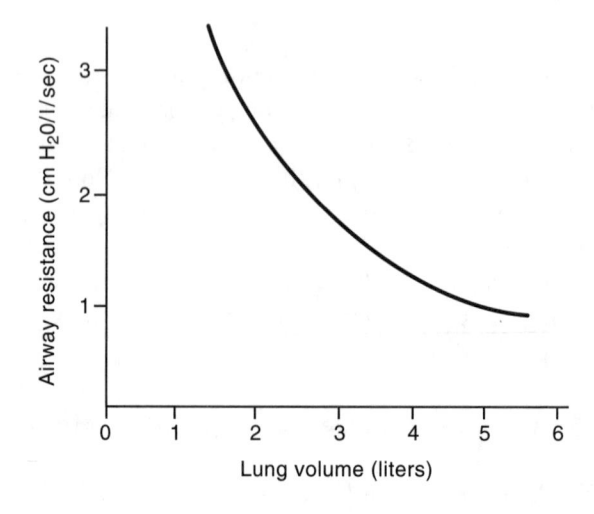

As lung volume increases, airway resistance falls.

73. The answer is E. *(Pathology; Pulmonary/respiratory)*
Most solitary coin lesions are granulomas. Histoplasma capsulatum is commonly found in the Ohio and Mississippi River valleys, so previous histoplasmosis would be the most common cause in a patient from the Midwest. Overall, greater than 75% of solitary coin lesions are benign, with the majority of these representing granulomas. Less than 25% are malignant, with most of these representing primary squamous carcinoma or adenocarcinoma. Metastatic cancer is not commonly solitary (< 25%). The majority of solitary coin lesions are benign if the patient is less than 35 years of age (1% malignant). However, the chances for malignancy increase to 50% to 60% if the patient is over 50 years of age. A smoking history also increases the risk for cancer. Calcifications that suggest a benign lesion are those that have a popcorn configuration (most commonly a bronchial hamartoma), concentric rings of calcification, or flecks of calcium in the center or at the periphery (usually a granuloma). The growth rate is also important. Calcifications that show no increase in size in 2 years are usually benign, whereas those that increase in size during this interval are often malignant. If they grow rapidly in less than 2 weeks, they are probably inflammatory lesions.

74. The answer is C. *(Pathology; Pulmonary/respiratory)*
The estimated occurrence of pulmonary metastases by primary malignancy is greatest for choriocarcinoma (80%), which is a trophoblastic tumor composed of cytotrophoblast and syncytiotrophoblast. Human chorionic gonadotropin (hCG) is a marker for this tumor. Gestationally derived choriocarcinomas portend a good prognosis even when metastasis is present because they respond extremely well to chemotherapeutic agents (e.g., methotrexate). However, this is not true for nongestationally derived choriocarcinoma.

The percent occurrence of pulmonary metastases for the other cancers is breast cancer, 55%; prostate cancer, 45%; malignant melanoma, 60%; and gastrointestinal carcinoma, 20%.

75. The answer is A. *(Pathology; Pulmonary/respiratory)*
The patient has interstitial pneumonitis secondary to cytomegalovirus (CMV). This is the most common pneumonia in bone marrow transplant patients. Note the 'owl's eye' intranuclear inclusion. The virus lives in lymphocytes; thus, the risk for infection in bone marrow transplant patients. It is a common infection in immunocompromised patients (e.g., AIDS patients). Ganciclovir is used as prophylaxis and for the treatment of CMV infections. The antibiotic enters cells infected by CMV and is phosphorylated into a triphosphate that preferentially inhibits CMV DNA polymerase. Its major side effects include leukopenia, thrombocytopenia, impairment of renal function, and seizures.

76. The answer is E. *(Pharmacology; Pulmonary/respiratory)*
Community-acquired, drug-resistant *Streptococcus pneumoniae* (DRSP) have emerged as a widespread clinical problem. Many strains are resistant to penicillins, cephalosporins, erythromycin, and trimethoprim-sulfamethoxazole. All strains of DRSP were initially sensitive to vancomycin and rifampin; however, some vancomycin-resistant strains have been reported recently, and resistance to rifampin often develops rapidly after exposure to the antibiotic. Vancomycin is a peptide inhibitor of cell-wall synthesis and is effective against many methicillin-resistant staphylococci and some enterococci.

77. The answer is D. *(Pharmacology; Pulmonary/respiratory)*
Pneumocystis carinii pneumonia (PCP) is the most common infection in immunocompromised AIDS patients and should be treated aggressively with prednisone and appropriate antimicrobial therapy. Trimethoprim-sulfamethoxazole (oral or intravenous) is the drug of choice unless contraindicated. Suitable alternatives include pentamidine, administered as an aerosol, or intravenous trimetrexate. Trimetrexate is a recent dihydrofolate reductase inhibitor related to methotrexate and must be followed by leucovorin three days after trimetrexate administration (leucovorin rescue) because trimetrexate inhibits both human and microbial dihydrofolate reductase. Leucovorin is an active form of folic acid that bypasses the inhibition of dihydrofolate reductase. Rifampin is an antitubercular drug

with some activity against gram-positive cocci and is not indicated for PCP.

78. The answer is C. *(Pharmacology; Pulmonary/respiratory)*
Isoniazid administration for 6 to 12 months is the only documented prophylaxis for individuals with a positive tuberculin test without evidence of disease. There is no evidence that a combination of drugs is more effective. Prophylaxis is most important for patients with risk factors, including illicit drug use, diabetes mellitus, prolonged glucocorticoid therapy, end-stage renal disease, hematologic disease and others. However, studies indicate that prophylaxis is also appropriate for individuals who are otherwise healthy. Patients with previous tuberculous infections that were not adequately treated with a combination of drugs should also receive isoniazid preventive therapy.

79. The answer is C. *(Pharmacology; Pulmonary/respiratory)*
Multidrug-resistant (MDR) tuberculosis is becoming a major problem and has been reported in at least 14 hospitals in the United States. Some organisms are resistant to as many as seven antitubercular drugs, and mortality with these strains is high. Any patient with suspected tuberculosis should be isolated. Areas with the highest prevalence of MDR tuberculosis include Korea and Southeast Asia. MDR tuberculosis has been reported in at least 11 states, with the highest occurrence in New York and New Jersey. People with AIDS are particularly susceptible to MDR tuberculosis.

80. The answer is A. *(Pharmacology; Pulmonary/respiratory)*
Both amantadine and rimantadine act by inhibiting viral uncoating through preventing fusion of the viral envelope and endosome membrane. Essentially, rimantadine is identical to amantadine in efficacy, frequency of administration, and use; however, rimantadine is a more polar compound and is less likely to cause central nervous systenm side effects than amantadine. Adverse effects of amantadine include dizziness, insomnia, and nervousness, and it must be used cautiously in patients with epilepsy or psychosis. There is no evidence that viral resistance to rimantadine is less than with amantadine. Both of these drugs are indicated for prophylaxis of influenza A and treatment of influenza A in patients within 20 hours of the onset of illness.

81. The answer is B. *(Gross anatomy; Cardiovascular)*
The right border of the heart is comprised entirely of the right atrium. The right border of the radiographic heart shadow includes the right atrium and the superior vena cava. A small portion of the inferior vena cava may also be included. The right border of the heart shadow is obliterated with right middle lobe pneumonia because this lobe is in contact with the right border of the heart. The left border of the heart shadow is comprised of the left ventricle, the left auricular appendage, the left pulmonary artery, and the aortic arch.

82. The answer is D. *(Gross anatomy; Cardiovascular)*
The brachiocephalic veins are in the superior mediastinum. They converge to form the superior vena cava. The separation between the superior mediastinum and the posterior mediastinum is an imaginary horizontal plane at the level of the fourth thoracic vertebra. The superior mediastinum contains both brachiocephalic veins and the superior vena cava, the aortic arch and the brachiocephalic, left common carotid and left subclavian arteries, the trachea, esophagus, and part of the thymus. The posterior mediastinum is the region of the thorax below the plane at the fourth thoracic vertebra, between the pleural cavities and behind the pericardial sac.

83. The answer is C. *(Gross anatomy; Pulmonary/respiratory)*
In a tension pneumothorax, the mediastinum shifts from the side with higher pressure toward the side with lower pressure. Because the pleural space with a tension pneumothorax has a higher than normal pressure, the mediastinum shifts toward the normal side. A tension pneumothorax may cause compression of the heart and reduce cardiac filling during diastole.

A pneumothorax may occur without any chest wall injury because of leakage of air from the lung into the pleural space caused by rupture of a bleb on the lung. Without the negative pressure of the pleural space to keep the lung expanded, the recoil of the elastic fibers of the lung wall cause the lung to collapse.

84. The answer is C. *(Gross anatomy; Cardiovascular)*
Accumulation of fluid in the pericardial space increases the pressure around the heart. The presence of a nonelastic fibrous pericardium prevents the pericardial cavity from expanding outward. The increased pressure compresses the heart and thereby reduces the ability of the heart to fill during diastole (cardiac tamponade). This results in a reduced end-diastolic volume and a reduced stroke volume; thus, cardiac output is reduced and blood pressure is reduced. This condition is a medical emergency.

85. The answer is B. *(Pathology; Cardiovascular)*
Ischemic heart disease is least likely to be associated with valvular vegetations. Most cases progress into chronic left heart failure with congestive cardiomyopathy. Marantic vegetations (nonbacterial thrombotic endocarditis) occur in malignancies that are associated with a hypercoagulable state, such as mucinous carcinomas of the colon or pancreas. This syndrome is considered paraneoplastic. Embolization is common and, in rare circumstances, the vegetations can become a nidus for seeding in a septicemia. Systemic lupus erythematosus is associated with warty vegetations on both sides of the valve leaflets in 10% to 20% of cases. This is called Libman-Sacks endocarditis. Ventricular septal defects are the most common congenital heart disease associated with infective endocarditis. Intravenous drug abuse using contaminated needles is associated with acute endocarditis of the tricuspid valve (50%) and aortic valve (50%).

86. The answer is D. *(Pathology; Cardiovascular)*
Right ventricular hypertrophy (RVH) would least likely be secondary to tricuspid stenosis, because stenosis refers to a problem with opening the valve, which reduces the workload of the right ventricle (RV).
 RVH refers to increased thickness of the RV. It is a compensatory reaction against increased pressure resistance (increased afterload) or volume overload (increased preload). Increased resistance is noted in pulmonary hypertension and pulmonic stenosis (part of the tetralogy of Fallot complex). Volume overload occurs in pulmonic or tricuspid valve incompetence or volume overload from left-to-right shunting in congenital heart disease (e.g., ventricular septal defect, atrial septal defect, patent ductus arteriosus). The term cor pulmonale refers to RVH secondary to pulmonary hypertension (PH) of lung origin (e.g., chronic obstructive pulmonary disease) or primary pulmonary vessel disease. It does not include PH from any cardiac disease, such as mitral stenosis or congenital heart disease. Biventricular hypertropy with a heart weighing in excess of 600 g to 1000 g (normal weight is 300 g to 350 g) is called cor bovinum.

87. The answer is B. *(Pathology; Cardiovascular)*
High output failure is least likely to be associated with Addison disease, which is characterized by massive salt wasting and hypotension from mineralocorticoid deficiency. Patients are more likely to experience hypovolemic shock than high output failure. High output failure is the outcome of any condition that increases blood volume, increases positive inotropism (contractility), increases blood flow by decreasing blood viscosity (severe anemia), or decreases peripheral vascular resistance with subsequent increase in venous return to the heart.

88. The answer is B. *(Pathology; Cardiovascular)*
A right ventricular infarct does not produce left heart failure, because blood builds up behind the failed heart. They occur in one-third of patients with inferior myocardial infarctions involving thrombosis of the right coronary artery. It is clinically significant in less than 50% of cases. A right ventricular infarct is recognized by the presence of hypotension in association with preserved left ventricular function. Patients have a normal pulmonary capillary wedge pressure (representing left ventricular end-diastolic pressure), an increased central venous pressure, and a low cardiac output. The return of blood to the damaged right heart is impaired (increased central venous pressure), which reduces the output from the right ventricle to the lungs with subsequent decrease of blood returning to the left heart.
 Severe anemia initially produces a high output failure because of a reduction in blood viscosity and drop in the total peripheral resistance. If not corrected, high output failure can progress into left heart failure.

Myocarditis, most commonly caused by Coxsackievirus B, usually involves the left and right ventricles. Contractility of cardiac muscle is decreased because of inflammation. Congestion cardiomyopathy may occur as a sequela to myocarditis.

Aortic regurgitation, most commonly caused by old rheumatic heart disease, predisposes to volume overload of the left heart, which can progress to left heart failure.

Chronic ischemic heart disease results in the gradual replacement of cardiac tissue by noncontractile scar tissue. Congestive cardiomyopathy involving both ventricles frequently occurs.

89. The answer is D. *(Pathology; Cardiovascular)*
Prinzmetal angina refers to chest pain at rest or pain that awakens patients during their sleep. It is characterized by ST elevation (transmural ischemia) on a stress test. The mechanism for the pain is vasospasm of the coronary vessels, with or without fixed stenosing coronary artery disease. Vasospasm is most likely caused by the release of thromboxane A_2 (TXA_2) from small platelet thrombi in the vessel.

Atherosclerosis develops in areas of damage and turbulence, particularly at vessel bifurcations. In descending order of frequency, it involves the abdominal aorta, coronary arteries, popliteal arteries, descending thoracic aorta, internal carotid arteries, and the vessels in the circle of Willis.

General complications of atherosclerosis include thrombosis leading to infarction and/or embolization and aneurysm formation from weakening of a vessel. Involvement of the coronary arteries results in angina (the most common manifestation), acute myocardial infarction, sudden cardiac death, or chronic ischemic heart disease.

Complications involving the peripheral vascular system include gangrene of the digits and intermittent claudication (calf pain on walking). Leriche syndrome refers to aortoiliac atherosclerosis in a male with impotence caused by involvement of the hypogastric arteries. Claudication when walking, atrophy of the calf muscles, and diminished or absent femoral pulses are also present.

Involvement of vessels in the circle of Willis may result in cerebral infarction, laminar necrosis (loss of neurons in layers 3, 5, and 6 of the cerebral cortex) with cerebral atrophy, and aneurysm formation. Internal carotid artery disease is frequently associated with transient ischemic attacks (neurologic deficits last less than 24 hours), atherosclerotic stroke, or embolic stroke (embolism of atheromatous plaques material). Vertebrobasilar system involvement produces transient ischemic attacks (dizziness, vertigo) or brain stem infarctions.

Atherosclerosis of the renal artery produces renovascular hypertension. Narrowing of the renal artery increases renin release, which increases angiotensin II and aldosterone production. Renal infarction and atrophy are also possible complications.

Involvement of the celiac, superior, and inferior mesenteric system often results in mesenteric angina (abdominal pain 30 minutes after eating), bowel infarction (particularly the small bowel), and ischemic colitis, usually located at the splenic flexure, where the superior and inferior mesenteric arteries overlap.

90. The answer is E. *(Pathology; Cardiovascular)*
Erysipelas is a diffuse cellulitis with a brawny, nonpitting edema associated with group A streptococcal infection. An acute lymphangitis manifested as a 'red streak' often extends from the area of cellulitis to the regional lymph nodes. Recall that lymphatics have an incomplete basement membrane, which predisposes them to infection. Drainage of infected material to regional lymph nodes results in reactive hyperplasia with enlarged, tender lymph nodes. From the lymph nodes, infected material has the potential for passing into the systemic circulation, resulting in septicemia. Fever and throbbing pain in the area of inflammation complete the clinical presentation of cellulitis.

Lymphedema refers to an abnormal interstitial collection of lymphatic fluid caused by congenital disease or blockage of the lymphatics. A primary type of lymphedema presents at birth (congenital lymphedema) or is delayed and first presents in adolescence (lymphedema precox). Milroy disease is a hereditary type of lymphedema caused by faulty development of lymphatics. Turner syndrome is associated with lymphedema of the hands and feet in newborns.

The secondary type of lymphedema is secondary to lymphatic obstruction by tumor (e.g., the peau d'orange appearance in inflammatory carcinoma of the breast), radiation (postradical mastectomy with axillary dissection; most common cause in the United States), infestation by filarial organisms *(Wuchereria bancrofti)* or a localized form of lymphedema in the vulva or scrotum associated with the granulomatous inflammation in patients with lymphogranuloma venereum (a chlamydial disease).

Clinically, lymphedema is nonpitting edema and does not resolve with overnight elevation. This differentiates it from the pitting edema of heart failure or deep venous insufficiency. A complication of longstanding lymphedema is lymphangiosarcoma.

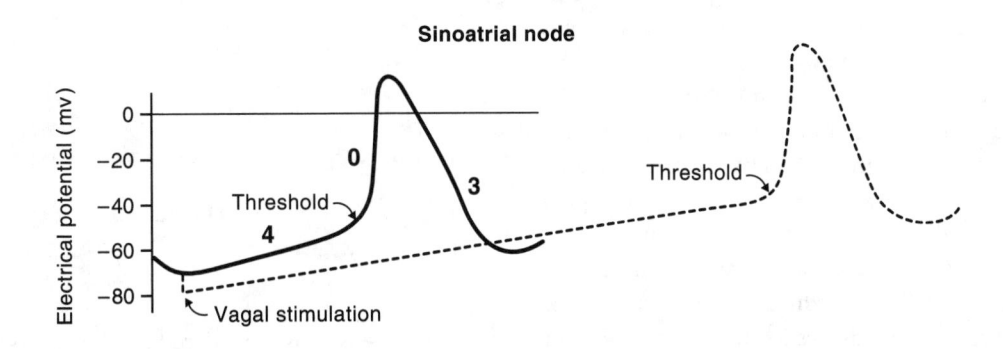

Sinoatrial node

91. The answer is E. *(Physiology; Cardiovascular)*
The electrical potential across a membrane is determined by the ion species to which the membrane is permeable and its concentration gradient across the membrane. For example, if a membrane were permeable only to K+, K+ would diffuse out of the cell in response to its concentration gradient and deposit positive charges on the outside of the membrane. These positive charges would generate an electrical potential across the membrane that exactly balances the concentration gradient; thus, no further net movement for K+ would occur. This potential is called the equilibrium potential for K+ and in cardiac tissue is -95 mV.

If the membrane were only permeable to Na+, Na+ would diffuse into the cell in response to its concentration gradient, deposit positive charges on the inside of the membrane, and the membrane potential would rest at the equilibrium potential for Na+, +60 mV. The equilibrium potential for Ca++ is greater than 200 mV.

The diagram depicts the change in the "resting" potential at the sinoatrial node as it spontaneously depolarizes (phase 4). The low, resting membrane potential (-65 mV) and its upward drift implies that during phase 4, the membrane is not only permeable to K+ but also has a considerable and increasing permeability to Na+. Threshold is reached at approximately -45 mV when a Ca++ action potential is elicited (phase

0). Phase 3 represents the repolarization of the membrane towards the resting potential of -65 mV.

Stimulating the vagus nerve to the heart releases acetylcholine at the sinoatrial node and increases its conductance to K+. This will result in the movement of the membrane potential from -65 mV towards the equilibrium potential for K+, -95 mV [i.e., the potential across the membrane will become more negative, (hyperpolarized) (dashed line)]. The increased conductance to K+ slows the rate of phase 4 depolarization. Thus, threshold is reached later and heart rate is reduced (increased P-P interval). Acetylcholine does not alter threshold.

92. The answer is E. *(Physiology; Pulmonary/respiratory)*
Asthma is characterized by a hyperreactivity of the airways to various stimuli, which leads to widespread narrowing of the bronchioles. During an asthma attack, airway resistance increases, and the expiratory flow rate is reduced. These actions lead to a large decrease in the forced expiratory volume in 1 second (FEV_1) and a smaller fall in the forced vital capacity (FVC), as airways close prematurely. Thus, the FEV_1 to FVC ratio also decreases. There is also a decrease in the forced expiratory flow of 25% to 75%. The early closure of the airways upon a full expiration also

leads to a high residual volume. Bronchodilators can reverse these effects in asthmatics.

93. The answer is E. *(Physiology; Pulmonary/respiratory)*
The partial pressure of a gas in liquid is a reflection of the amount of gas dissolved in a given volume of fluid. As blood passes through the lungs, O_2 comes into equilibrium across the alveolocapillary membrane with respect to the oxygen tension (P_{O_2}). Because the concentration of hemoglobin has no effect on this equilibrium, an anemic individual has a normal arterial P_{O_2}. At any given P_{O_2}, the percent of sites occupied by an O_2 molecule is the same for anemic and normal individuals, but because there are fewer sites in anemia, the O_2 content (vol%) is reduced. The same rate of O_2 consumption by the tissues would result in a lower venous P_{O_2}. Alveolar ventilation depends on the tidal volume, frequency of breathing, and dead space; these are not different from normal in anemia.

94. The answer is A. *(Physiology; Pulmonary/respiratory)*
CO_2 is formed in the tissues and diffuses into the blood. A small fraction dissolves in plasma, where it exerts a partial pressure. The remainder diffuses into the red blood cell, where it can either bind directly to the free NH_2 groups on hemoglobin (a carbamino compound) or become hydrated in the presence of the enzyme, carbonic anhydrase, to form carbonic acid. This acid is in equilibrium with its dissociated form of H^+ and HCO_3^-. This diagram describes the fate of the H^+ and HCO_3^- generated by the CO_2.

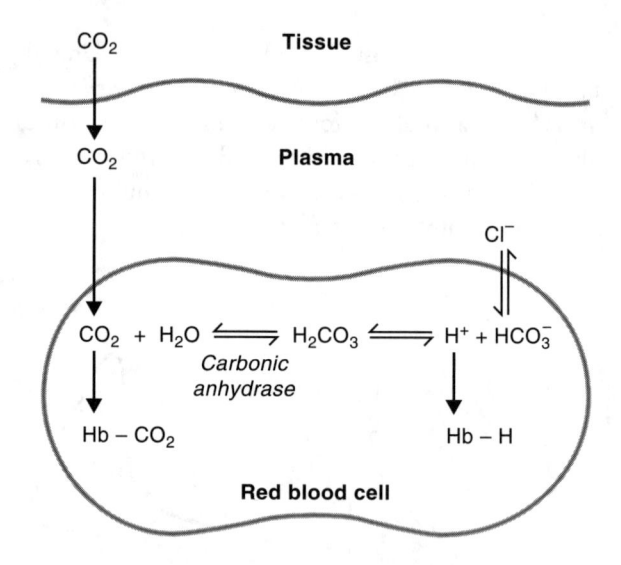

The H^+ is mostly buffered by the free NH_2 groups on hemoglobin. As hemoglobin gives up its O_2, it changes shape and exposes more binding sites. Thus, deoxyhemoglobin is a better buffer than oxyhemoglobin because as the tissues consume more O_2, hemoglobin changes to a form that is better able to buffer the products of metabolism (Haldane effect). The HCO_3^- formed by the hydration of CO_2 diffuses out of the red blood cell (RBC) in exchange for Cl^-; this action is called a chloride shift. In the plasma, most of the CO_2 produced by metabolism is carried in the form of HCO_3^-.

95. The answer is E. *(Pathology; Pulmonary/respiratory)*
The schematic below represents the compounds in the figure.

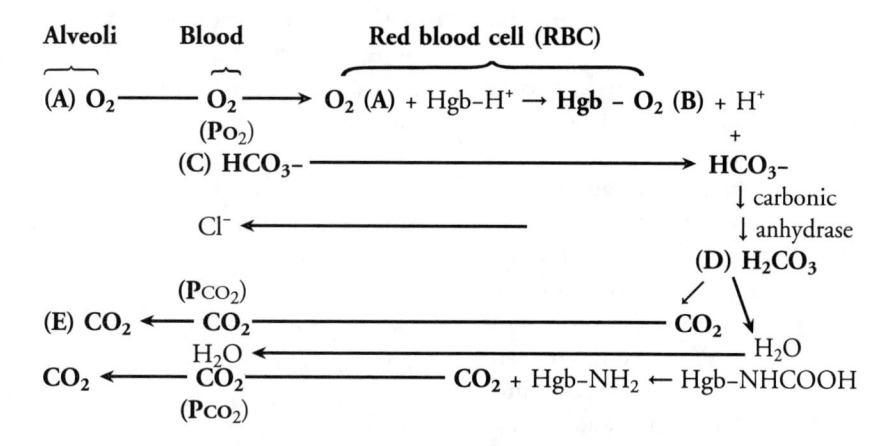

```
                Alveoli      Blood         Red blood cell (RBC)
                (A) O₂ ──────── O₂ ──────→ O₂ (A) + Hgb-H⁺ → Hgb - O₂ (B) + H⁺
                              (Po₂)                                           +
                (C) HCO₃- ─────────────────────────────→ HCO₃-
                                                                    ↓ carbonic
                         Cl⁻ ←─────────────────────────            ↓ anhydrase
                                                               (D) H₂CO₃
                              (Pco₂)                              ↙    ↘
                (E) CO₂ ←──── CO₂ ─────────────────────── CO₂
                              H₂O ←─────────────────────────── H₂O
                    CO₂ ←──── CO₂ ─────────────────────── CO₂ + Hgb-NH₂ ← Hgb-NHCOOH
                              (Pco₂)
```

Gas exchange in the lungs is the reverse of what happens in the tissue. Because the partial pressure of oxygen (P_{O_2}) [A] is higher in the alveoli (P_{O_2} 100 torr) than in the pulmonary capillary (P_{O_2} 40 torr), oxygen (O_2) diffuses into the plasma.

When O_2 enters the red blood cell (RBC), it combines with deoxygenated hemoglobin to form oxyHgb (Hgb-O_2)[B] and H⁺. The H⁺ combines with HCO_3^- (C), the main carrier of carbon dioxide, when entering the red blood cell (RBC) to form H_2CO_3 (D). Chloride (Cl⁻) leaves the RBC to counterbalance the entry of HCO_3^-. Carbonic acid (H_2CO_3) dissociates into CO_2 and H_2O.

Carbon dioxide (CO_2) [E] leaves the RBC and dissolves in the plasma, increasing the P_{CO_2}. Because the partial pressure of CO_2 in the plasma (46 +/- 4 mm Hg) is greater than that in the alveoli (40 mm Hg), CO_2 passes into the alveoli and is eliminated. CO_2 is far more soluble than O_2, and it passes directly through the interface into the alveoli even in the presence of interstitial fluid and fibrosis. If alveolar P_{CO_2} increases (respiratory acidosis), this automatically lowers the alveolar P_{O_2} because the sum of the partial pressures of these gases along with nitrogen must equal 760 mm Hg. Therefore, as alveolar P_{CO_2} increases, there is a proportionate decrease in alveolar P_{O_2}.

96. The answer is D. *(Pharmacology; Pulmonary/respiratory)*

Pyrazinamide is associated with arthralgia, hyperuricemia, and hepatitis. Other antitubercular agents that may cause hepatitis include isoniazid and rifampin. Ethambutol may affect visual acuity and red/green color discrimination. Rifampin induces, rather than inhibits, cytochrome P450 enzymes and may accelerate the metabolism of other drugs, such as warfarin, corticosteroids, and oral contraceptives, thereby reducing their plasma concentration and effectiveness. Kanamycin, an aminoglycoside, may cause otic and renal toxicity.

97. The answer is A. *(Pharmacology; Pulmonary/respiratory)*

Trimethoprim-sulfamethoxazole (TMP-SMX) is often effective against several pathogens that cause upper and lower respiratory tract infections, including *Haemophilus influenzae, Streptococcus pneumoniae,* and *Pneumocystis carinii*. However, TMP-SMX is not active against *Chlamydia, Legionella,* or *Mycoplasma*. Infections caused by these bacteria are effectively treated with a macrolide, such as erythromycin, azithromycin, or clarithromycin. Alternatively, doxycycline may be used to treat chlamydial and mycoplasma infections. The newer macrolides, azithromycin and clarithromycin, offer somewhat greater antimicrobial activity and favorable pharmacokinetic properties, including better bioavailability and longer half-lives, thereby decreasing their dosing frequency and duration of treatment.

98. The answer is E. *(Pharmacology; Pulmonary/respiratory)*

Cefuroxime and other oral cephalosporins may be safely used in pregnant women to treat minor respiratory, urinary tract, and soft tissue infections. Parenteral penicillins, cephalosporins, and vancomycin may be used to treat more serious infections in these patients. Macrolides, such as erythromycin or azithromycin, may be used to treat chlamydial and mycoplasma infections in pregnant patients.

Antimicrobial agents that lack specific toxicity to the mother and fetus include penicillins, cephalosporins (e.g., cefuroxime), and vancomycin. Clarithromycin is the only macrolide that should not be given to pregnant women because it produces fetal toxicity in primates. Fluoroquinolones, such as ciprofloxacin, may damage cartilage and cause arthropathy in children. Sulfonamides may cause neonatal kernicterus and hemolytic anemia when administered to pregnant women. Tetracyclines may stain teeth when administered during pregnancy, and large intravenous doses have caused fatal liver toxicity in pregnant women.

99. The answer is B. *(Pharmacology; Pulmonary/respiratory)*
Clindamycin is not associated with pulmonary reactions, but it causes diarrhea and pseudomembranous colitis. Several drugs can produce pulmonary toxicity that often begins with alveolitis or pneumonitis, then progresses to fibrosis, especially with higher doses or chronic administration. The anticancer drugs (e.g., bleomycin, busulfan, methotrexate, cyclophosphamide) cause diffuse interstitial pulmonary disease with fibrosis. Other drugs causing pulmonary fibrosis include amiodarone (a class III antiarrhythmic agent) and methysergide (a serotonin antagonist/partial agonist used to prevent migraine headaches). Methysergide causes retroperitoneal, pleuropulmonary, and cardiac fibrosis, all of which are potentially fatal.

Nitrofurantoin may cause pulmonary-sensitivity reactions, including interstitial pneumonia with fibrosis, and should be discontinued at the earliest sign of pulmonary reactions.

100. The answer is C. *(Pharmacology; Pulmonary/respiratory)*
Exacerbations of chronic pulmonary infections in patients with cystic fibrosis are usually caused by *Pseudomonas aeruginosa* and are effectively treated with antipseudomonal aminoglycosides (tobramycin and others), penicillins (piperacillin and others), third-generation cephalosporins (ceftazidime and cefoperazone), or fluoroquinolones (ciprofloxacin or ofloxacin). Other third-generation cephalosporins, such as ceftriaxone and cefotaxime, lack sufficient antipseudomonal activity for use in this infection. Fluoroquinolones are not approved for use in children under 14 years old in the United States because of concerns about possible damage to cartilage. However, European studies of large numbers of children receiving fluoroquinolones have failed to reveal cartilage damage or arthropathy.

101-102. The answers are: 101-B, 102-B. *(Pathology; Skin/connective tissue)*
The following is the schematic with the components of the diagram provided.

Tissue	Blood	Red blood cell (RBC)
(A) CO_2	CO_2 P_{CO_2}	CO_2 + Hemoglobin (Hgb) → (B) **Carbamino** compounds (HgbNHCOOH)
		(1) Carbonic anhydrase
CO_2	CO_2	CO_2 + H_2O → (C) H_2CO_3 ↓
	(D) HCO_3 ←	HCO_3^- + H^+
		+
	(E) Cl^- → Cl^-	Hgb-O_2 oxyHgb
	Chloride shift	↓ **Bohr effect**
		H^+-Hgb Deoxy Hgb
		+
O_2 ←	O_2 P_{O_2}	O_2 (F)
	(G) H_2O →	H_2O (RBC swells)

Carbon dioxide (CO_2; A) is carried in blood in three forms: dissolved, bicarbonate (D; most important), and a combination with other proteins as carbamino compounds (B).

CO_2 from tissue enters the plasma, and most of this CO_2 enters the red blood cell (RBC). In the RBC, the majority of CO_2 combines with H_2O via carbonic anhydrase (1) to form H_2CO_3 (C). Recall that carbonic anhydrase is also located in the brush border of the proximal tubules and the cytosol of the distal tubule in the kidneys. H_2CO_3 dissociates into H^+ and HCO_3^- (D), the latter leaving the RBC and entering the plasma. HCO_3 is the primary extracellular buffer for neutralizing excess hydrogen ions. Chloride (Cl^-; E) shifts into the RBC to counterbalance the loss of the negatively charged HCO_3. H^+ combines with oxygenated Hgb (Hgb-O_2), which releases oxygen (O_2). The Bohr effect is when H^+ in the RBC enhances the release of O_2 (F).

O_2 leaves the RBC, enters the plasma, and dissolves, increasing the capillary partial pressure of oxygen (P_{O_2}). As long as the capillary P_{O_2} is higher than the tissue P_{O_2}, O_2 diffuses into the tissue. Therefore, it is the capillary P_{O_2} that is the main force that drives O_2 into tissue. However, the O_2 carried by hemoglobin (Hgb) determines the amount of O_2 that passes into the tissue. When the partial pressure of O_2 in tissue is the same as capillary P_{O_2}, O_2 diffusion ceases.

Carbon monoxide occupies the heme group of Hgb and decreases oxygen saturation in arterial blood (Sa_{O_2}). Methemoglobin also reduces Sa_{O_2} because iron is in the ferric rather than the ferrous state. Respiratory acidosis reduces the Pa_{O_2} and the Sa_{O_2}. Therefore, carbon monoxide poisoning, methemoglobinemia, and respiratory acidosis all reduce both the O_2 content of blood and the total amount of O_2 that is available to tissue.

103-105. The answers are: 103-D, 104-E, 105-D. *(Gross anatomy; Human development and genetics)* The branchial arches are the mesodermal structures in the lateral wall of the pharnyx. The branchial pouches are the endodermal evaginations of the pharynx between the arches. The branchial clefts are the ectodermal invaginations of the skin ectoderm between the arches. The branchial arches give rise to cartilaginous, bony, and muscular structures (first arch: Meckel cartilage, malleus, incus, tensor tympani, masticatory muscles; second arch: part of hyoid bone, stapes, stapedius muscle, facial muscles; third arch: part of hyoid bone, stylohyoid ligament, stylopharyngeus muscle; fourth arch: thyroid cartilage, arytenoid cartilages, laryngeal and pharyngeal muscles; sixth arch: cricoid cartilage, laryngeal muscles). The branchial pouch endoderm gives rise to epithelial structures (first pouch: tympanic cavity mucosa, eustachian tube mucosa; second pouch: palatine tonsils; third pouch: inferior parathyroid gland and thymus; fourth pouch: superior parathyroid gland and ultimobranchial body). The branchial cleft ectoderm gives rise to epithelial structures (first cleft: external ear canal epithelium and external surface of tympanic membrane).

106-111. The answers are: 106-B, 107-C, 108-E, 109-A, 110-D, 111-C. *(Pathology; Cardiovascular)*

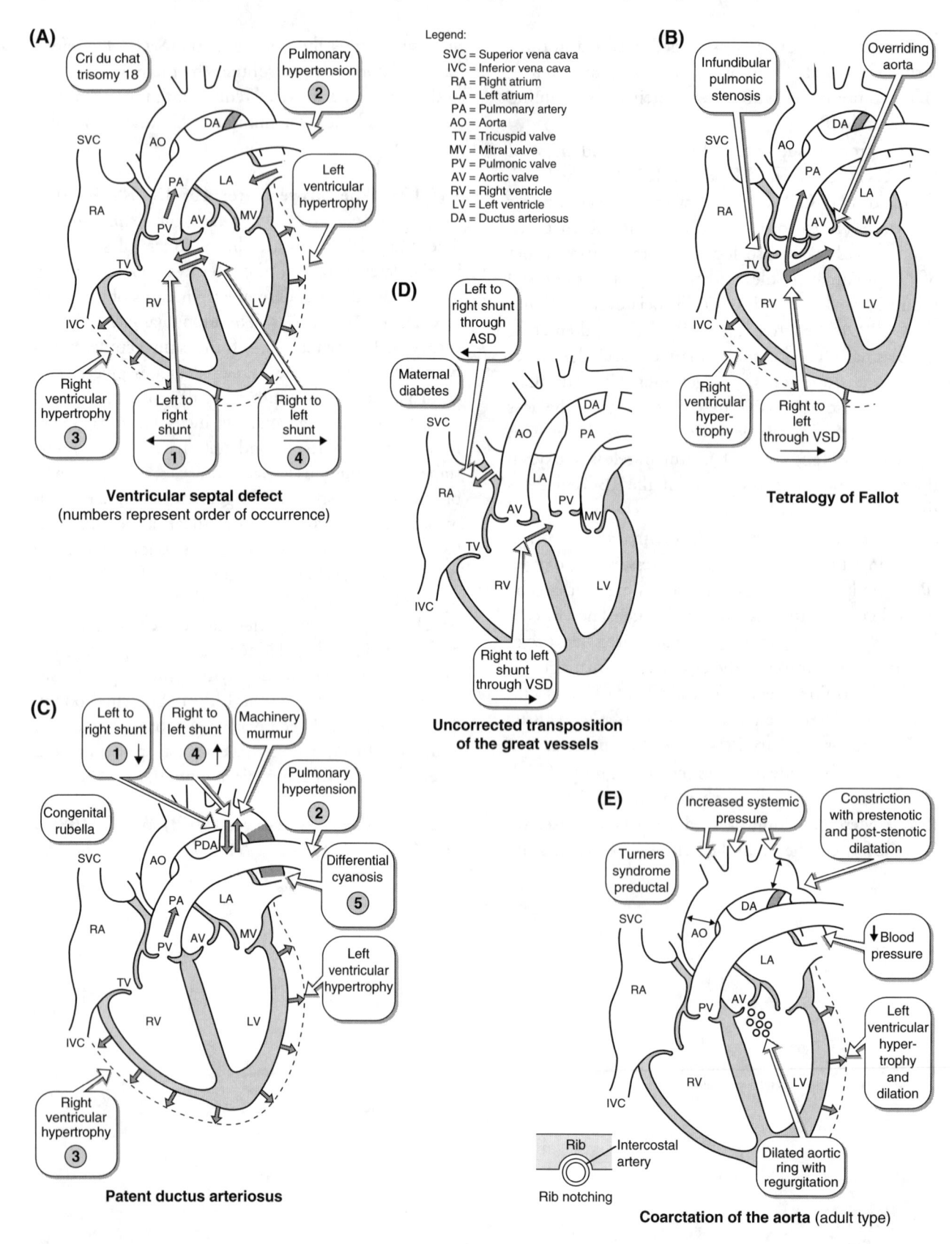

240 Body Systems Review I: Hematopoietic/Lymphoreticular, Respiratory, Cardiovascular

(A) Ventricular septal defect
(numbers represent order of occurrence)

- Cri du chat trisomy 18
- Pulmonary hypertension ②
- Left ventricular hypertrophy
- Right ventricular hypertrophy ③
- Left to right shunt ①
- Right to left shunt ④

Legend:
- SVC = Superior vena cava
- IVC = Inferior vena cava
- RA = Right atrium
- LA = Left atrium
- PA = Pulmonary artery
- AO = Aorta
- TV = Tricuspid valve
- MV = Mitral valve
- PV = Pulmonic valve
- AV = Aortic valve
- RV = Right ventricle
- LV = Left ventricle
- DA = Ductus arteriosus

(B) Tetralogy of Fallot

- Infundibular pulmonic stenosis
- Overriding aorta
- Right ventricular hypertrophy
- Right to left through VSD

(C) Patent ductus arteriosus

- Left to right shunt ① ↓
- Right to left shunt ④ ↑
- Machinery murmur
- Congenital rubella
- Pulmonary hypertension ②
- Differential cyanosis ⑤
- Left ventricular hypertrophy
- Right ventricular hypertrophy ③

(D) Uncorrected transposition of the great vessels

- Left to right shunt through ASD
- Maternal diabetes
- Right to left shunt through VSD

(E) Coarctation of the aorta (adult type)

- Increased systemic pressure
- Constriction with prestenotic and post-stenotic dilatation
- Turners syndrome preductal
- ↓ Blood pressure
- Left ventricular hypertrophy and dilation
- Dilated aortic ring with regurgitation
- Rib / Intercostal artery
- Rib notching

Tetralogy of Fallot (diagram B) is the most common congenital heart disease (CHD) that presents without cyanosis if the degree of pulmonic stenosis is not severe. Recall that tetralogy consists of right ventricular hypertrophy (RVH secondary to pulmonic stenosis), infundibular (below the valve) pulmonic stenosis, a ventricular septal defect (VSD), and an overriding aorta (25% of cases). The magnitude of the right-to-left shunt is primarily dependent on the degree of pulmonic stenosis, which determines how much blood goes into the pulmonary artery (PA) or through the VSD. If the degree of pulmonic stenosis is significant, right-to-left shunting is worse, because most of the unoxygenated blood goes into the left ventricle (LV) through the VSD; however, mild pulmonic stenosis allows most of the blood to go into the PA.

Reversal of a left-to-right shunt that results in differential cyanosis between the upper and lower extremities is characteristic of a patent ductus arteriosus (PDA, diagram C). A PDA refers to a persistence of the ductus arteriosus (DA), which normally connects the PA to the aorta. Delayed closure may occur in patients with prolonged hypoxemia (e.g., respiratory distress syndrome), acidosis, and/or prematurity. There is an association of PDAs with congenital rubella, but the incidence of CHD has markedly decreased with improved adherence to an immunization schedule. Most cases of PDA represent isolated defects rather than one of many cardiac defects. Initially, there is a left-to-right shunt (aorta-PDA-PA), which increases blood flow in the PA and volume overloads the LV, resulting in LV hypertrophy and dilatation. Increased PA blood flow eventually leads to PH and RVH with reversal of the shunt (PA-PDA-aorta). Shunting of unoxygenated blood into the aorta dumps the blood below the subclavian artery; the upper extremities are pink and the lower extremities are cyanotic. This is particularly true if a preductal coarctation is also present and blood has difficulty in getting through the specific portion of the aorta. In this situation, the major vessels above the constriction supply oxygenated blood to the upper body, while a patent PDA with a right-to-left shunt (PA-PDA-aorta) supplies unoxygenated blood to the lower extremities. Whatever the case, there is a pressure gradient between the aorta and PA during both parts of the cardiac cycle, resulting in a classic 'machinery murmur' throughout systole and diastole.

A coarctation of the aorta (diagram E) is the most common CHD associated with diastolic hypertension and an increased incidence of congenital bicuspid aortic valves (75%) and berry aneurysms in the central nervous system. A coarctation refers to a constriction of the aorta, either proximal to the ligamentum arteriosum (preductal, or infantile type) or distal to the ligamentum arteriosum (postductal, or adult type). The preductal type is commonly associated with a hypoplastic left ventricle, congestive heart failure at birth, Turner syndrome, and cyanosis of the lower extremities (see previous discussion under PDA). The postductal type usually develops over a period of time and does not present at birth. Obstruction of blood flow in the aorta is a stimulus for the development of a prominent collateral circulation involving the intercostal arteries (producing rib notching) and connections between the internal mammary arteries and the superficial epigastric arteries. Distal aortic dilatation is caused by the impact of the jet stream of blood through the area of stenosis. Patients have hypertension involving the upper extremities and hypotension in the lower extremities. The diminished flow of blood distally reduces renal artery perfusion, which is a stimulus for renin release with subsequent development of a renovascular type of diastolic hypertension. The increased pressure in the proximal portion of the aorta predisposes to dissecting aortic aneurysms, berry aneurysms in the central nervous system, and dilatation of the aortic ring with subsequent aortic regurgitation. Clinical exam reveals diminished or absent femoral pulses, a blood pressure disparity greater than a 10 mm Hg difference between the arms and legs, and a systolic ejection murmur heard best between the shoulder blades.

The most common CHD associated with Eisenmenger syndrome is a VSD (diagram A). Associations with VSDs include trisomy 13 (Patau syndrome) and 18 (Edward syndrome) and the cri du chat syndrome (deletion of chromosome 5, mental retardation, and crying like a cat). The most common location of the defect is in the membranous portion (90%) of the interventricular septum. A defect less than 0.5 cm is called maladie de Roger disease. These small defects spontaneously close in 50% of cases. In larger defects, the blood initially shunts from the left to right ventricle. These defects are more serious and require surgery before the left-to-right shunt produces PH and RVH with reversal of the shunt (Eisenmenger syndrome, or cyanosis tardive). A VSD is also the most common CHD that is associated with infective endocarditis.

The CHD that is fatal without left-to-right and right-to-left shunting is a transposition of the great vessels (diagram D). There is an increased incidence in the offspring of diabetic mothers. In the usual type of transposition, the aorta empties the right ventricle and the pulmonary artery empties the left ventricle. The right atrium still receives unoxygenated venous blood and the left atrium receives oxygenated blood from the pulmonary veins. For survival to occur, a PDA, atrial septal defect (ASD), or VSD is necessary (an ASD and VSD are most commonly present). Oxygenated blood from the left atrium goes through the ASD and mixes with unoxygenated blood coming from the venous system. The aorta pumps some of this mixture into the systemic circulation. In addition, some of this blood is shunted from the right to left ventricle through the VSD, which is emptied by the PA. The PA brings the blood to the lungs for oxygenation. A corrected transposition is where the atria are also switched; the left atrium receives the venous return and the right atrium receives the oxygenated blood from the pulmonary veins.

112-122. The answers are: 112-D, 113-E, 114-B, 115-H, 116-P, 117-J, 118-K, 119-O, 120-F, 121-C, 122-A. *(Pathology; Hematopoietic/lymphoreticular)*

A high-grade B-cell malignant lymphoma that most often develops in the setting of abnormal immune states, such as systemic lupus erythematosus, Sjogren's syndrome, renal transplants, and AIDS, is a large cell, immunoblastic lymphoma. In general, both B-cell and T-cell immunoblastic lymphomas have very poor prognoses.

A common low-grade, adult malignant lymphoma, which frequently has a leukemic phase and a propensity for metastasis to the bone marrow, is a follicular lymphoma (small-cleaved, poorly differentiated lymphocytic lymphoma). This is one of the most common non-Hodgkin disease malignant lymphomas. This lymphoma grows slowly but is difficult to treat. Like all malignant lymphomas that metastasize to the bone marrow, the neoplastic cells are in juxtaposition to the bony trabecula. The same cells also get into the peripheral blood. Many of these lymphomas have a t14;18 translocation.

Cat-scratch disease produces granulomatous microabscesses, which require silver stains to confirm the presence of the organisms. This disease is self limited and is caused by Gram-negative cell wall defective rods, which are best visualized with Warthin-Starry silver stains of infected tissue (e.g., lymph nodes). Inoculation of the organism by a cat scratch is followed by variable systemic signs and regional lymphadenopathy. The granulomatous reaction is noncaseating and most commonly involves lymph nodes in the axilla and/or neck.

A high-grade T-cell malignant lymphoma with mediastinal involvement that is most commonly seen in children is a lymphoblastic lymphoma. These lymphomas have an increased mitotic rate and convoluted-appearing nuclei resembling the imprint of a chicken foot. They commonly metastasize to the bone marrow and have a leukemic phase. Like the majority of immature T-cell malignancies, lymphoblastic lymphomas are positive for terminal deoxynucleotidyl transferase (Tdt). They have a poor prognosis.

A collection of cells in bone or the upper-respiratory tract that frequently evolves into a malignancy associated with a monoclonal gammopathy is a plasmacytoma. Plasmacytomas of bone usually present as a single, lytic lesion in the spine, pelvis, or ribs. Approximately 50% of cases are associated with an M spike on a serum protein electrophoresis. Most patients eventually develop multiple myeloma. Plasmacytomas located outside the bone are most commonly found in the upper-respiratory tract and rarely evolve into multiple myeloma.

A monoclonal proliferation of lymphoplasmacytoid cells, generalized lymphadenopathy, and the hyperviscosity syndrome is characteristic of Waldenström macroglobulinemia. The lymphoplasmacytoid cells secrete excessive amounts of immunoglobulin M (IgM), which produce the hyperviscosity syndrome in 85% to 95% of all cases. This syndrome is characterized by neurologic problems (e.g., dizziness, headaches), visual disturbances associated with retinal hemorrhages and exudate, Raynaud phenomenon because IgM forms cold-reacting cryoglobulins that gel in cold temperatures, and bleeding abnormalities caused by the effect of IgM interfering with platelet aggregation.

Unlike multiple myeloma, Waldenström macroglobinemia is associated with generalized lymphadenopathy, splenomegaly, and hepatomegaly.

A neoplastic infiltration of the small bowel leading to malabsorption and the presence of a monoclonal spike on a serum protein electrophoresis is IgA heavy chain disease. Heavy chain diseases are characterized by the presence of monoclonal proteins composed of a portion of the heavy chain in the serum and/or urine. Bence Jones protein is not a feature of the disease. The diseases are named according to the heavy chains: α (most common), γ, and μ (least common). α-Heavy chain disease has two clinical patterns, mainly a neoplastic infiltration of the small bowel leading to malabsorption (Mediterranean lymphoma) or a localized disease in the upper-respiratory tract.

A skin disease associated with a dermal infiltrate of mast cells is called urticaria pigmentosa. Mast cell diseases may be localized to the skin where they produce urticaria (hives) on stroking of the skin. In some cases, these diseases are associated with dissemination to multiple organs (systemic mastocytosis). Urticaria pigmentosa is primarily limited to the skin and is diagnosed by demonstrating an increased number of mast cells on toluidine blue or Giemsa staining on a skin biopsy.

Well differentiated lymphocytic lymphomas (small lymphocytic lymphomas) are the most common malignant lymphoma associated with a leukemic phase that is identical with chronic lymphocytic leukemia (CLL). These lymphomas are most commonly found in elderly patients where they produce painless swelling or peripheral lymph nodes often indistinguishable from metastatic CLL. Most observers consider the distinction between the two diseases slight because they both are low-grade, indolent B-cell neoplastic processes. In rare cases, these diseases may develop into a large cell lymphoma, which is called Richter syndrome.

Sinus histiocytosis is a good prognostic sign in ipsilateral axillary lymph nodes because it is thought to represent an immunologic response against the tumor, draining breast cancer.

When pigmented macrophages in lymph nodes are confused with metastatic malignant melanoma, this is called dermatopathic lymphadenopathy. It is associated with nodes draining a chronic dermatitis, usually of the exfoliative type (e.g., psoriasis). The macrophages phagocytize melanin from the exfoliated squamous cells, giving them the appearance of malignant melanoma cells. It is considered to be a type of reactive hyperplasia.

123-125. The answers are: 123-C, 124-D, 125-B. *(Pathology; Tissue biology and associated response to disease)*
Congenital asplenia is associated with malformations of the heart in greater than 80% of cases.

Hyperplastic arteriolosclerosis ("onion skinning") of the arterioles in the spleen is a characteristic finding in systemic lupus erythematosus.

Portal hypertension, most commonly secondary to cirrhosis, results in congestive splenomegaly. The spleen is frequently surfaced by a "sugar-coated" capsule from perisplenitis. Calcium and iron concretions called Gamna-Gandy bodies are present in the collagen tissue.

Felty syndrome is associated with rheumatoid arthritis, splenomegaly, and neutropenia. Hairy cell leukemia is a B-cell leukemia that commonly metastasizes to the red pulp of the spleen. Sago and lardaceous spleens refer to splenic involvement in systemic amyloidosis.

126-127. The answers are: 126-B, 127-A. *(Pathology, Hematology; Hematopoietic/lymphoreticular)*
In clinical practice, it is frequently necessary to evaluate the patient's hemostatic mechanisms for potential problems. The four tests that are useful in screening for hemostasis disorders are the platelet count (PC), bleed time (BT), prothrombin time (PT), and partial thromboplastin time (PTT).

The patient with menorrhagia, whose platelets do not aggregate when exposed to ristocetin, has von Willebrand disease (VWD). In classic VWD, an autosomal dominant disease, the patient lacks the von Willebrand factor (necessary for platelet adhesion), factor VIII antigen (a carrier protein for factor VII coagulant), and factor VIII coagulant (in the intrinsic coagulation pathway). This patient has a prolonged BT because of the defect in platelet adhesion, a normal platelet count, a normal PT because factor VIII is not in the extrinsic pathway, and a prolonged PTT. The PTT is useful in identifying XII, XI, IX, and VIII factor deficiencies. Both the PTT and the PT share the final common pathway factors [X, II (prothrombin), V, I (fibrinogen)] up to the formation of the clot.

A patient with uremia with a hemorrhagic diathesis most likely has a qualitative platelet defect, which would only affect the BT. A toxin in the plasma interferes with platelet factor III, which is the phospholipid substrate on the platelet membrane upon which the coagulation process occurs. Because phospholipid is added to the test tube, the PT and PTT are normal.

128-132. The answers are: 128-A, 129-B, 130-C, 131-D, 132-D. *(Physiology; Cardiovascular)*
The dashed loop in graph A indicates an increase in preload. Moving from the erect to the supine position results in an increase in venous return. The increased end-diastolic volume (preload) results in an increased force of contraction and stroke volume (i.e., stroke volume goes from 70 ml to 90 ml). Note that end-systolic volume does not change.

The dashed loop in graph B indicates an increase in contractility. In the presence of a positive inotropic agent, the ventricle develops a greater force and ejects a larger stroke volume while end-diastolic volume remains the same. The increased stroke volume (from 70 ml to 90 ml) comes from the end-systolic reserve volume (i.e., end-systolic volume is lower).

The dashed loop in graph C is an example of moderate exercise in which both preload and contractility have increased. The left ventricle has a higher end-diastolic volume and a lower end-systolic volume. Stroke volume has increased from 70 ml to 110 ml.

The dashed loop in graph D is an example of reduced ventricular filling, either as a result of a severely stenotic mitral valve that limits inflow into the ventricle or through the use of venodilators, which decrease peripheral venous pressure and thus decrease venous return, end-diastolic volume, and stroke volume.

133-135. The answers are: 133-B, 134-C, 135-D. *(Physiology; Cardiovascular)*
Loop A represents a normal left ventricular pressure–volume relationship.

Loop B represents an instance of increased afterload possibly caused by aortic stenosis. End-diastolic volume and end-systolic volume increase somewhat, but stroke volume remains normal. The left ventricle develops a high systolic pressure. Hypertension produces a similar right shift in the position of the loop.

In loop C, a volume overload results in dilatation of the left ventricle. This is typical in aortic regurgitation, where end-diastolic volume, end-systolic volume, and stroke volume are all exceedingly higher than normal. Ventricular systolic pressure is elevated.

Chronic heart failure, represented by loop D, results in a dilated heart (high end-diastolic volume) that is not capable of developing an adequate pressure or stroke volume.

136-139. The answers are: 136-C, 137-D, 138-A, 139-A. *(Physiology; Pulmonary/respiratory)*
At functional residual capacity (FRC), the intrapleural pressure is negative relative to atmospheric pressure because the elastic recoil of the lung is opposite to the elastic recoil of the chest wall. When a fluid filled "space" is subjected to a distending force, the pressure in that space becomes subatomspheric. The greater the distending force, the more negative the pressure. When the thorax is in the erect position, the lungs are subjected to the pull of gravity. Thus, at the apex of the lungs, in addition to the elastic recoil is the weight of the lungs pulling away from the thorax. At the apex the intrapleural space is subjected to a greater distending force than at the base of the lungs (where in fact, the intrapleural space is being compressed by the weight of the lungs). At the apex of the lungs when the thorax is in the erect position, the intrapleural pressure at FRC (before inspiration begins) is -10 cm H_2O. At the base of the lungs, the intrapleural pressure is -2 cm H_2O. At the end of inspiration, intrapleural pressure falls to a lower number. The only possible choices are either -15 or -30 cm H_2O. Two individuals, one normal and the other with stiffer, less compliant lungs, take the same size breath. Intrapleural pressure must fall to a lower value (-30 cm H_2O) in the individual with fibrotic lungs to enable the lungs to expand to the same volume as the normal person.

Intraalveolar pressure at the apex of the lungs at FRC is 0 cm H_2O (atmospheric). The pressure is negative during inspiration, but returns to 0 cm H_2O at the end of inspiration. This is the case in both normal and fibrotic lungs.

140-143. The answers are: 140-A, 141-D, 142-B, 143-C. *(Pathology; Pulmonary/respiratory)*
Dimorphic means that the fungus is in yeast form in a tissue but in hyphal form in a culture. Dimorphic fungi include *Coccidioides, Histoplasma, Blastomyces, Sporothrix,* and *Paracoccidioides.*

The dimorphic fungus that causes lung disease characterized by multiple granulomas with extensive dystrophic calcification is *Histoplasma capsulatum* (A). Note that the yeast forms are in the cytoplasm of a macrophage. Histoplasmosis, acquired by inhalation of spores, is the most common systemic fungal infection in the Midwest. As with all of the systemic fungi, *Histoplasma capsulatum* is characterized by granulomatous inflammation and the formation of caseous granulomas. This fungus has a dormant phase, similar to tuberculosis, and can reactivate. Other similarities include cavitation, solitary coin lesions in the lung, miliary spread, and phagocytosis by macrophages. Usually, old infection sites (e.g., spleen, liver, lung) contain multiple calcified granulomas. Amphotericin B is the treatment of choice.

The systemic fungus that lives in old tuberculous cavities and is associated with massive hemoptysis, bronchial asthma, and invasive vasculitis is Aspergillus fumigatus (D). Note the corona with the fruiting body in figure D. The narrow angled septae are not depicted in the photograph. Aspergillosis that involves old pulmonary cavities is called an aspergilloma, or fungus ball, and is one of the few causes of massive hemoptysis. As with the *Mucor* and *Candida* species, *Aspergillus* invades vessels and is associated with hemorrhagic infarctions in the lungs, external otitis, and dissemination throughout the body. Amphotericin B is used for invasive and disseminated disease.

The dimorphic fungus that is most commonly seen along the Southeast coast, is predominantly in males, and produces both skin and lung disease is *Blastomyces dermatitidis* (B). Note the broad-based bud of *Blastomyces* in figure B. Blastomycosis produces a mixed suppurative and granulomatous infection. Amphotericin B is the treatment of choice.

The dimorphic fungus that was a prevalent pulmonary pathogen after the recent Los Angeles earthquake was *Coccidioides immitis*, which causes coccidioidomycosis (C). Note the round spherules filled with endospores in figure C. The disease ("valley fever") is acquired by inhaling arthrospores while living or passing through the Southwest or the San Joaquin Valley in California. These arthrospores can also be introduced into the atmosphere by dust caused by an earthquake. Coccidioidomycosis frequently presents with flu-like symptoms and erythema nodosum, which are painful nodules most commonly located on the lower legs. The cavities are thin (similar to an egg shell) and prefer the lower rather than the upper lobes. Amphotericin B is the treatment of choice.

144-157. The answers are: 144-L, 145-G, 146-Y, 147-F, 148-R, 149-D, 150-J, 151-A, 152-H, 153-S, 154-K, 155-I, 156-T, 157-N. *(Pathology; Pulmonary/respiratory)*
The pathogen that commonly produces lung abscesses and is a common secondary invader in the lung in patients with rubeola (regular measles) or influenza is *Staphylococcus aureus*. The high mortality of influenza A epidemics is most frequently the result of secondary bacterial pneumonias, as with pneumonia associated with rubeola.

When the infant passes through the birth canal, the pathogen that is contracted and produces a pneumonia characterized by an abrupt onset of tachypnea, wheezing, hyperaeration, eosinophilia, and a lack of fever is *Chlamydia trachomatis*. Approximately 10% to 20% of infants who pass through an infected endocervical canal will develop this pneumonia. The hyperaeration and wheezing results from an inflammation of the nonrespiratory bronchioles and trapping of air behind these airways. Inclusion conjunctivitis, also caused by *Chlamydia trachomatis*, occurs in 50% of these cases. Chlamydia is the most common cause of ophthalmia neonatorum. Erythromycin is the treatment of choice for both pneumonia and the conjunctivitis.

The systemic pathogen that is often associated with indwelling venous/arterial catheters and immunodeficiency states is *Candida albicans*. In the lungs, this pathogen produces a pneumonia characterized by diffuse nodular infiltrates and evidence of vessel invasion. Metastatic abscesses are most commonly found in the brain and the kidneys. Amphotericin B is the treatment of choice.

The childhood pathogen with a pneumonia associated with characteristic Warthin-Finkeldey multinucleated giant cells is rubeola. Symptoms of rubeola begin to appear after the 7 to 14 day incubation period.

Fever, cough, conjunctivitis, and coryza (excessive nasal mucous production) occur initially, followed by the pathognomonic Koplik spots (red with white centers) in the mouth. Posterior cervical lymphadenopathy and a generalized blanching maculopapular rash (viral vasculitis) appear. Warthin-Finkeldey giant cells are pathognomonic for rubeola and are found in the lungs and lymphoid tissue.

The water-loving pathogen that is most commonly seen in men more than age 40 years who smoke is *Legionella pneumophila*. In the lungs, this pathogen produces a confluent, macrophage dominant bronchopneumonia, with fever, nonproductive cough, hemoptysis, and other systemic symptoms. It is a weak-staining, gram-negative rod that favors water reservoirs and cooling units. Next to *Streptococcus pneumoniae*, it is a common cause of community-acquired pneumonia. However, this organism also produces nosocomial infections, particularly in immunocompromised patients. *Legionella* is best visualized in clinical material by direct immunofluorescence or Dieterle silver staining. Mental confusion, diarrhea, and disturbances in the liver and kidneys are common. *Legionella* infections respond to erythromycin and rifampin.

The respiratory pathogen with a significant mortality in those more than age 55 years, particularly in those who have underlying renal, cardiac, or lung problems, is the influenza virus. In the lungs, this virus produces a severe, exudative pneumonia with a propensity for secondary bacterial invasion. Infection peaks in the late fall and winter and is transmitted by hand-to-hand transfer of infected material and by respiratory droplet infection. Influenza type A, which produces pandemics and epidemics, is considered the most severe form of the disease. Type B also produces epidemics, whereas type C is associated with sporadic disease. Clinically, respiratory disease ranges from a mild cold, to bronchitis, to a severe pneumonia with exudate, similar to a bacterial pneumonia. The vaccine is effective in preventing disease in 70% to 90% of healthy young people, whereas it is 50% effective in older people. The vaccine is egg-based, so there is a danger of anaphylaxis in patients who are allergic to eggs. There is an association in children with Reye syndrome, particularly if the child is taking salicylates. Amantadine is useful in treating influenza if given early in the infection. It inhibits viral uncoating of influenza A (not B).

The respiratory pathogen that is transmitted primarily by inhalation and is commonly contracted by dairy farmers or individuals who work with the birthing process in sheep, cows, and goats is *Coxiella burnetii*. This is the only rickettsial disease unassociated with a vector. *Coxiella burnetii* produces an interstitial pneumonitis associated with high fevers, chest pain, and myalgias. Granulomatous hepatitis with characteristic ring granulomas develops in 50% of patients. It is a rare cause of infective endocarditis. Tetracycline is the treatment of choice.

The respiratory pathogen that is transmitted by direct hand-to-hand transfer (usually by school children) of infected material and by respiratory droplet infection is the rhinovirus. Rhinoviruses are the most common cause of the common cold (25-30%). Because there are at least 100 serotypes of the rhinovirus, the development of a vaccine is unlikely. One unique characteristic is that it is acid labile.

Chlamydia pneumoniae (TWAR) is transmitted by droplet infection, accounts for approximately 10% of community-acquired atypical pneumonias and a smaller percentage of cases of bronchitis, and responds well to tetracycline as with other members of its family. Pharyngitis and hoarseness are also common findings. It can be confused with *Mycoplasma pneumonia*, but the cold agglutinin titer is negative.

Actinomyces israelii, a gram-positive filamentous bacteria, can produce an empyema with a sinus tract draining pus with yellow flecks of material that has a characteristic Gram-stain appearance. Actinomycosis localizes to the cervicofacial area and the thoracoabdominal cavities. A common presentation is a draining sinus in the jaw or neck area after a recent history of extraction of an abscessed tooth. Actinomyces is also the most common infection associated with intrauterine devices. Penicillin G is the treatment of choice.

The pathogen that is commonly contracted in military stations and in crowded situations is *Mycoplasma pneumoniae*. This pathogen is the most common cause of interstitial pneumonitis, or atypical pneumonia ("walking pneumonia"). The pneumonia presents with fever and upper respiratory tract signs of pharyngitis, nasal congestion, otitis media (bullous myringitis), and a nonproductive cough (most common symptom). The chest radiograph reveals segmental interstitial infiltrates. There is also an association with erythema multiforme, which is characterized by targetoid lesions on the skin. Cold agglutinin titers are elevated in 40% to 70% of cases and is not recommended as a screening test. Cultures reveal colonies with a characteristic "fried-egg" appearance. Erythromycin is the treatment of choice.

The incidence of *Chlamydia psittaci*, which is associated with an interstitial pneumonia, has been declining due to the addition of tetracycline to animal feed. Inhalation of the organism by individuals working with psittacine birds (e.g., parrots, parakeets, pigeons, or turkeys) is the usual mode of transmission. The organism produces an interstitial pneumonitis that responds well to erythromycin.

Nocardia asteroides is a strict aerobe most commonly seen in patients with defects in cellular immunity, particularly in the setting of heart transplantation. This pathogen produces microabscesses in the lungs, often with granuloma formation. A characteristic feature, aside from its being a gram-positive filamentous bacteria, is that it is partially acid fast. From a primary site in the lungs, this organism frequently disseminates to the central nervous system and the kidneys. *Nocardia* species are also associated with mycetomas, which are traumatically induced, localized subcutaneous infections that are very resistant to therapy. Pulmonary disease is treated with trimethoprim sulfamethoxazole.

Hemophilus influenzae (HIB), which is recovered in the sputum of patients with cystic fibrosis, is second to *Streptococcus pneumoniae* in causing septicemia in children in the absence of a localized site of infection. HIB also produces septicemia in patients with sickle-cell anemia who have splenic dysfunction. Children have a 30% to 90% carrier rate for this gram-negative coccobacillus, thus the susceptibility for infection in the 1 month to 5 year age bracket. However, the peak incidence of infection is in children under 2 years of age (75%). Rifampin is the drug of choice for eradicating the chronic carrier state. HIB conjugate vaccines given as a primary series in infancy have a high efficacy (> 90%) in preventing disease. The series is usually started at age 2 months. There are no contraindications for administering the vaccine.

158-162. The answers are: 158-D, 159-C, 160-A, 161-E, 162-B. *(Pharmacology; Pharmacodynamic and pharmacokinetic processes)*
Meclizine and cyclizine are often used in preventing motion sickness and for short-term treatment of vertigo because they are less sedating than diphenhydramine and promethazine.

Promethazine also causes antiemetic activity and is available as a suppository or for injection. Chlorpheniramine and brompheniramine, which are less sedating than diphenhydramine and promethazine, are often used alone or in combination with decongestants for allergies and colds. Cyproheptadine has antiserotonin activity and may be useful in treating patients with carcinoid tumor, where it blocks the effects of serotonin and histamine on smooth muscle. Aztemizole, loratadine, and terfenadine are nonsedating antihistamines that do not cross the blood–brain barrier.

Test 4

QUESTIONS

DIRECTIONS:
Each of the numbered items or incomplete statements in this section is followed by answers or by completions of the statement. Select the ONE lettered answer or completion that is BEST in each case.

1. Which of the following structures are attached directly to the free margins of the mitral and tricuspid valves?

   (A) Papillary muscles
   (B) Chordae tendineae
   (C) Pectinate muscles
   (D) Trabeculae carneae
   (E) Semilunar nodules

2. A patient presents with hoarseness and is subsequently diagnosed as having an aneurysm of the aortic arch. The hoarseness is caused by compression of which of the following nerves?

   (A) Left phrenic nerve
   (B) Right vagus nerve
   (C) Left recurrent laryngeal nerve
   (D) Left internal laryngeal nerve
   (E) Left external laryngeal nerve

3. Which of the following statements regarding the thorax is correct?

   (A) In the root of the right lung, the pulmonary veins lie posterior to the pulmonary artery
   (B) The right upper lobar bronchus passes inferior to the right pulmonary artery
   (C) The bronchial arteries typically arise from the segmental branches of the pulmonary arteries
   (D) The bifurcation of the trachea is at approximately the sixth thoracic vertebra
   (E) At the bifurcation of the trachea, the left bronchus diverges from the midline more than the right bronchus

4. A major cause of respiratory distress syndrome (hyaline membrane disease) in premature infants is

   (A) failure of the respiratory diverticulum to develop from the foregut
   (B) failure of the walls of the terminal sacs to become thin
   (C) failure of the splanchnic mesoderm to develop into pulmonary vasculature
   (D) failure of the coelom to develop into the pleural space
   (E) failure of alveolar type II cells to produce sufficient surfactant

5. As a result of trauma, a patient is bleeding into the pericardial cavity. To aspirate this blood and minimize risk to the pleura, a needle should be introduced into

   (A) the right fourth intercostal space
   (B) the left fourth intercostal space
   (C) the suprasternal notch
   (D) the left xiphocostal angle
   (E) the right xiphocostal angle

6. An afebrile 28-year-old woman who is 6 weeks postpartum complains of fatigue and swelling of her feet at the end of the day. A chest radiograph shows generalized cardiomegaly and pulmonary edema. The patient most likely has a

   (A) toxic cardiopathy
   (B) restrictive cardiomyopathy
   (C) primary valvular disease
   (D) congestive cardiomyopathy
   (E) hypertrophic cardiomyopathy

7. Which of the following screening tests for primary prevention of coronary artery disease is recommended in the United States?

(A) Chest radiograph
(B) Electrocardiogram
(C) Sputum cytology
(D) Exercise stress test
(E) Serum cholesterol and high-density lipoproteins

8. Compared with skeletal muscle, cardiac muscle

(A) has less resting tension
(B) is less difficult to stretch
(C) has a greater force of contraction when stretched
(D) has peripherally located nuclei
(E) is repaired by hyperplasia of subjacent muscles

9. Under normal physiologic conditions, changes in stroke volume are most often secondary to changes in

(A) afterload
(B) heart rate
(C) contractility
(D) preload

10. If a person has a heart rate of 70 beats/min, a left ventricular end-diastolic volume (LVEDV) of 100 ml, and an ejection fraction of 0.50, then the cardiac output is

(A) 3.0 liters/min
(B) 3.5 liters/min
(C) 4.0 liters/min
(D) 4.5 liters/min
(E) 5.0 liters/ min

11. A young patient with a rhabdomyoma of the heart most likely has

(A) mental retardation
(B) a trisomy chromosomal abnormality
(C) a single umbilical artery at birth
(D) a primary rhabdomyosarcoma located in the thigh
(E) café au lait spots

12. A 64-year-old man with a prosthetic aortic valve presents with a sudden onset of left upper quadrant abdominal pain, and a friction rub is heard over this area. The pain radiates to his left shoulder. He has fever and a neutrophilic leukocytosis. The pathologic process that is most likely causing the patient's pain is

(A) a left lower lobe pulmonary infarction
(B) thromboembolism to the spleen with infarction
(C) acute pancreatitis
(D) perforation of a peptic ulcer with air under the diaphragm
(E) an acute myocardial infarction with fibrinous pericarditis

13. Two months following an anterior wall myocardial infarction, a 62-year-old man presents with fever; continuous substernal chest pain, including pain when leaning forward in inspriation; and generalized myaglia and arthralgia. The pain is not relieved by nitroglycerin. On physical examination, a substernal, scratchy, three-component sound in atrial systole, ventricular systole, and ventricular diastole is heard. A complete blood cell count shows an absolute neutrophilic leukocytosis. An electrocardiogram exhibits a Q wave similar to that from his previous infarction as well as diffuse ST segment elevation that is concave upward. The etiology of the disorder is most closely associated with

(A) an infection
(B) neoplasia
(C) autoimmune disease
(D) an acute myocardial infarction
(E) a cardiomyopathy

14. Which of the following pathophysiologic sequences of development of right ventricular hypertrophy in a patient with mitral stenosis is correct?

(A) Increased left atrial pressure, increased pulmonary venous pressure, pulmonary hypertension, pulmonary edema, chronic pulmonary congestion
(B) Increased pulmonary venous pressure, increased left atrial pressure, pulmonary edema, chronic pulmonary congestion, pulmonary hypertension
(C) Increased pulmonary venous pressure, pulmonary edema, increased left atrial pressure, pulmonary hypertension, chronic pulmonary congestion
(D) Increased left atrial pressure, increased pulmonary venous pressure, pulmonary edema, chronic pulmonary congestion, pulmonary hypertension
(E) Increased left atrial pressure, pulmonary edema, increased pulmonary venous pressure, chronic pulmonary congestion, pulmonary hypertension

15. The following figure shows pressure–volume loops for the left heart. The normal loop is represented by the *solid lines*, and the abnormal loop is represented by *dashed lines*. The numbers represent events in the cardiac cycle.

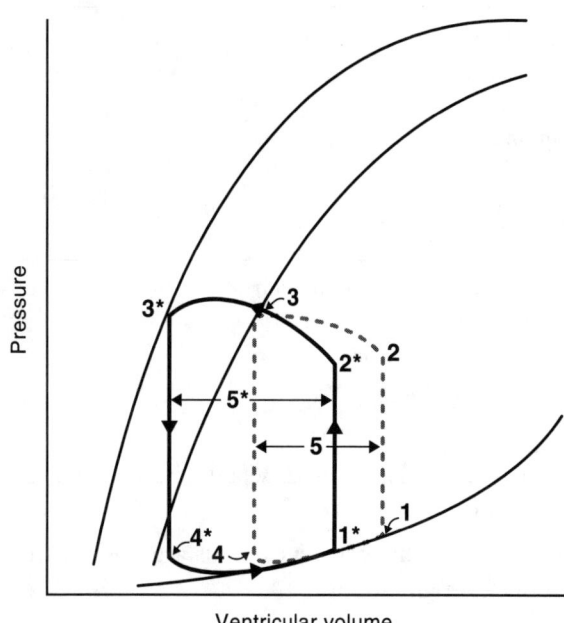

The abnormal pressure–volume loop is most likely associated with

(A) essential hypertension
(B) constrictive pericarditis
(C) congestive heart failure
(D) a patient who is taking digitalis
(E) a patient who has hypovolemia

16. The following table shows the arterial oxygen saturation (SaO₂) concentration for five patients who have a congenital heart defect. Which of the following statements is correct?

	Normal SaO₂ (%)	Patient A	Patient B	Patient C	Patient D	Patient E
Right atrium	75	↔	↑	↔	↔	↑
Right ventricle	75	↔	↑	↔	↑	↑
Pulmonary artery	75	↑	↑	↔	↔	↑
Pulmonary vein	95	↔	↔	↔	↔	↔
Left atrium	95	↔	↔	↔	↔	↔
Left ventricle	95	↔	↔	↓	↔	↓
Aorta	95	↔	↔	↓	↔	↓

SaO₂ = arterial oxygen saturation
↔ = normal
↑ = greater than normal
↓ = less than normal

(A) Patient A has an atrial septal defect with a left-to-right shunt

(B) Patient B has a patent ductus arteriosus with a left-to-right shunt

(C) Patient C has a ventricular septal defect with a left-to-right shunt

(D) Patient D has a tetralogy of Fallot with a right-to-left shunt

(E) Patient E has a complete transposition with an atrial left-to-right shunt and a ventricular right-to-left shunt

17. Which of the following combination of events would most likely increase venous return to the right heart and cardiac output?

	Right atrial pressure ↑	Right atrial pressure ↓	Arteriolar Vasodilation	Arteriolar Vasoconstriction	Venous Vasodilation	Venous Vasoconstriction	Ventricular relaxation ↑	Ventricular relaxation ↓	Blood volume ↑	Blood volume ↓
(A)	X			X	X			X		X
(B)		X	X		X			X	X	
(C)		X	X			X	X		X	
(D)	X		X			X		X		X
(E)		X		X		X	X		X	

18. The following figure shows the effect of contractility on the stroke volume. Which of the following statements is correct?

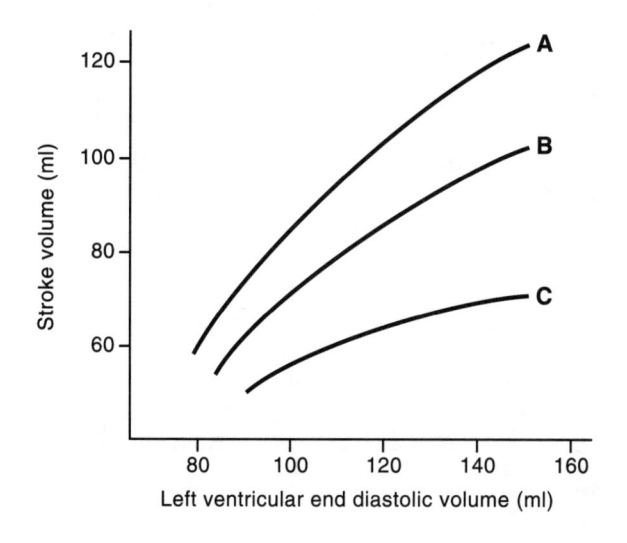

(A) At a left ventricular end-diastolic volume (LVEDV) of 120 ml, the ejection fraction remains the same as contractility increases
(B) Physiologic hypertrophy of the left ventricle in a well-trained athlete is best represented by *curve A*, because it has a higher stroke volume and ejection fraction than *curves B* and *C*
(C) In *curve B*, the ejection fraction at a LVEDV of 120 ml is less than that at 150 ml
(D) *Curve C* most likely represents a patient who is receiving digitalis
(E) An increase in contractility is expected when there is an increase in LVEDV

19. The following diagram shows the effect of heart rate on cardiac output. The change in the cardiac output that occurs when the heart rate is greater than 180 beats/min is most closely associated with

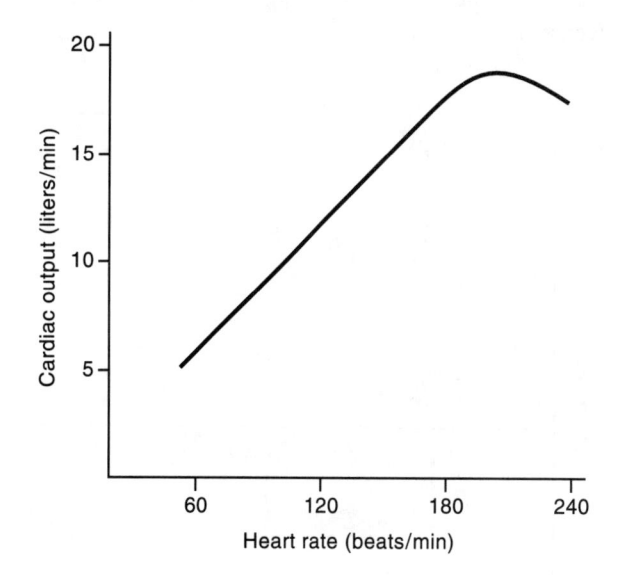

(A) a reduction in the ejection fraction
(B) a decreased myocardial contractility
(C) a decreased stroke volume
(D) myocardial ischemia
(E) a decreased compliance

Body Systems Review I: Hematopoietic/Lymphoreticular, Respiratory, Cardiovascular

Questions 20-21

The following figure shows the pressure–volume loop for the left heart. The normal loop is represented by the *solid line*, and the abnormal loop is represented by the *dotted line*. Letters A through I represent different events in the cardiac cycle.

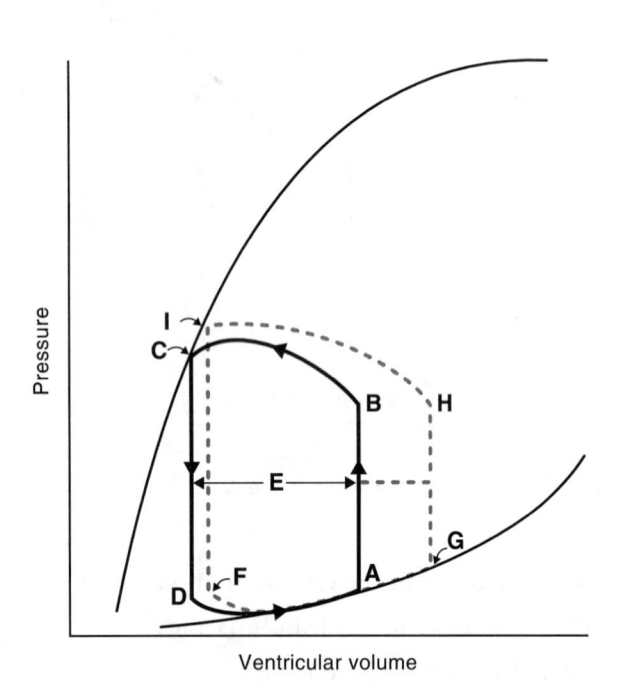

20. Which of the following statements concerning the normal loop (*solid line*) is correct?

(A) Isovolumic contraction occurs from *point A* to *point B* and encompasses ventricular diastole
(B) Isovolumic relaxation occurs from *point C* to *point D* and encompasses ventricular systole
(C) *Point A* represents the preload of the left ventricle, and *point B* represents the afterload of the left ventricle
(D) Atrial contraction would occur between *points C* and *D*
(E) *Point B* is one of the components of the S_1 heart sound, and *point D* is one of the components of the S_2 heart sound

21. Which of the following statements concerning the abnormal loop (*dotted line*) is correct?

(A) The preload is decreased
(B) The stroke volume is increased
(C) Contractility is increased
(D) Afterload is increased
(E) The heart would be a normal size on a chest radiograph

22. This following is an autopsy finding in the heart of an afebrile 45-year-old woman with a history of repeated group A streptococcal sore throats. She presented with an acute onset of a stroke in the distribution of the middle cerebral artery and died within 6 hours of admission to the hospital. Physical examination showed an irregularly irregular pulse, and no *a* wave was present in the jugular venous pulse. Carotid artery bruits were not present bilaterally. A diastolic heart murmur that was best heard on expiration at the apex was noted. The most likely cause of her stroke was

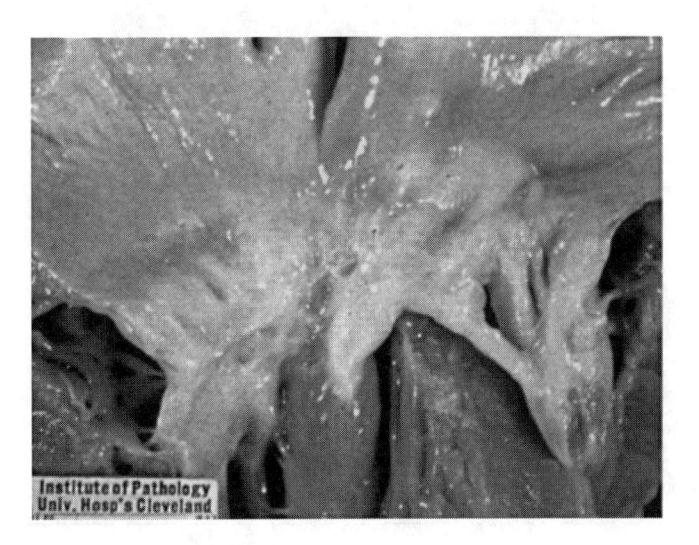

(A) atherosclerosis of the carotid artery
(B) embolization of a bacterial vegetation from the mitral valve
(C) rupture of a berry aneurysm in the central nervous system
(D) mitral valve prolapse
(E) embolism of a clot from the left atrium

23. The following autopsy finding is from the heart of a 30-year-old homeless man. He presented to the emergency room with fever, a sudden onset of hyperdynamic, bounding pulses in all extremities, and bilateral rales in his lungs. A grade III diastolic blowing murmur was noted in the right second intercostal space following the second heart sound. His pulse pressure was widened. Which of the following associations best describes this patient's clinical presentation?

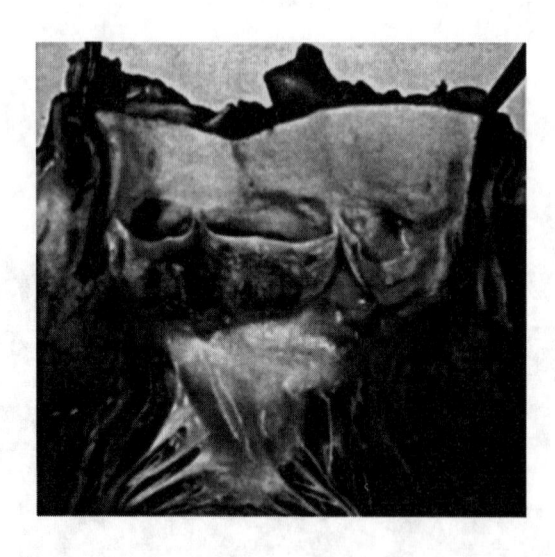

	Probable Past History	Type of Valvular Lesion	Associated Physical Findings
(A)	Rheumatic fever	Rheumatic vegetations	Aortic stenosis
(B)	Rheumatic fever	Subacute endocarditis	Aortic insufficiency
(C)	Rheumatic fever	Acute endocarditis	Aortic insufficiency
(D)	Intravenous drug abuser	Subacute endocarditis	Aortic stenosis
(E)	Intravenous drug abuser	Acute endocarditis	Aortic insufficiency

24. Which one of the following electrocardiographic findings is associated with first-degree heart block?

(A) Widening of the QRS complex
(B) Dropped ventricular beats
(C) A prolonged PR interval
(D) Sinus tachycardia
(E) A dropped P wave

25. Which combination would cause the largest increase in cerebral blood flow?

(A) Arterial pH: 7.4, arterial carbon dioxide tension (Pa_{CO_2}): 40 mm Hg
(B) Arterial pH: 7.3, Pa_{CO_2}: 40 mm Hg
(C) Arterial pH: 7.3, Pa_{CO_2}: 50 mm Hg
(D) Arterial pH: 7.5, Pa_{CO_2}: 40 mm Hg
(E) Arterial pH: 7.5, Pa_{CO_2}: 30 mm Hg

26. The triad of Howell-Jolly bodies, target cells, and reticulocytosis is most compatible with which clinical finding?

(A) Splenectomy
(B) Thrombotic thrombocytopenic purpura (TTP)
(C) Disseminated intravascular coagulation (DIC)
(D) Sickle cell trait
(E) Autoimmune hemolytic anemia

27. A patient with a prolonged prothrombin time (PT) is administered an intramuscular injection of vitamin K, but the PT does not improve. This finding is most consistent with

(A) a malabsorptive state
(B) hemophilia A
(C) von Willebrand's disease
(D) cirrhosis
(E) circulating factor VIII

28. A patient with venous thrombosis of the deep saphenous veins in both legs is placed on warfarin and develops hemorrhagic skin necrosis. What is the most likely cause of the skin necrosis?

(A) Protein C deficiency
(B) Antithrombin III deficiency
(C) Increased platelet aggregation
(D) A hypersensitivity reaction
(E) Disseminated intravascular coagulation (DIC)

29. Full anticoagulation in a patient on warfarin does not occur during the first 24 hours because

(A) warfarin is poorly absorbed in the gastrointestinal tract
(B) coadministration of heparin inhibits warfarin absorption
(C) the previously activated vitamin K–dependent factors have different half-lives
(D) the liver is still synthesizing functional vitamin K–dependent factors
(E) the inactivation of epoxide reductase by warfarin is not complete for at least 3 days

30. Patients with glucose 6-phosphate dehydrogenase (G6PD) deficiency often have an increased susceptibility to infection, which, in turn may precipitate a hemolytic episode. What is the most likely underlying mechanism for the increased risk of infection?

(A) Decreased amount of glutathione in the white blood cells
(B) Decreased amount of reduced nicotinamide adenine dinucleotide (NADPH) in the white blood cells
(C) Defective phagocytosis
(D) Deficiency of myeloperoxidase
(E) Inhibition of adenosine triphosphate (ATP) production by the white blood cells

31. Which one of the following occurs in mature erythrocytes?

(A) Pyruvate, via pyruvate dehydrogenase, is converted into acetyl coenzyme A (acetyl CoA), producing two adenosine triphosphate (ATP) molecules
(B) Glycolysis generates two reduced nicotinamide adenine dinucleotide (NADH) molecules that are shuttled to the oxidative phosphorylation pathway
(C) Reduced nicotinamide adenine dinucleotide phosphate (NADPH) is formed via the pentose phosphate pathway
(D) Lactate is converted into pyruvate for use as a fuel for gluconeogenesis

32. Muscle and red blood cells are able to obtain more glucose for use as fuel by

(A) using ketone bodies as an alternate fuel
(B) directly converting lactate into pyruvate as a substrate for gluconeogenesis
(C) degrading glycogen into glucose
(D) blocking oxidation of fatty acids
(E) liver conversion of lactate into pyruvate as a substrate for gluconeogenesis

33. The normal increase in hemoglobin concentration in the newborn is most closely attributed to

(A) a right shift in the oxygen dissociation curve
(B) left-to-right shunting of blood in the heart
(C) a patent ductus arteriosus
(D) erythropoietin release stimulated by ABO incompatibility
(E) the increased concentration of hemoglobin F in fetal red blood cells

34. The symptoms of anemia associated with pyruvate kinase deficiency are mild primarily because

(A) enhanced oxidative phosphorylation offsets the decreased adenosine triphosphate (ATP) production by glycolysis
(B) no abnormal red blood cells are formed, so there is no extravascular hemolysis by macrophages in the spleen
(C) increased production of 2,3 bisphosphoglycerate increases oxygen delivery to tissue
(D) decreased synthesis of ATP produced by glycolysis results in increased β-oxidation of fatty acids
(E) osmotic fragility is decreased

35. Which of the following conditions is associated with a left shift of the oxygen dissociation curve?

(A) Chronic obstructive pulmonary disease (COPD)
(B) Being employed as a toll taker on the turnpike
(C) Tetralogy of Fallot
(D) A 103°F fever
(E) Diabetic ketoacidosis

36. When compared to the reference intervals for serum iron and serum ferritin in a normal adult man, the reference intervals for serum iron and serum ferritin in a normal adult woman are

	Serum Iron	Serum Ferritin
(A)	Less	Same
(B)	Same	Same
(C)	Less	Less
(D)	Same	Less
(E)	Less	Greater

37. Most of the carbon dioxide produced in metabolism is transported as

(A) carbamate bound to the α amino groups of hemoglobin
(B) dissolved carbon dioxide in plasma
(C) bicarbonate ions
(D) carbonic acid
(E) glutamine

38. A patient with generalized cyanosis and a normal arterial oxygen tension (PaO_2) and a low arterial oxygen saturation (SaO_2) does not respond to oxygen with an increase in SaO_2. The patient most likely has

(A) respiratory acidosis
(B) methemoglobinemia
(C) an intrapulmonary shunt
(D) a right-to left-cardiac shunt
(E) a diffusion abnormality in the lungs

39. A pregnant woman has been a strict vegetarian for 2 years. She is not taking any vitamin supplements or consuming dairy products. What nutrient is she most likely to be deficient in?

(A) Folate
(B) Vitamin B_{12}
(C) Vitamin C
(D) Iron
(E) Zinc

40. Which one of the following is derived from something other than hemoglobin?

(A) Bilirubin
(B) Hemosiderin
(C) Haptoglobin
(D) Ferritin
(E) Hematin

41. Which one of the following, when added to stored blood, has the capacity to improve oxygen delivery to tissue?

(A) Citrate
(B) Dextrose
(C) Inosine
(D) Phosphate
(E) Fructose

42. A patient with δ β-thalassemia would most likely have the greatest increase in hemoglobin

(A) A
(B) A_2
(C) F
(D) H
(E) C

43. Which one of the following statements describes mast cells and basophils?

(A) They are the primary effector cells in type I hypersensitivity reactions
(B) They contain material that forms Charcot-Leyden crystals in sputum
(C) They contain enzymes that neutralize histamine and luekotrienes
(D) Interleukin-5 is important in the synthesis of eosinophils
(E) They kill parasites via a type II hypersensitivity mechanism

44. Which of the following enzymes is present in the specific granules of neutrophils and serves as a marker for distinguishing a leukemoid reaction from chronic myelogenous leukemia (CML)?

(A) Myeloperoxidase
(B) Lysozyme
(C) Alkaline phosphatase
(D) Elastase
(E) Acid hydrolase

45. Which one of the following antibody associations is correct?

(A) Anti-D antibody— most common cause of hemolytic disease of the newborn
(B) Anti-Kell antibody— most common cause of febrile transfusion reactions
(C) Anti-I antibody— associated with cold autoimmune hemolytic anemia in infectious mononucleosis
(D) Anti-i antibody—associated with cold autoimmune hemolytic anemias in *Mycoplasma pneumoniae* infections
(E) Anti-Lewis antibody— generally harmless; present in blood and body secretions

46. Which one of the following anemias requires a bone marrow examination for confirmation of the diagnosis?

(A) Iron deficiency
(B) Pernicious anemia
(C) Anemia associated with renal failure
(D) Anemia of chronic inflammation
(E) Pyridoxine-responsive sideroblastic anemia

47. α Globulin chain synthesis is controlled by four genes (each chromosome 16 contains two genes). Which of the following parentage combinations would most likely result in deletion of the α-globulin gene (Bart's hemoglobin disease)?

	Mother	Father
(A)	—α/αα	—α/—α
(B)	—α/αα	—α/αα
(C)	—α/αα	—α/αα
(D)	——/αα	αα/αα
(E)	——/αα	——/αα

48. Which one of the following is the most common cause of anemia in malignancy?

(A) Blood loss
(B) Bone marrow metastasis
(C) Anemia of inflammation
(D) Vitamin B_{12} deficiency
(E) Folate deficiency

49. Which one of the following is unlikely to be associated with leukemia?

(A) Hyperuricemia
(B) Enlargement of the liver, spleen, and lymph nodes
(C) Bleeding
(D) Infection
(E) Microcytic anemia

50. The bone marrow biopsy shown below was taken from a patient with pancytopenia.

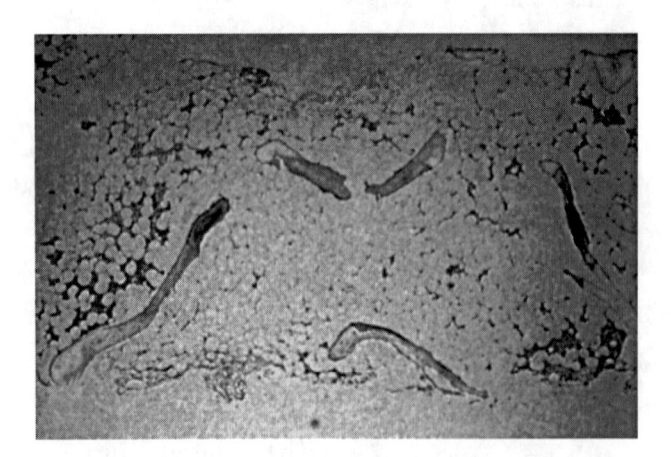

What is the most likely acquired cause of this patient's disease?

(A) Radiation
(B) Infection
(C) Another anemia
(D) Drugs
(E) Industrial agents

51. In a vascular bed that autoregulates its flow, an increase in intravascular pressure will result in

(A) contraction of vascular smooth muscle
(B) a reflex increase in sympathetic nerve activity
(C) relaxation of vascular smooth muscle
(D) a decrease in the vessel wall tension
(E) an axon reflex

52. Which one of the following statements regarding increases in the oxygen demands of the myocardium is true?

(A) Increased demands are met by increasing oxygen extraction
(B) Increased demands can be met only if the perfusion pressure is increased
(C) Increased demands are met by increasing coronary blood flow
(D) Sympathetic nerves mediate the response to increased oxygen demands
(E) Increased oxygen demands are met when local vasoconstriction redistributes blood to the working portions of the heart

53. Hypoxic vasoconstriction is a feature of which one of the following circulations?

(A) Coronary
(B) Cerebral
(C) Cutaneous
(D) Skeletal
(E) Pulmonary

Questions 54-55

The following questions refer to an individual whose mean arterial blood pressure has fallen below 40 mm Hg as a result of severe blood loss.

54. Which one of the following is primarily responsible for maintaining peripheral resistance and blood pressure in this patient?

(A) Arterial baroreceptors
(B) Central chemoreceptors
(C) Stretch receptors in the right atrium
(D) Peripheral chemoreceptors
(E) Stretch receptors in the left atrium

55. Which of the following effects would result from the administration of 100% oxygen to this patient?

(A) The patient's blood pressure would increase
(B) Oxygenation of the peripheral organs would improve
(C) The patient's blood pressure would decrease
(D) Firing of the central chemoreceptors would decrease
(E) Firing of the arterial baroreceptors would increase

56. A Valsalva maneuver is most likely to result in which one of the following events?

(A) A decrease in intrapleural pressure
(B) An initial decrease, followed by a more prolonged increase, in mean arterial blood pressure
(C) An increase in venous return to the right atrium
(D) A decrease in arterial baroreceptor afferent activity
(E) Collapse of the jugular veins

57. An increased heart rate is associated with

(A) a shortened QT interval
(B) a prolonged PR interval
(C) a smaller QRS complex
(D) a longer ST segment
(E) a wider QRS complex

58. A middle-aged woman complains of an uncomfortable feeling in her chest. Her radial pulse is fast and irregular, and an electrocardiogram (EKG) shows a normal QRS complex that is not accompanied by distinct P waves. What is the most likely diagnosis?

(A) Third-degree heart block
(B) Second-degree heart block
(C) First-degree heart block
(D) Atrial fibrillation
(E) Ventricular fibrillation

59. The mean electrical axis (MEA) of a patient is +150. In which one of the following leads would you expect the QRS complex to be inverted?

(A) Lead I
(B) Lead aVR
(C) Lead II
(D) Lead aVF
(E) Lead III

60. A normal β chain gene is acted on by MST II endonuclease:

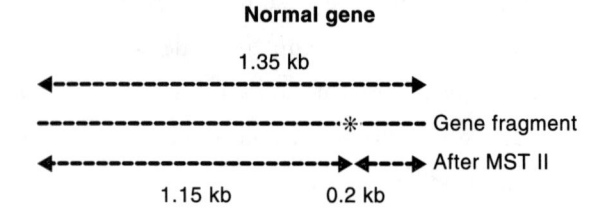

Normal gene

1.35 kb

────────────────────────── * ───── Gene fragment
────────────────────────── ──── After MST II
1.15 kb   0.2 kb

* = MST II cleavage site

In normal genes, the enzyme cleaves the gene on the β chain at the same site where the point mutation involved in sickle cell anemia occurs. What effect would you expect MST II endonuclease to have on the β chain gene in a patient with sickle cell disease?

(A) One 1.35-kb fragment, one 1.15-kb fragment, and one 0.20-kb fragment
(B) Two 1.35-kb fragments
(C) Two 1.15-kb fragments and two 0.20-kb fragments
(D) One 1.35-kb fragment, one 1.15-kb fragment, and two 0.10-kb fragments
(E) One 1.15-kb fragment

61. Which one of the following statements concerning cerebral blood flow is correct?

(A) Cerebral vascular resistance is higher at a mean arterial pressure of 140 mm Hg than at a mean arterial pressure of 60 mm Hg
(B) Changing the mean arterial blood pressure from 70 mm Hg to 110 mm Hg will increase cerebral blood flow in proportion to the pressure change
(C) Cerebral vascular resistance is higher at a mean arterial pressure of 60 mm Hg than at a mean arterial pressure of 140 mm Hg
(D) Flow does not increase at an arterial blood pressure of 130 mm Hg because the cerebrum is contained by the cranium
(E) Cerebral blood flow would be maintained during a hemorrhage that caused the mean arterial blood pressure to drop to 40 mm Hg

62. What type of arrhythmia is shown in the electrocardiogram (EKG) below?

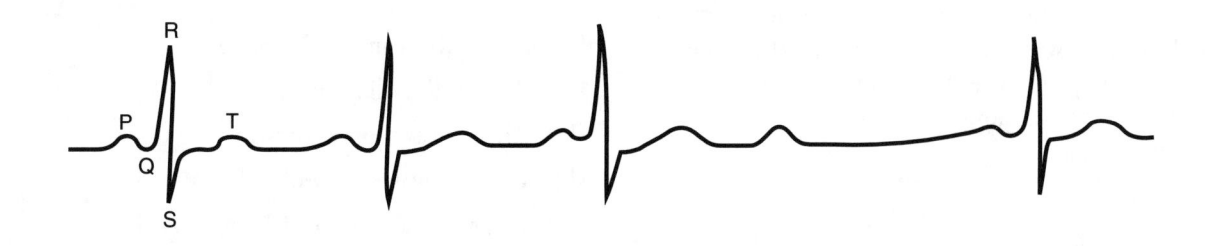

(A) Premature ventricular contraction
(B) Second-degree heart block
(C) Third-degree heart block
(D) Sinus bradycardia
(E) Atrial fibrillation

63. Which interval best approximates the duration of a ventricular myocyte action potential?

(A) PR interval
(B) RR interval
(C) QRS interval
(D) QT interval
(E) PP interval

64. What is the most likely effect of hemorrhage?

(A) An increase in the discharge rate of the carotid baroreceptors
(B) An increase in the firing rate of the vagal efferent fibers to the heart
(C) An increase in total peripheral resistance
(D) A decrease in myocardial contractility
(E) An increase in the rate of fluid filtration into the capillaries

65. The graph below is the result of an experiment concerning the effect of various procedures on the femoral arterial blood pressure. The femoral arterial blood pressure was measured directly. Which one of the following procedures would result in the changes that occur between point X and point Y?

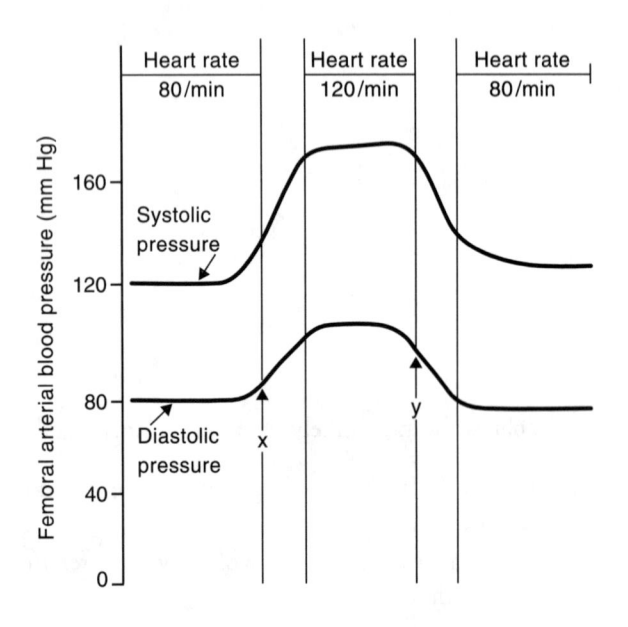

(A) Intravenous injection of an arteriolar vasoconstrictor
(B) Intravenous injection of a venoconstrictor
(C) Occlusion of the femoral artery
(D) Occlusion of both common carotid arteries
(E) Intravenous injection of a venodilator

66. Myocardial ischemia frequently occurs in the subendocardial muscle of patients with aortic regurgitation because

(A) during systole, epicardial blood flow is reduced to a greater extent than subendocardial blood flow
(B) during systole, subendocardial and epicardial blood flow are reduced to the same extent
(C) during diastole, subendocardial blood flow is less than epicardial blood flow
(D) subendocardial blood flow is more affected than epicardial blood flow by the diastolic pressure
(E) increases in myocardial oxygen demand are met by increases in oxygen extraction from blood

67. Which of the following clinical findings would be observed in a patient with aortic valve stenosis?

(A) A rapidly rising radial pulse
(B) A mid-diastolic murmur
(C) An increased pulse pressure
(D) Large R waves in lead I
(E) Large R waves in lead III

68. The graph below depicts the change in lung volume that results when the intrapleural pressure decreases. Curves A, B, and C represent three patients.

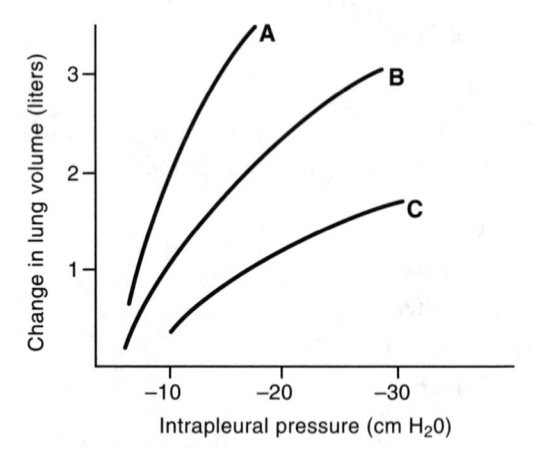

Curve C was most likely obtained from someone with

(A) asthma
(B) emphysema
(C) normal pulmonary mechanics
(D) age-related changes in pulmonary compliance
(E) interstitial fibrosis

69. The relaxation pressure–volume curves shown below were obtained from two individuals, one with normal pulmonary function and the other with fibrotic lung disease. Chest wall compliance is the same for both individuals. What is the functional residual capacity (FRC) in the patient with restrictive disease?

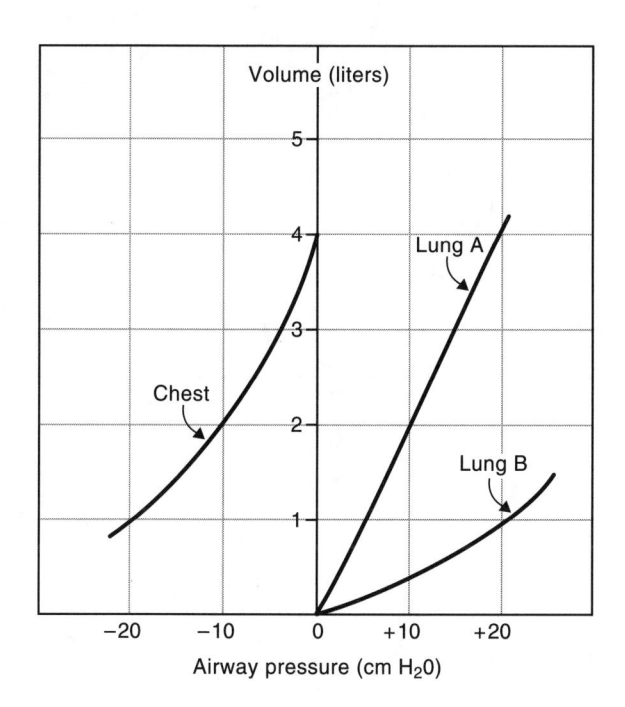

(A) 200 ml
(B) 500 ml
(C) 800 ml
(D) 1000 ml
(E) 2000 ml

70. The flowmeter tracing shown below was obtained from an individual during a forced vital capacity (FVC) maneuver. Which one of the following statements is correct?

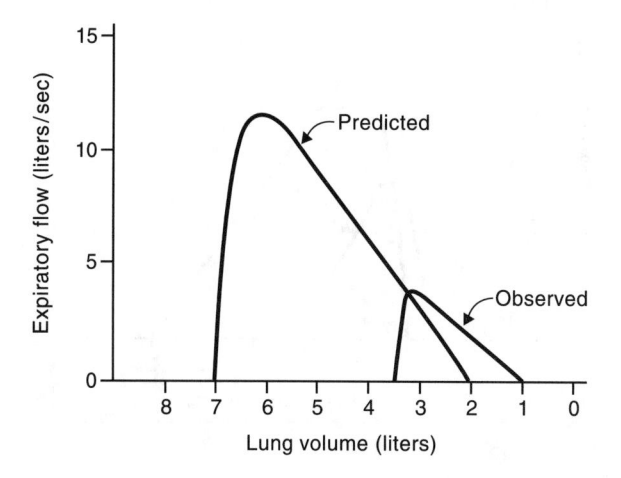

(A) Had the inspiratory capacity (IC) been measured, it would have been less than predicted
(B) Had the functional residual capacity (FRC) been measured, it would have been greater than predicted
(C) At comparable lung volumes, the observed flow is lower than predicted
(D) Had the forced expiratory volume in 1 second (FEV_1/forced vital capacity (FVC)) ratio been measured, it would have been less than predicted
(E) Had the residual volume (RV) been measured, it would have been greater than predicted

71. The flowmeter tracing shown below was obtained during a forced vital capacity (FVC) maneuver. Which one of the following statements is correct?

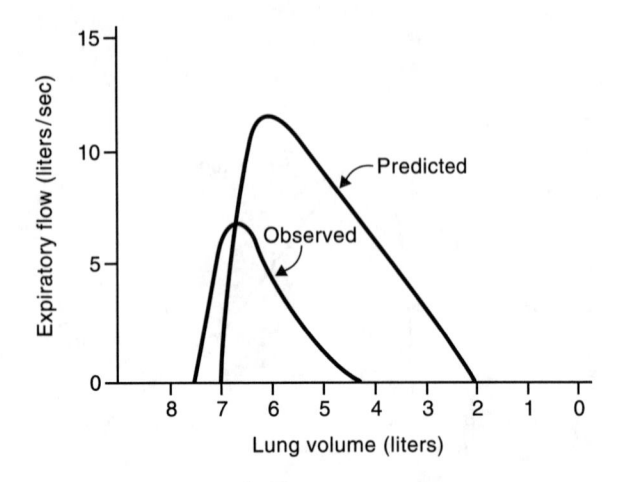

(A) Had the forced expiratory volume in 1 second (FEV$_1$)/forced vital capacity (FVC) ratio been measured, it would have been lower than predicted
(B) Had the FEV$_1$ been measured, it would have been higher than predicted
(C) Had the functional residual capacity (FRC) been measured, it would have been lower than predicted
(D) Had the forced expiratory flow$_{25\%-75\%}$ (FEF$_{25\%-75\%}$) been measured, it would have been greater than normal
(E) Had the maximal inspiratory flow been measured, it would have been reduced to a similar or greater extent than the maximal expiratory flow

72. The spirogram observed in a patient is compared with the spirogram that is predicted for this individual's size, sex, and age:

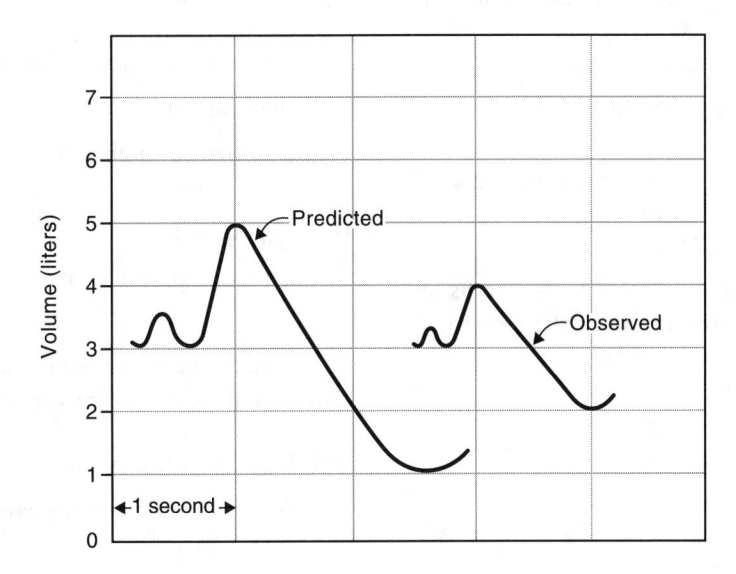

The findings suggest that

(A) the functional residual capacity (FRC) is normal
(B) the residual volume (RV) is higher than normal
(C) there is an obstruction in the trachea
(D) pulmonary compliance may be reduced
(E) the patient would improve following administration of a bronchodilator

73. In an effort to counteract the unpleasant effects of hypoxia at the top of a montain, a beginning climber is given 40% oxygen to breathe. Given the following data,

Total barometric pressure = 427 mm Hg
Alveolar carbon dioxide tension (P_{ACO_2}) = 28 mm Hg
Vapor pressure of water at body temperature = 47 mm Hg
Respiratory exchange ratio = 0.7

what will the climber's alveolar oxygen tension (P_{AO_2}) be following administration of the oxygen?

(A) 40 mm Hg
(B) 60 mm Hg
(C) 80 mm Hg
(D) 112 mm Hg
(E) 152 mm Hg

74. Which one of the following statements about carbon dioxide is correct?

(A) At their respective partial pressures in arterial blood, less carbon dioxide is dissolved than oxygen
(B) The ability of hemoglobin to carry carbon dioxide is greater within the capillaries of metabolically active tissues than in resting tissues
(C) Carbon dioxide produced by metabolism is transported in the blood primarily as a carbamino compound
(D) The reduced ability of hemoglobin to bind carbon dioxide in venous blood demonstrates that oxyhemoglobin is a weaker acid than deoxyhemoglobin
(E) The sigmoidal shape of the carbon dioxide dissociation curve prevents hyperventilation from significantly reducing the carbon dioxide content of the blood

75. A consistent finding in a patient with moderate ventilation/perfusion (V_A/Q_c) mismatching (right-to-left shunting) would be

(A) retention of carbon dioxide
(B) hypocapnia
(C) a somewhat higher than normal arterial oxygen tension (P_{aO_2})
(D) hypoxemia
(E) a normal arterial carbon dioxide tension (P_{aCO_2})

76. Which one of the following disease—hypersensitivity reaction associations is correct?

(A) Goodpasture's syndrome—type III
(B) Allergic bronchopulmonary aspergillosis—type I, type II
(C) Berylliosis—type III
(D) Farmer's lung—type III, type IV
(E) Histoplasmosis—type II, type IV

77. Alveolitis is the precursor lesion for

(A) bronchopneumonia
(B) honeycomb (Hamman-Rich) lung
(C) sarcoidosis
(D) bronchioloalveolar carcinoma
(E) Goodpasture's syndrome

78. Which one of the following is an oxygen-related injury associated with respiratory distress syndrome (RDS) in newborns?

(A) Patent ductus arteriosus
(B) Bronchopulmonary dysplasia
(C) Necrotizing enterocolitis
(D) Intraventricular hemorrhage
(E) Atelectasis

79. A 65-year-old woman with abdominal pain and weight loss has a right-sided hemorrhagic pleural effusion. A pleural tap is performed and cytologic evaluation exhibits clumps of malignant cells, which stain positive for carcinoembryonic antigen (CA)-125. The patient most likely has

(A) metastatic pancreatic carcinoma
(B) a primary mesothelioma
(C) Meig's syndrome
(D) metastatic serous cystadenocarcinoma of the ovary
(E) metastatic breast cancer

80. Which one of the following disorders would most likely produce a pleural effusion that contains less than 3 g/dl of protein and very few inflammatory cells?

(A) Congestive heart failure
(B) Acute pancreatitis
(C) Metastatic breast cancer
(D) Pulmonary infarction
(E) Lobar pneumonia

81. Which one of the following lung tumors would most likely contain neurosecretory granules on electron microscopy?

(A) Bronchioloalveolar carcinoma
(B) Squamous carcinoma
(C) Large cell undifferentiated carcinoma
(D) Bronchial carcinoid
(E) Scar carcinoma

82. Which one of the following is most likely to represent a systemic finding in primary lung cancer rather than a finding related to localized spread of the cancer?

(A) Hoarseness
(B) Pericardial effusion
(C) Endogenous lipoid pneumonia
(D) Dysphagia for solids
(E) Hypercalcemia

83. An executive in an office in Philadelphia has a window air conditioning unit that is a favorite roost for pigeons. She and her secretary both develop lung disease. What is the most likely causative agent?

(A) *Candida albicans*
(B) *Coccidioides immitis*
(C) *Cryptococcus neoformans*
(D) *Chlamydia trachomatis*
(E) *Histoplasma capsulatum*

84. Which of the following lung lesion–etiology associations is correct?

(A) Farmer's lung—inhalation of nitrous oxide
(B) Byssinosis—inhalation of moldy sugar cane
(C) Silo filler's disease—inhalation of thermophilic actinomycetes
(D) Bagassosis—exposure to hemp, cotton, or other fibers
(E) Silicosis—inhalation of sand

85. A 25-year-old man with bilateral gynecomastia and multiple metastatic tumor nodules in both lung fields would most likely have primary cancer of the

(A) bone marrow
(B) testicle
(C) kidney
(D) islet cells of the pancreas
(E) anterior pituitary

86. Lung disease produced by bleomycin and amiodarone is most closely associated with the type of lung disease seen in

(A) progressive systemic sclerosis
(B) bronchial asthma
(C) aspergillosis
(D) Wegener's granulomatosis
(E) pulmonary alveolar proteinosis

87. Select the preferred and alternative drugs for the prevention of malaria caused by the chloroquine-resistant organism *Plasmodium falciparum*.

(A) Mefloquine, doxycycline
(B) Primaquine, pyrimethamine
(C) Quinine, mefloquine
(D) Primaquine, doxycycline
(E) Quinine, sulfadoxine

88. Which drug is used to eradicate the persistent hepatic infection that characterizes the advanced stages of malaria caused by *Plasmodium vivax* or *Plasmodium ovale*?

(A) Mefloquine
(B) Quinine
(C) Chloroquine
(D) Primaquine
(E) Pyrimethamine-sulfadoxine

89. The combination of mechlorethamine, vincristine, prednisone, and procarbazine is used for the treatment of patients with

(A) acute lymphocytic leukemia
(B) chronic lymphocytic leukemia
(C) Hodgkin's disease
(D) acute myelocytic leukemia
(E) chronic myelocytic leukemia

90. What type of toxicity is associated with didanosine (dideoxyinosine) and zalcitabine (dideoxycytidine), two drugs used in the treatment of human immunodeficiency virus (HIV)-positive patients?

(A) Cardiomyopathy
(B) Sensorimotor neuropathy
(C) Hepatocellular hepatitis
(D) Pulmonary fibrosis
(E) Intestinal ulceration

91. Selective toxicity of zidovudine for human immunodeficiency virus (HIV) is based on inhibition of

(A) viral genomic RNA replication
(B) complementary DNA synthesis
(C) incorporation of complementary DNA into the host genome
(D) transcription of viral DNA by host RNA polymerase
(E) translation of viral messenger RNA into structural and regulatory proteins

92. Which one of the following immunosuppressants is often preferred because it causes less bone marrow suppression than other agents?

(A) Cyclophosphamide
(B) Azathioprine
(C) Mercaptopurine
(D) Cyclosporine
(E) Methotrexate

93. Which one of the following is most likely to precipitate digoxin toxicity in a patient if digoxin dosage is not reduced?

(A) Administration of gastric antacids or cholestyramine
(B) Administration of cimetidine
(C) Administration of probenecid
(D) Reduced creatinine clearance
(E) Reduced hepatic blood flow

94. A patient with normal renal function has received a daily maintenance dose of digoxin for 2 weeks. If the dosage is changed, the new steady-state plasma digoxin concentration would be achieved in approximately

(A) 2 days
(B) 1 week
(C) 2 weeks
(D) 4 weeks
(E) 8 weeks

Test 4

DIRECTIONS:
Each of the numbered items or incomplete statements in this section is negatively phrased, as indicated by a capitalized word such as NOT, LEAST, or EXCEPT. Select the ONE lettered answer or completion that is BEST in each case.

95. All of the following statements regarding the right coronary artery are true EXCEPT

(A) it is a branch of the ascending aorta
(B) it anastomoses with the circumflex branch of the left coronary artery
(C) it provides blood supply to the right ventricle
(D) it provides blood supply to the sinoatrial node
(E) it lies in the interventricular sulcus

96. All of the following surface projections of the lungs and pleura are correct EXCEPT

(A) the inferior limit of the lung is at the eighth rib in the midaxillary line
(B) the inferior limit of the parietal pleura is at the eighth rib in the midclavicular line
(C) the inferior limit of the parietal pleura is at the tenth rib in the paravertebral line
(D) the inferior limit of the lung is at the sixth rib in the midclavicular line
(E) the superior limit of the parietal pleura is above the first rib

97. All of the following events typically occur at or shortly after birth EXCEPT

(A) the ductus venosus closes, allowing blood to bypass the liver
(B) the ductus arteriosus closes, ending the shunting of blood from the aorta to the lungs
(C) the foramen ovale closes, ending the shunting of blood from the right to the left atrium
(D) blood pressure in the left atrium increases
(E) blood pressure in the right atrium decreases

98. Infection, either directly or indirectly, is LEAST likely to be primarily involved in the pathogenesis of

(A) a berry aneurysm in the central nervous system
(B) Osler nodes
(C) "spots" in Rocky Mountain spotted fever
(D) myocarditis
(E) pericarditis

99. All of the following would most likely occur in a well-trained athlete EXCEPT

(A) a greater stroke volume during exercise caused by increased preload in the left ventricle
(B) a slower resting heart rate
(C) a greater left ventricular ejection fraction during exercise
(D) less increase in heart rate during exercise
(E) greater contractility of the left ventricle

100. The cardiac index is increased by all of the following factors EXCEPT

(A) stroke volume
(B) venoconstriction
(C) arteriolar vasodilatation
(D) sinus tachycardia
(E) decreased preload in the left ventricle

101. A 16-year-old girl with a history of repeated sore throats presents with fever, migratory polyarthritis, and a grade III pansystolic murmur that is heard best at the apex. The murmur radiates into the axilla. She has wet inspiratory rales, neck vein distention, and a positive hepatojugular reflux. Based on these findings, the patient LEAST likely has

(A) a history of exudative tonsillitis in the recent past
(B) a positive blood culture for group A streptococcus
(C) an increased antistreptolysin O titer
(D) small sterile vegetations on the lines of closure of the mitral valve
(E) left- and right-sided heart failure

102. All of the following statements about fetal circulation are true EXCEPT

(A) a single umbilical vein brings oxygenated blood from the placenta to the fetal heart
(B) oxygenated blood from the right atrium enters primarily the pulmonary artery, which directs most of the blood through the lungs into the left atrium
(C) the ductus arteriosus is kept open by hypoxemia and prostaglandin E_2 derived from the placenta
(D) following the newborn's first breath, the rising partial pressure of oxygen (PaO_2) lowers the pulmonary artery resistance, thus blood flows through the lungs, returns to the left atrium, and functionally closes the foramen ovale
(E) indomethacin is commonly used to close the ductus arteriosus

103. In constrictive pericarditis encircling the entire heart, which of the following is LEAST likely to occur?

(A) A decrease in stroke volume
(B) Reduced filling of the ventricle during both early and late diastole
(C) A pericardial (diastolic) knock
(D) A decrease in blood pressure during inspiration
(E) Neck vein distention on inspiration

Questions 104-105

The following figure illustrates the cardiac cycle.

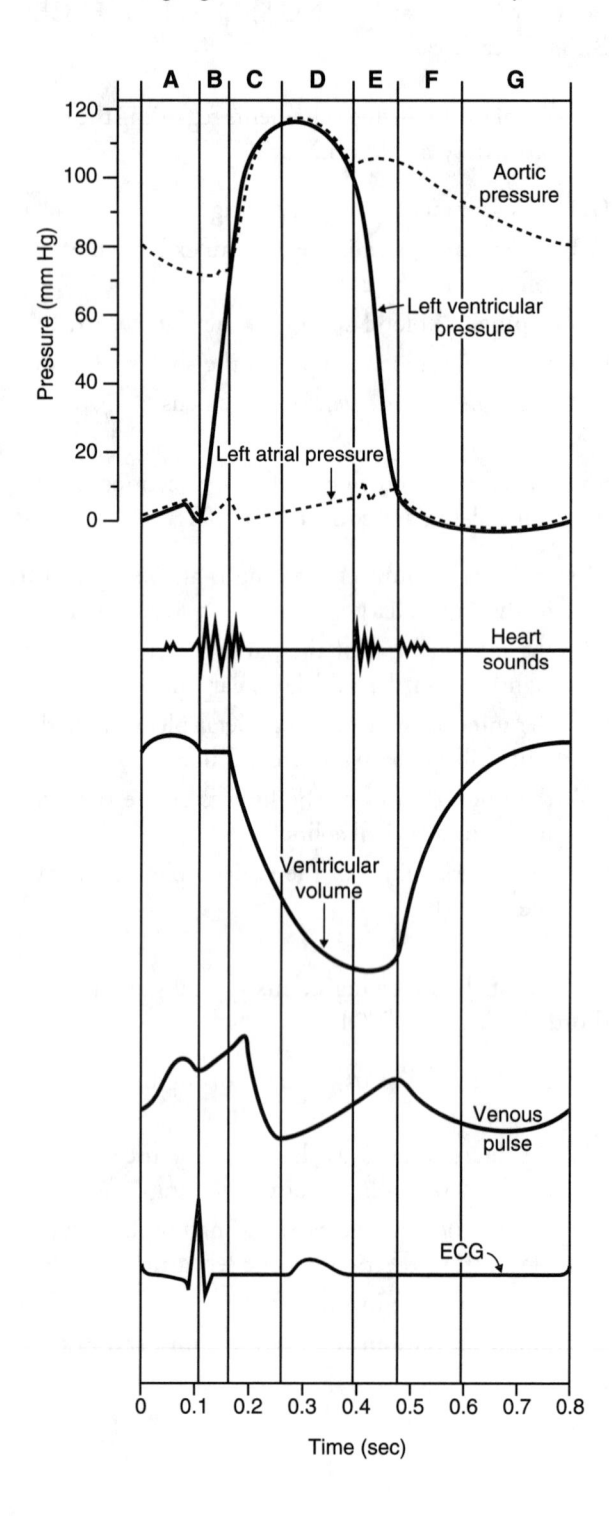

104. Which of the following statements about the cardiac cycle is NOT correct?

(A) Systole occurs from the beginning of *section B* to the end of *section D*

(B) Diastole occurs from the beginning of *section E* to the end of *section A*

(C) In *section A*, the upward deflection of the wave in the jugular venous pulse is caused by closure of the tricuspid valve

(D) In *section B*, the heart sound is an S_1 and is caused by closure of the mitral valve followed closely by closure of the tricuspid valve

(E) In *section F*, the increase in ventricular volume is associated with the descent of the wave in the jugular venous pulse

105. Which of the following statements about abnormalities associated with the cardiac cycle is NOT correct?

(A) In *section A*, the *a* wave in the left atrial pressure curve has a greater amplitude in mitral stenosis

(B) In aortic stenosis, the aortic pressure in the ejection phase of ventricular systole increases to the same degree as does the left ventricular pressure

(C) In mitral insufficiency, the *v* wave in the left atrial pressure curve is accentuated

(D) In aortic insufficiency, there is a greater-than-normal increase in ventricular volume from the beginning of *section E* to the end of *section A*

(E) In cardiac tamponade, there is no descent in the jugular venous pulse wave in *section F*, because the ventricles cannot fill properly during early and late diastole

106. The following figure of a ventricular function curve shows the relationships between changes in stroke volume and changes in the left ventricular end-diastolic volume. All of the following statements are true EXCEPT

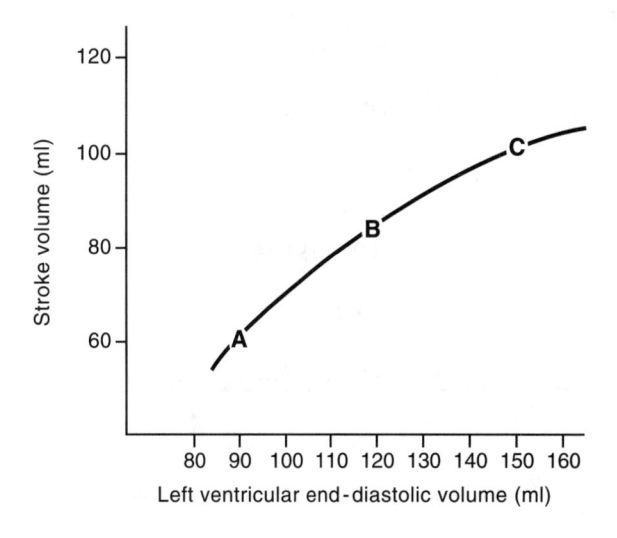

(A) *point A* most likely represents a patient in hypovolemic shock caused by blood loss

(B) *point B* most likely represents a healthy person in the standing position

(C) *point C* most likely represents a patient who is lying down or who has been infused with an excessive amount of isotonic saline

(D) the ejection fractions are different for *points A, B* and *C*, indicating the role that contractility plays in increasing stroke volume

(E) the ventricular function curve shows that an increase in preload increases stroke volume

Questions 107-108

The following series of biochemical reactions occurs in the developing red blood cell. The letters represent substrates and the numbers represent enzymes.

$$\text{(A)} + \text{Glycine}$$
$$\downarrow \underline{1}$$
$$\delta - \text{Aminolevulinic acid}$$
$$\downarrow \underline{2}$$
$$\text{Porphobilinogen}$$
$$\downarrow \underline{3}$$
$$\text{Uroporphyrinogen III}$$
$$\downarrow \underline{4}$$
$$\text{Coproporphyrinogen III}$$
$$\downarrow$$
$$\text{(B)} + \text{Iron}$$
$$\downarrow \underline{5}$$
$$\text{(C)} + \text{(D)} \rightarrow \text{Hemoglobin}$$

107. Which one of the following statements is NOT correct?

(A) Compound A is a water-soluble vitamin
(B) δ-Aminolevulinic acid is increased in the urine in patients with lead poisoning
(C) Compound B is increased in patients with iron deficiency, lead poisoning, and anemia of chronic inflammation
(D) Compoud C exerts negative feedback on enzyme 1
(E) Compound D is decreased in patients with thalassemia syndromes

108. Which one of the following statements is NOT correct?

(A) Enzymes 1, 2, and 5 are inhibited by lead
(B) Enzyme 1 is enhanced in acute intermittent porphyria
(C) Enzyme 3 is deficient in acute intermittent porphyria
(D) Enzyme 4 is deficient in porphyria cutanea tarda
(E) Consumption of compound C during drug metabolism can precipitate an attack of acute intermittent porphyria

109. Which one of the following statements about mast cells is NOT true?

(A) The granules contain biogenic amines
(B) The granules contain chemotactic factors
(C) The mast cells are activated by anaphylatoxins, immunoglobulin E (IgE), and physical stimuli
(D) Radioimmunosorbent and radioallergosorbent testing detect the late phase reaction
(E) Interleukins 3 and 4 are important in mast cell synthesis

110. All of the following statements accurately describe platelets EXCEPT

(A) They contain human leukocyte antigen (HLA) and platelet antigen-1 (PLA-1) antigens
(B) They contain histamine, adenosine diphosphate (ADP), and calcium
(C) They lack ABO and Rh antigens
(D) They synthesize thromboxane A_2
(E) They contain glycoprotein IIb and glycoprotein IIIa (GpIIb and GpIIIa) for binding fibrinogen and glycoprotein Ib (GpIb) for binding von Willebrand's factor

111. All of the following statements regarding endothelial cells are correct EXCEPT

(A) They contain nitric oxide
(B) They contain prostacyclin (PGI_2)
(C) They contain tissue plasminogen activator, von Willebrand's factor, and clotting factors
(D) Weibel-Palade bodies are noted on electron microscopy
(E) They are positive for S100 antigen in tissues

112. All of the following statements regarding monocytes are true EXCEPT

(A) They act as scavenger cells in the atherosclerotic process
(B) They process antigens for presentation to B and T lymphocytes
(C) They kill microbes via the oxygen-dependent myeloperoxidase system
(D) They release interleukin-2 (IL-2) to enhance CD4 T helper cell activity
(E) They serve as reservoirs for human immunodeficiency virus (HIV)

113. Which one of the following blood components is LEAST likely to pose a hepatitis risk when infused into a patient?

(A) Hepatitis B immune globulin
(B) Cryoprecipitate
(C) Packed red blood cells
(D) Platelet concentrates
(E) Fresh frozen plasma

114. A 25-year-old woman is involved in a head-on collision just outside the hospital. In the emergency room, the initial examination reveals a few minor scratches and point tenderness over the tenth rib on her left side. The patient's supine blood pressure is 120/80 mm Hg and her pulse rate is 100 beats/minute. When the patient sits up, her blood pressure drops to 80/60 mm Hg and her pulse rate increases to 130 beats/minute. Prior to the intravenous infusion of normal saline, the patient's hemoglobin is 13.5 gms/dl with a hematocrit of 41%. Which of the following statements is NOT correct?

(A) The blood pressure drop is consistent with hypovolemia
(B) The hemoglobin and hematocrit do not accurately reflect the patient's current hematologic status
(C) The hemoglobin and hematocrit should remain the same following administration of normal saline or Ringer's lactate solution
(D) The most likely diagnosis is splenic rupture
(E) Sinus tachycardia after sitting up is a compensatory reaction for reduced cardiac output

115. Which one of the following factors is LEAST likely to predispose to leukemia?

(A) Fanconi's syndrome
(B) Wiskott-Aldrich syndrome
(C) Increased exposure to ultraviolet light
(D) Alkylating agents
(E) Bloom's syndrome

116. Autoregulation of blood flow occurs in all of the following circulations EXCEPT the

(A) pulmonary circulation
(B) renal circulation
(C) cerebral circulation
(D) coronary circulation

117. All of the following would predispose to the development of a reentrant signal EXCEPT

(A) cardiac dilatation resulting from aortic valve insufficiency
(B) reduced plasma potassium levels
(C) epinephrine
(D) ischemia of the Purkinje system
(E) electrical pacing of the heart at a high frequency

118. The flowmeter tracing below was obtained from a normal individual during a forced vital capacity maneuver. Which one of the following statements is NOT correct?

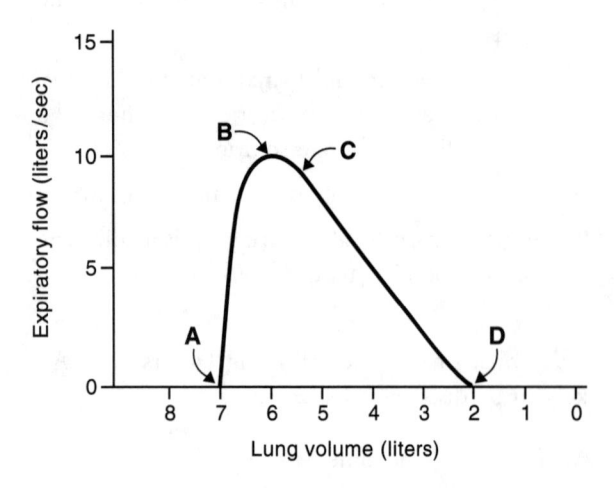

(A) Point A represents the total lung capacity (TLC)
(B) Point D represents the residual volume (RV)
(C) The expiratory flow rate from point A to point B depends on the effort made by the patient
(D) The expiratory flow rate from point C to point D is independent of the effort made by the patient
(E) If the volume at point A was 9 liters and the volume at point D was 3 liters, the patient would most likely have restrictive lung disease

119. All of the following statements regarding pulmonary surfactant are true EXCEPT

(A) in its absence, the change in intrapleural pressure required to achieve a given tidal volume decreases
(B) in its absence, the functional residual capacity (FRC) is reduced
(C) maternal diabetes is a risk factor for respiratory distress syndrome (RDS) in the newborn
(D) it is produced by the alveolar type II cells
(E) in its absence, progressive atelectasis will occur

120. Reduced lung compliance is LEAST likely to be associated with which one of the following conditions?

(A) Mesothelioma
(B) Thoracic kyphoscoliosis
(C) Bronchial asthma
(D) Sarcoidosis
(E) Aging

121. All of the following drugs are useful in the treatment of acute lymphocytic leukemia (all) EXCEPT

(A) Prednisone
(B) Vincristine
(C) Chlorambucil
(D) Methotrexate
(E) Mercaptopurine

122. Which one of the following treatments is LEAST likely to cause acute myelogenous leukemia in a patient who has been treated for another malignancy?

(A) Procarbazine therapy
(B) Mechlorethamine therapy
(C) Cyclophosphamide therapy
(D) Tamoxifen therapy
(E) Ionizing radiation

123. In a patient with human immunodeficiency virus (HIV) infection, all of the following may result from treatment with zidovudine EXCEPT

(A) decreased incidence of opportunistic infections
(B) decreased anemia and thrombocytopenia
(C) increased CD4 lymphocyte count
(D) improved neurologic and mental symptoms
(E) prolonged life

124. All of the following immunomodulating drugs inhibit the maturation and proliferation of lymphocytes EXCEPT

(A) cyclosporine
(B) azathioprine
(C) prednisone
(D) levamisole
(E) cyclophosphamide

125. In a patient with low-output congestive heart failure, the administration of digoxin may increase all of the following EXCEPT

(A) cardiac output
(B) stroke volume
(C) ejection fraction
(D) venous pressure
(E) cardiac contractility

126. All of the following drugs may increase cardiac output in the patient with congestive heart failure EXCEPT

(A) amrinone
(B) digoxin
(C) furosemide
(D) captopril
(E) nitroprusside sodium

127. The effects of digoxin on cardiac muscle include all of the following EXCEPT

(A) increased intracellular calcium
(B) inhibition of sodium–potassium–adenosine triphosphatase (ATPase)
(C) increased cardiac contractility
(D) increased peak systolic tension
(E) increased duration of systole

128. The electrophysiologic and electrocardiographic effects of digoxin include all of the following EXCEPT

(A) decreased atrioventricular conduction velocity
(B) increased atrioventricular refractory period
(C) prolongation of the PR interval
(D) widened QRS complex
(E) shortened QT interval

129. All of the following may be useful in the treatment of cardiac arrhythmias caused by digoxin toxicity EXCEPT

(A) administration of potassium
(B) administration of magnesium
(C) administration of calcium
(D) administration of lidocaine
(E) administration of digoxin antibody

DIRECTIONS:

Each set of matching questions in this section consists of a list of four to twenty-six lettered options (some of which may be in figures) followed by several numbered items. For each numbered item, select the ONE lettered option that is most closely associated with it. To avoid spending too much time on matching sets with a large number of options, it is generally advisable to begin each set by reading the list of options. Then for each item in the set, try to generate the correct answer and locate it in the option list, rather than evaluating each option individually. Each lettered option may be selected once, more than once, or not at all.

Questions 130-132

For each factor listed below, select the effect on preload, afterload, or both that is most likely associated with it.

(A) Increase in preload
(B) Increase in afterload
(C) Increase in preload and afterload
(D) Increase in preload and decrease in afterload
(E) Decrease in preload and afterload
(F) Decrease in preload
(G) Decrease in afterload
(H) No change in preload or afterload

130. Venoconstriction

131. Restricted salt and water intake in a patient with congestive heart failure

132. Increased contractility

Questions 133-134

Match each cell with the stain that best identifies it.

(A) Periodic acid—Schiff (PAS) stain
(B) Specific esterase stain
(C) Tartrate-resistant acid phosphatase stain
(D) Nonspecific esterase stain
(E) Leukocyte alkaline phosphatase stain
(F) Platelet peroxidase stain

133. Monocytes in acute monocytic leukemia

134. Neoplastic erythroid precursor cells in Diguglielmo's erythroleukemia, lymphoblasts in acute lymphoblastic leukemia, and Sezary cells

Questions 135-137

For each description, choose the appropriate factor.

(A) Hemoglobin concentration
(B) Arterial oxygen tension (Pa_{O_2})
(C) Hemoglobin concentration and the Pa_{O_2}
(D) Arterial carbon dioxide tension (Pa_{CO_2})

135. Determines the total oxygen content of arterial blood

136. Determines the percent saturation of hemoglobin

137. Reduced in anemia

Questions 138-140

The mediastinum compartments are represented by circled letters in the diagram below. Match each clinical description with the appropriate mediastinum compartment.

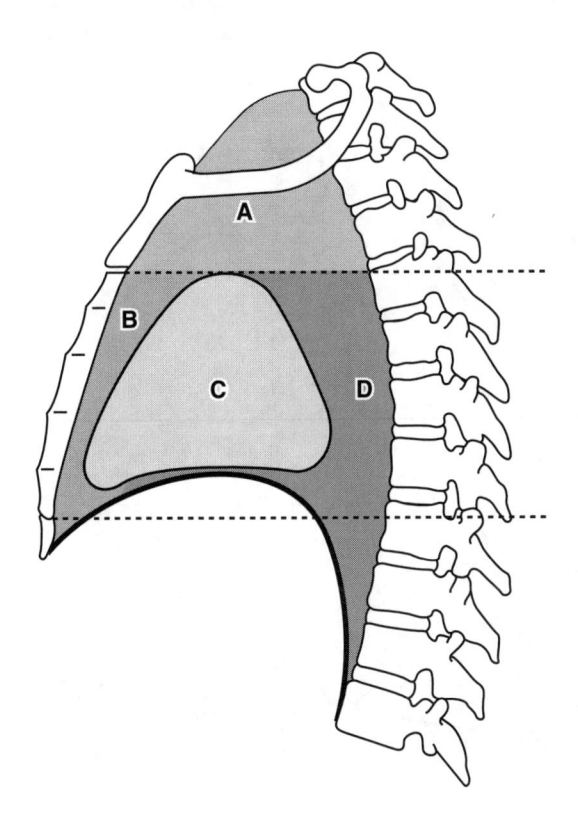

138. Neurogenic tumors that are usually benign in adults and malignant in children are most commonly located in this compartment

139. A tumor associated with myasthenia gravis and pure red blood cell aplasia is most commonly located in this compartment

140. The most common cystic lesion in the mediastinum is located in this compartment

Questions 141-146

For each hematologic disorder, select the appropriate pharmacologic therapeutic regimen.

(A) Ferrous sulfate
(B) Filgrastim or sargramostim
(C) Chromium picolinate
(D) Cyanocobalamin
(E) Folic acid
(F) Erythropoietin
(G) Zinc sulfate
(H) Erythropoietin plus filgrastim or sargramostim

141. Anemia associated with aplastic bone marrow

142. Leukopenia following cancer chemotherapy

143. Macrocytic anemia accompanied by gastric hypochlorhydria and paresthesias of the fingers and toes

144. Megaloblastic anemia in a patient receiving phenytoin

145. Anemia characterized by a decreased mean corpuscular volume and mean corpuscular hemoglobin concentration

146. Anemia in a patient with end-stage renal disease

Questions 147-151

Match each indication with the appropriate immunomodulating drug.

(A) Prednisone
(B) Cyclosporine
(C) Levamisole
(D) Muromonab-CD3
(E) Aldesleukin

147. Renal cell carcinoma

148. Autoimmune hemolytic anemia

149. Prevention of cardiac transplant rejection

150. Reversal of renal allograph rejection crisis

151. Colorectal cancer

Questions 152-162

The letters on the following diagram of the anterior view of the lungs represent different sites in the lungs (for example, central or peripheral location) and structures in the lungs. G represents the hilar lymph nodes. K is in a segment of lung that is posteriorly located. Match each clinical description with the appropriate letter.

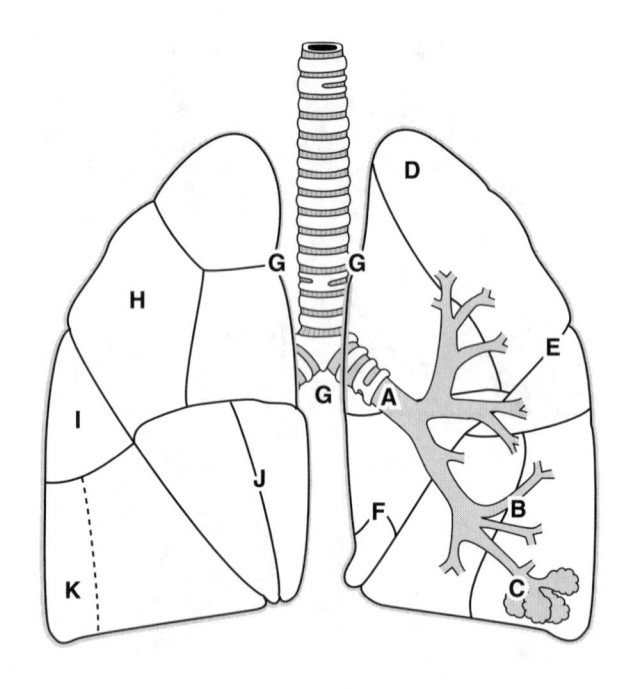

152. In a patient who is sitting or standing up, a foreign body or aspirated material would most commonly localize to this primary site in the lung

153. Panacinar emphysema most commonly involves this area of the bronchial tree

154. In a patient who is lying on his right side, a foreign body or aspirated material would most commonly localize to this primary site in the lung

155. Small cell carcinoma is most commonly located in this area of the lung (use left lung)

156. This is the most common primary site for the Pancoast tumor and for patients who present with hoarseness (use left lung)

157. The most common primary cancer of the lung, scar carcinoma, Ghon foci, and pulmonary infarction occur most commonly in this portion of the lungs (use left lung)

158. In a patient who is lying on her back, aspiration of a foreign body or aspirated material would most commonly localize to this site

159. Squamous carcinoma is most commonly located in this part of the lung (use left lung)

160. This is the most common site for metastasis of a primary lung cancer

161. In a patient who is lying on his left side, aspiration of a foreign body or aspirated material would most commonly localize to this site

162. Involvement of this part of lung with a consolidation (e.g., lobar pneumonia) can obliterate a border of the heart on a postero-anterior radiograph

ANSWER KEY

1. B	28. A	55. C	82. E	109. D	136. B
2. C	29. C	56. D	83. C	110. C	137. A
3. E	30. B	57. A	84. E	111. E	138. D
4. E	31. C	58. D	85. B	112. D	139. B
5. D	32. E	59. A	86. A	113. A	140. C
6. D	33. E	60. B	87. A	114. C	141. H
7. E	34. C	61. A	88. D	115. C	142. B
8. C	35. B	62. B	89. C	116. A	143. D
9. C	36. C	63. D	90. B	117. B	144. E
10. B	37. C	64. C	91. B	118. E	145. A
11. A	38. B	65. D	92. D	119. A	146. F
12. B	39. D	66. D	93. D	120. E	147. E
13. C	40. C	67. D	94. B	121. C	148. A
14. D	41. C	68. E	95. E	122. D	149. B
15. C	42. C	69. D	96. C	123. B	150. D
16. E	43. A	70. A	97. A	124. D	151. C
17. C	44. C	71. A	98. A	125. D	152. K
18. B	45. E	72. D	99. A	126. C	153. C
19. C	46. E	73. D	100. E	127. E	154. H
20. C	47. E	74. B	101. B	128. D	155. A
21. B	48. C	75. D	102. B	129. C	156. D
22. E	49. E	76. D	103. B	130. A	157. E
23. E	50. D	77. B	104. C	131. F	158. I
24. C	51. A	78. B	105. B	132. H	159. A
25. C	52. C	79. D	106. D	133. D	160. G
26. A	53. E	80. A	107. A	134. A	161. F
27. D	54. D	81. D	108. A	135. C	162. J

ANSWERS AND EXPLANATIONS

1. The answer is B. *(Gross anatomy; Cardiovascular)*
The chordae tendineae connect the papillary muscles to the free margins of the atrioventricular valves. The pectinate muscles and the trabeculae carneae are the ridges of muscle in the walls of the atria and ventricles, respectively. The semilunar nodules are the thickenings of the edges of the semilunar valves of the aorta and the pulmonary trunk. Only the mitral and tricuspid valves have chordae tendineae attached to them. Rupture of the chordae tendineae is associated with mitral or tricuspid valve prolapse and valvular insufficiency. Rheumatic valvular heart disease most commonly affects the mitral and aortic valves.

2. The answer is C. *(Gross anatomy; Pulmonary/respiratory)*
The left recurrent laryngeal nerve passes under the aortic arch, and an aneurysm of this arch may compress the nerve. This nerve innervates all the muscles of the left side of the larynx except the cricothyroid muscle. The left phrenic nerve innervates the diaphragm, not the larynx. The right vagus nerve is not related to the aortic arch. The internal and external laryngeal nerves are branches of the superior laryngeal nerve, which is not related to the aortic arch. The internal laryngeal nerve provides innervation to the mucosa of the larynx and does not innervate any muscles. The external laryngeal nerve innervates only the cricothyroid muscle.

3. The answer is E. *(Gross anatomy; Pulmonary/respiratory)*
The pulmonary veins are anterior in the hilus in both lungs. The right upper lobar bronchus is the eparterial bronchus and passes superior to the pulmonary artery. The bronchial arteries are systemic arteries and are branches of the aorta or its branches. The trachea bifurcates at the fourth thoracic vertebra. The right bronchus is more vertical than the left and therefore is more likely to receive a foreign body. These aspirated foreign bodies follow gravity. They are most likely to enter the posterior basal segment of the right lower lobe in the erect position and the apical segment of the right lower lobe in the supine position.

4. The answer is E. *(Gross anatomy; Pulmonary/respiratory)*
Alveolar type II cells increase their production of surfactant during the terminal period of gestation. Premature infants may not yet be producing sufficient surfactant. Thyroxine and glucocorticoids enhance surfactant synthesis. Surfactant reduces surface tension and facilitates the expansion of the alveoli, particularly during expiration.

5. The answer is D. *(Gross anatomy; Pulmonary/respiratory)*
At the left xiphocostal angle, the inferior end of the pericardium contacts the anterior chest wall, and the pleura is displaced laterally because of the cardiac notch. In pericardiocentesis, the objective is to enter the lower portion of the pericardial space where the fluid would accumulate, without passing through the pleura. This procedure is performed to remove fluid from the pericardial space and reduce pressure on the heart.

6. The answer is D. *(Pathology; Cardiovascular)*
The patient most likely has postpartum congestive (dilated) cardiomyopathy. Cardiomyopathies are either primary or secondary, and they are subclassified into congestive, hypertrophic, and restrictive types. Patients present with signs and symptoms associated with myocardial dysfunction. Congestive cardiomyopathy is characterized by a dilated heart with impaired contractility, resulting in both left- and right-sided heart failure. Arrhythmias and sudden death also occur. Congestive cardiomyopathy primarily occurs in young adults and is usually idopathic. In 20% of cases, there is an autosomal or sex-linked recessive or autosomal dominant inheritance pattern.

Acquired etiologies include alcoholism (thiamine deficiency or direct cardiotoxicity); myxedema heart; neuromuscular disease (e.g., Friedreich ataxia); previous viral myocarditis, usually coxsackievirus B; postpartum state; and drug toxicities (e.g., tricyclic antidepressants, lithium, adriamycin, cobalt, and cylclophosphamide).

Diagnosis is made by endomyocardial biopsy. A massively dilated heart is noted on chest radiograph. Echocardiogram confirms left ventricular dilatation, thinning, and global dysfunction. The prognosis of the disease is poor, and most patients require cardiac transplantation if medical therapy is not effective.

7. The answer is E. *(Pathology; Cardiovascular)*
Serum cholesterol and high-density lipoproteins are recommended by most major authoritites as a screening test for primary prevention of coronary artery disease. Screening should begin at 20 years of age and every 5 years thereafter. Scientific evidence has not documented that electrocardiogram, chest radiograph, sputum cytology, and an exercise stress test are useful screening tests in asymptomatic, normal-risk individuals.

8. The answer is C. *(Physiology; Pulmonary/respiratory)*
Cardiac muscle obeys Frank-Starling length–tension relationships; when the muscle is stretched (increased preload), the force of contraction is increased. This is not true of skeletal muscle.

Cardiac muscle has a greater resting tension than skeletal muscle, is much more difficult to stretch than skeletal muscle, has centrally located nuclei, and repairs damaged tissue by fibrosis. Hyperplasia does not occur in either skeletal or cardiac muscles.

9. The answer is C. *(Physiology; Pulmonary/respiratory)*
The three most important factors that control stroke volume, or the volume of blood that is normally ejected during ventricular systole, are contractility, preload, and afterload. Overall, changes in contractility caused by increased sympathetic stimulation are most responsible for changes in the stroke volume under normal physiologic conditions.

10. The answer is B. *(Physiology; Cardiovascular)*
Cardiac output = heart rate × stroke volume. The ejection fraction = stroke volume/left ventricular end-diastolic volume (LVEDV). Therefore, stroke volume = ejection fraction × LVEDV, or 0.50 × 100 ml = 50 ml. Cardiac output = 50 ml × 70 beats/min = 3500 ml/minute, or 3.5 liters/min.

11. The answer is A. *(Pathology; Cardiovascular)*
Rhabdomyoma is the second most common benign tumor of heart muscle. These tumors are usually associated with tuberous sclerosis, which is an autosomal dominant disease characterized by mental retardation, calcified tubercles in the brain, angiomyolipomas of the kidney, and skin tumors (adenoma sebaceum). Rhabdomyomas are not associated with trisomy syndromes, single umbilical artery, rhabdomyosarcoma, or café au lait spots in neurofibromatosis.

12. The answer is B. *(Pathology; Cardiovascular)*
The patient most likely has an infective endocarditis involving his prosthetic aortic valve and thromboembolization to the spleen, which produces a splenic infarct, which in turn produces fibrinous inflammation of the serosal surface and results in a friction rub. Irritation of the diaphragm sends pain to the left shoulder through the phrenic nerve.

Complications associated with prosthetic heart valves include thromboembolism, infective endocarditis caused by *Staphylococcus epidermidis*, microangiopathic hemolytic anemia caused by trauma to red blood cells passing through the valve, and mechanical dysfunction with heart failure.

13. The answer is C. *(Pathology; Cardiovascular)*
The patient has an autoimmune pericarditis, or Dressler syndrome, which usually occurs several weeks to months following an acute myocardial infarction. Dressler syndrome is characterized by fever, pericarditis, pleuritis, myalgia, arthralgia, and a neutrophilic leukocytosis. Nonsteroidal anti-inflammatory agents or systemic corticosteroids are used for treatment.

14. The answer is D. *(Pathology; Cardiovascular, Pulmonary/respiratory)*
Mitral stenosis is usually caused by long-standing rheumatic heart disease. The valve opening becomes stenotic as the leaflets and chordae thicken and fuse. Dystrophic calcification is also present in most cases. Left atrial pressure increases because of the difficulty emptying blood into the left ventricle, which in turn increases the hydrostatic pressure in the pulmonary veins, eventually overriding the oncotic pressure in the pulmonary capillaries. Then, pulmonary edema occurs, followed by chronic pulmonary congestion as

the pressures continue to increase. Over time, the increased pulmonary pressures cause pulmonary hypertension, right ventricular hypertrophy, and right-sided heart failure.

15. The answer is C. *(Pathology, Physiology; Cardiovascular)*
The abnormal pressure–volume loop illustrates a heart with an increased preload (*point 1* is farther along the diastolic volume curve than *point 1**); a decrease in contractility (*point 3* intercepts a different pressure curve than *point 3**); a decrease in the stroke volume caused by a decrease in contractility (*point 5* is less than *point 5**); and a normal afterload (*point 2* is the same as *point 2**). These findings are associated with congestive heart failure.

In essential hypertension, there is an increased afterload without change in preload. In constrictive pericarditis, there is a decrease in preload and stroke volume but normal contractility. A patient who is taking digitalis has an increase in contractility with an increase in stroke volume. A patient who has hypovolemia has a decreased preload and stroke volume but normal contractility.

16. The answer is E. *(Pathology, Physiology; Cardiovascular, Pulmonary/respiratory)*
Patient E has a complete transposition, in which the aorta empties the right ventricle and the pulmonary artery empties the left ventricle. The left and right atria have a normal relationship with the venous return from the lungs and the tissue, respectively. In most cases, there is an atrial septal defect for left-to-right shunting and a ventricular septal defect for right-to-left shunting. There is an increase in arterial oxygen saturation (SaO_2) in the right atrium and right ventricle from the left-to-right shunt through the foramen ovale, and there is a decrease in the SaO_2 in the left ventricle from right-to-left shunting of blood through a ventricular septal defect.

Patient A has a patent ductus arteriosus with a left-to-right shunt from the aorta to the pulmonary artery. Not the increase in oxygen in the pulmonary artery.

Patient B has an atrial septal defect with a left-to-right shunt. Note the increase in oxygen in the right atrium, right ventricle, and pulmonary artery.

Patient C has a tetralogy of Fallot with a right-to-left shunt through a ventricular septal defect. This occurs because the infundibular pulmonic stenosis thwarts emptying of the right ventricle, and coupled with right ventricular hypertrophy, a greater amount of unoxygenated blood shunts into the left ventricle. Note the decrease in oxygen in the left ventricle and aorta.

Patient D has a ventricular septal defect with a left-to-right shunt. Note the increase of oxygen in the right ventricle.

17. The answer is C. *(Pathology, Physiology; Cardiovascular)*
Venous return to the heart influences cardiac output, because whatever blood enters the right heart leaves the left heart. An increase in venous return increases preload, which automatically increases stroke volume and cardiac output in both the right and left ventricles. Likewise, a decrease in venous return decreases preload, which decreases stroke volume and cardiac output in both ventricles.

Right atrial pressure, which represents the gradient between the capillary pressure (normally 15 mm Hg) and the right atrium (0 mm Hg), is the primary factor influencing venous return. The lower the right atrial pressure, the greater the gradient and thus the greater the venous return. However, if the right atrial pressure has a negative value (e.g. -4 mm Hg), a negative pressure is exerted on the inside of the large veins and causes the collapse of these veins and a decrease in venous return.

Arteriolar vasodilation increases venous return by increasing the pressure in the capillary circuit, thus increasing the gradient between the capillaries and the right atrium. Drugs that decrease afterload by arteriolar vasodilation increase both venous return and preload.

Venous vasoconstriction increases venous pressure and the pressure gradient between the capillaries and right atrium, thus increasing venous return.

Ventricular relaxation following systole (i.e., isovolumic relaxation) produces a negative pressure, which draws more blood into the right side of the heart, thus increasing venous return. A decrease in ventricular relaxation, as occurs in congestive heart failure, decreases venous return to the heart.

An increase in blood volume increases the pressure gradient between the capillaries and the right atrium, thus increasing venous return to the heart. This occurs in mineralocorticoid excess states, such as primary aldosteronism, pregnancy, or overzealous infusion of crystalloid solutions.

18. The answer is B. *(Physiology; Cardiovascular, Pulmonary/respiratory)*
Contractility, or the vigor of contraction, is increased only by an increase in sympathetic stimulation of β_1 receptors in the cardiac muscle or an increase in size of the individual cardiac muscle fibers (i.e., hypertrophy). *Curve A* represents increased contractility; *curve B* represents normal contractility during the resting state; and *curve C* represents decreased contractility. Physiologic hypertrophy of the left ventricle in a well-trained athlete is best represented by *curve A*, because it shows an increased stroke volume of 100 ml at a left ventricular end-diastolic volume (LVEDV) of 120 ml, whereas *curve B* shows a stroke volume of 80 ml at an LVEDV of 120 ml. The ejection fraction is calculated by dividing the stroke volume by the LVEDV, thus the ejection fraction is 0.83 (100/120) for *curve A* and 0.67 (80/120) for *curve B*.

Contractility is independent of the preload, or LVEDV, in the left ventricle. Increasing contractility increases the stroke volume by increasing the ejection fraction. This is caused by an increased force of contraction, secondary to increased sympathetic stimulation of β_1-receptors in cardiac muscle or hypertrophy. For example, at a normal LVEDV of 120 ml, the normal resting stroke is 80 ml and the ejection fraction is 0.67 (80/120). If contractility is increased (e.g., if the patient is exercising), and assuming that 100 ml of blood is ejected (stroke volume), the ejection fraction increases to 0.83 (100/120). Increasing stroke volume by increasing preload (LVEDV) does not increase contractility as defined above, but utilizes Frank-Starling mechanisms for increasing the force of contraction by placing a stretch on the muscle.

19. The answer is C. *(Physiology, Pathology; Cardiovascular, Pulmonary/respiratory)*
The cardiac output is equal to the stroke volume multiplied by the heart rate. The figure shows that as the heart rate increases, the cardiac output increases until the heart rate is greater than 180 beats/min. This situation is most closely associated with a decrease in stroke volume caused by a decrease in the amount of time for diastole, or the time the left ventricle has to fill.

The ejection fraction remains essentially unchanged, because the fraction of blood ejected per total volume is the same. Myocardial contractility is not affected by an increase in heart rate. Myocardial ischemia is not a normal feature of an increase in heart rate unless the patient has an underlying coronary artery disease. Compliance is not affected by an increase in heart rate.

20-21. The answers are: 20-C, 21-B. *(Physiology, Pathology; Cardiovascular)*
Point A represents the left ventricle at the end of diastole, corresponding with the preload in the left ventricle. Isovolumic contraction occurs between *points A* and *B*. During this phase, the muscle contracts isometrically against the closed aortic valve, which eventually opens at *point B*, representing the afterload of the left ventricle. Opening of the aortic valve is not associated with the S_1 or S_2 heart sound. The S_1 heart sound is caused by closure of the mitral and tricuspid valves, and the S_2 heart sound is caused by closure of the aortic and pulmonic valves. Between *points B* and *C*, the left ventricle is emptying. The decrease in pressure at *point C* closes the aortic valve, marking the end of ventricular systole. In summary, ventricular systole extends from *point A* to *point C*.

Isovolumic relaxation occurs between *points C* and *D*. The left atrium fills against a closed mitral valve. The mitral valve opens at *point D*, and blood empties into the left ventricle from *point D* to *point A*. Atrial contraction occurs towards the tail end of the loop between *points D* (closure of the mitral valve) and *A*, representing the end of ventricular diastole. Therefore, ventricular diastole extends from *point C* to *point A*. The difference in volume between the *A–B* line (isovolumic contraction) and the *C–D* line (isovolumic relaxation) represents the stroke volume (*point E*).

The abnormal pressure volume loop represents an increase in preload. The amount of blood in the left ventricle is increased. This increased preload is shown in the figure by *point G* being greater than *point A*. The aortic valve opens at the same pressure (*point H* equals *point B*) and closes at the same pressure (*point*

C), marking the end of systole. The stroke volume is increased (*dotted line* at *point E*) due to a greater force of contraction through Frank-Starling mechanisms. Note that contractility did not increase, because *point I* touches the same curve as *point C*. Sympathetic stimulation, not an increase in preload, increases cardiac contractility. With an increase in left ventricular end-diastolic volume, the left ventricle of the heart would most likely show dilatation on a chest radiograph. Afterload, represented by *point H*, is not affected when the preload is increased. This type of curve may be seen in a patient who has a volume overload of isotonic saline or in a pregnant woman who has an increased plasma volume. This curve would not be associated with heart failure, because contractility would be decreased, resulting in a decreased stroke volume.

22. The answer is E. *(Pathology; Cardiovascular)*
The gross findings in the heart show thickening and fusion of the leaflets of the mitral valve and the attached chordae tendineae. Dystrophic calcification is also present in these thickened areas. All of these findings, in addition to a diastolic murmur at the apex, are indicative of mitral stenosis in a patient who most likely had long-standing rheumatic fever.

The irregularly irregular pulse and the absence of the *a* wave in the jugular venous pulse are characteristic of atrial fibrillation. In mitral stenosis, this common arrhythmia is a result of left atrial enlargement as blood backs up behind the stenotic valve. This patient's stroke was most likely embolic. Because the patient was not febrile and vegetations were not present on the mitral valve, the stroke was most likely caused by a clot that dislodged from a thrombus in the left atrium, which occurs in 20% of patients. The clots commonly embolize to the middle cerebral artery, producing an embolic hemorrhagic infarction. The vibratory effect of atrial fibrillation probably enhanced thromboembolization, which occurs in approximately one third of patients with atrial fibrillation. If the patient had been on an anticoagulant, this fatal complication may not have occurred.

23. The answer is E. *(Pathology; Cardiovascular)*
A large vegetation is located on the aortic valve immediately below the free margin of the valve cusps. The patient was most likely an intravenous drug abuser; he was febrile, his aortic valve was involved, and the onset was rapid, with early signs of aortic regurgitation (i.e., diastolic murmur, bounding pulses, and wide pulse pressure) that progressed to left-sided heart failure. This patient most likely had acute bacterial endocarditis caused by *Staphylococcus aureus*.

Generally, acute endocarditis is secondary to repeated contamination of the valves by virulent organisms from indwelling catheters or intravenous drug abuse. With intravenous drug abuse, vegetations are located on either the right side (50% involve the tricuspid valve) or the left side (50% involve the aortic valve). Valve destruction causing valve incompetency is common, as occurred in this patient. Preexisting heart disease (e.g., rheumatic heart disease, congenital heart disease, or mitral valve prolapse) increases the risk for bacterial seeding of the valves. Infarctions of the kidney, spleen, bowel, and digits is often caused by thromboembolization of parts of the vegetations. However, immune complex disease due to antibodies against bacterial antigens assumes a prominent role in this disease. Roth retinal hemorrhages, splinter hemorrhages under the nails, Osler nodes (painful nodules on the hands and feet), Janeway lesions (painless nodules on the hands and feet), and glomerulonephritis are more frequently caused by immune vasculitis than by embolization.

24. The answer is C. *(Physiology; Cardiovascular)*
First-degree heart block is manifested on the electrocardiogram (EKG) as a prolonged PR interval (i.e., one that lasts longer than 200 msec). First-degree heart block is usually indicative of an increase in atrioventricular (AV) nodal conduction time. In second-degree heart block, not every P wave elicits a QRS complex, so that the ventricle is said to "drop" a beat. Widening of the QRS complex is usually associated with premature ventricular contractions. The duration of the QRS complex is prolonged because the impulse is conducted through cardiac muscle rather than the Purkinje system. Though rare, it is possible for the impulse generated by the sinoatrial (SA) node to be blocked before it reaches the atrial muscle. In this case, the P wave is absent.

25. The answer is C. *(Physiology; Pulmonary/respiratory)*

An arterial pH of 7.3 and an arterial carbon dioxide tension ($PaCO_2$) of 50 mm Hg is representative of respiratory acidosis and would be most likely to cause the largest increase in cerebral blood flow. Moderate increases in $PaCO_2$ can double the normal cerebral blood flow. A high $PaCO_2$ leads to vasodilation: The carbon dioxide diffuses into the interstitial fluid, where it becomes hydrated to form carbonic acid (H_2CO_3). The carbonic acid dissociates into hydrogen ion (H^+) and bicarbonate (HCO_3^-). The receptor responsible for vasodilation is a H^+ receptor; therefore, blood flow and cerebral perfusion increase as the amount of H^+ increases.

26. The answer is A. *(Histology; Hematopoietic/lymphoreticular)*
The finding of Howell-Jolly bodies, target cells, and reticulocytosis is most compatible with splenectomy. The spleen has many functions, including the removal of defective and aged blood cells, hematopoiesis (in times of severe stress when the marrow is unable to meet the needs of the organism), iron conservation, blood cell storage, and host defense in the form of humoral immunity. Therefore, splenectomy can result in a number of hematologic abnormalities. Following splenectomy, Howell-Jolly bodies (leftover nuclear remnants) are not removed from the cytoplasm of erythrocytes, resulting in increased numbers of nucleated erythrocytes and erythrocytes containing Howell-Jolly bodies. The spleen plays an important role in the maturation of reticulocytes into erythrocytes; therefore, splenectomy can result in a reticulocytosis. Target cells are erythrocytes with too much membrane; normally, the spleen would remove the excess membrane. Thrombocytosis also occurs because normally, the spleen is a reservoir for one-third of the platelets in the body. Thrombotic thrombocytopenic purpura (TTP) is associated with increased numbers of reticulocytes (as a result of intravascular damage to erythrocytes); however, Howell-Jolly bodies and target cells are not normally seen in patients with TTP. Disseminated intravascular coagulation (DIC) is associated with an increase in reticulocytes due to destruction of erythrocytes by fibrin in the microvasculature, but again, target cells and Howell-Jolly bodies are not normally present. Sickle cell trait is not associated with any hematologic abnormalities. Autoimmune hemolytic anemias are associated with a pronounced reticulocytosis that is often accompanied by nucleated erythrocytes because the markedly stressed bone marrow is barely able to keep pace with the extravascular removal of erythrocytes. However, target cells are not usually present.

27. The answer is D. *(Pathology; Gastrointestinal)*
A prolonged prothrombin time (PT) that does not correct following an intramuscular injection of vitamin K is most consistent with hepatic cirrhosis because the liver synthesizes the nonfunctional precursors for the vitamin K–dependent coagulation factors, mainly prothrombin (factor II) and factors VII, IX, and X. These precursors must undergo γ-carboxylation by vitamin K in order to become functional when activated in the coagulation sequence. If an intramuscular injection of vitamin K does not correct the PT, then the problem must be in the liver. However, if the PT does correct itself, then the problem is vitamin K deficiency. A malabsorptive state (e.g., celiac disease) often results in vitamin K deficiency; therefore, the PT should correct if vitamin K is given intramuscularly. Hemophilia A and von Willebrand's disease involve deficiencies in the factor VIII molecule, which is not a vitamin K–dependent factor. Because factor VIII is in the intrinsic system, the partial thromboplastin time (PTT) would be prolonged and the PT would be normal (measures extrinsic system).

28. The answer is A. *(Pathology; Hematopoietic/lymphoreticular)*
Hemorrhagic skin necrosis is a complication associated with warfarin therapy in patients with an underlying protein C deficiency. Protein C normally inactivates factors V and VIII and enhances fibrinolysis. Therefore, a deficiency of protein C predisposes the patient to deep venous thrombosis, because factors V and VIII are not degraded and the fibrinolytic system is inactive. Patients who are heterozygotes for protein C deficiency (an autosomal dominant disease) have protein C levels that are approximately 50% lower than normal during the initial loading phase of warfarin. Warfarin inhibits epoxide reductase, which maintains vitamin K in an active form. Inactivation of vitamin K results in the inactivation of vitamin K–dependent factors [prothrombin (factor II), VII, IX, X, protein C, and protein

S] as well. Because protein C is a vitamin K–dependent factor, warfarin inhibits activation of protein C, so the factor level drops from 50% to 0%, resulting in a paradoxical thrombosis of vessels in the skin. A deficiency of antithrombin III also predisposes the patient to deep venous thrombosis, because antithrombin III normally inactivates coagulation factors that are serine proteases. However, antithrombin III deficiency is not associated with hemorrhagic skin necrosis, because it is not a vitamin K–dependent factor and is not inactivated by warfarin derivatives. Warfarin derivatives do not increase platelet aggregation or produce a hypersensitivity vasculitis. The patient history does not suggest disseminated intravascular coagulation (DIC), which is associated with thrombosis in the microvasculature, consumption of clotting factors, and bleeding, rather than deep venous thrombosis.

29. The answer is C. *(Hematology; Pharmacodynamic and pharmacokinetic processes)*
Warfarin derivatives inhibit epoxide reductase, which keeps vitamin K in an active form. By inactivating vitamin K, the vitamin K–dependent factors [factors II [prothrombin), VII, IX, X, and proteins C and S) are also inactivated. These factors are synthesized in a nonfunctional state by the liver and require vitamin K to render them functional. Once the factors are activated, levels decrease according to half-life: factors VII and protein C (4–6 hours), factor IX (20–25 hours), factor X (40–50 hours), and factor II (72 hours). Therefore, full anticoagulation does not occur within 24 hours of initiating therapy because warfarin does not inactivate those factors that have already been γ-carboxylated and are still operational.

30. The answer is B. *(Hematology; Biochemistry and molecular biology)*
Acute hemolytic episodes in patients with glucose 6-phosphate dehydrogenase (G6PD) deficiency are most commonly precipitated by infection. Increased susceptibility to infection in these patients is related to a decrease in reduced nicotinamide adenine dinucleotide phosphate (NADPH) synthesis in neutrophils and monocytes. NADPH is an essential component of the oxygen-dependent myeloperoxidase system, an important means of eliminating bacteria. NADPH acts with NADPH oxidase to convert molecular oxygen into a superoxide free radical, generating the respiratory burst reponsible for the destruction of bacteria. Absence of NADPH, as a result of G6PD deficiency, interferes with this important bactericidal mechanism, thus rendering the patient susceptible to infection. Infection increases the formation of peroxide, which cannot be degraded into water by glutathione because glutathione is lacking as well. Peroxide denatures hemoglobin, producing Heinz bodies. Macrophages in the spleen extravascularly remove the damaged erythrocytes, resulting in a brisk hemolytic anemia. A decreased amount of glutathione in the white blood cells does not predispose to infection, but it does allow peroxide to denature the hemoglobin and produce Heinz bodies. There is no specific defect in phagocytosis of bacteria, deficiency of myeloperoxidase, or deficiency of adenosine triphosphate (ATP) production in the white blood cells in patients with G6PD deficiency.

31. The answer is C. *(Biochemistry; Biochemistry and molecular biology)*
Mature erythrocytes depend on the pentose phosphate pathway for the formation of nicotinamide adenine dinucleotide phosphate (NADPH), which the cells require to maintain glutathione in a reduced state. Reduced glutathione is involved in maintaining the integrity of the cell membrane. Mature red blood cells do not have mitochondria. Therefore, they are unable to use fatty acids for fuel (β-oxidation), they do not have a tricarboxylic acid (TCA) cycle, and they do not employ oxidative phosphorylation for the generation of adenosine triphosphate (ATP). They do have a glycolytic cycle, and glucose is their only energy source. For each glucose molecule, the erythrocyte obtains two ATP molecules. The two reduced nicotinamide adenine dinucleotide (NADH) molecules generated during the glycolytic cycle are used to replenish the two nicotinamide adenine dinucleotide (NAD) molecules in the glycolytic pathway. In the process of replenishing NAD, pyruvate is converted into lactate, not acetyl coenzyme A (acetyl CoA). Because gluconeogenic enzymes are located only in the liver and kidney, the lactate is eliminated from the erythrocyte and converted by the liver into glucose (Cori cycle).

32. The answer is E. *(Biochemistry; Biochemistry and molecular biology)*

Neither muscle nor red blood cells have gluconeogenic enzymes to synthesize glucose for fuel. However, both tissues produce lactate, which is taken up by the liver and converted into pyruvate. The pyruvate is then used as a substrate for gluconeogenesis, which provides glucose for use by muscle and erythrocytes. Muscles are able to use glucose, ketone bodies, and fatty acids for fuel, whereas erythrocytes are restricted to using glucose. Because erythrocytes do not have mitochondria, they cannot metabolize fatty acids by β-oxidation. Muscles have glycogen, which can be degraded to yield glucose, but erythrocytes lack glycogen stores.

33. The answer is E. *(Hematology; Skin/Connective Tissue)*
The fetal hemoglobins include hemoglobin Gower (the first hemoglobin synthesized), hemoglobin F, and hemoglobin Portland. Newborn infants have 75% hemoglobin F, which is composed of two α and two γ chains. Hemoglobin F shifts the oxygen dissociation curve to the left, increases oxygen affinity, and stimulates the release of erythropoietin, which increases red blood cell production and the hemoglobin concentration. Left-to-right shunting of blood in the heart does not increase the hemoglobin concentration because oxygenated blood mixes with unoxygenated blood. This blood is then oxygenated in the lungs, so there is no hypoxemic stimulus for erythropoietin release. A patent ductus arteriosus starts out as a left-to-right shunt, with blood from the aorta passing through the ductus to the pulmonary artery. This shunt does not produce hypoxemia, so there is no stimulus for erythropoietin release. Anemia as a result of ABO incompatibility in the fetus is associated with the release of erythropoietin, but this is not the explanation for the normal physiologic increase in hemoglobin in all newborns.

34. The answer is C. *(Hematology; Biochemistry and molecular biology)*
Pyruvate kinase deficiency, an autosomal recessive disease, produces an intrinsic hemolytic anemia with extravascular hemolysis. Pyruvate kinase catalyzes the following reaction:

$$2 \text{ Phosphoenolpyruvate} + \text{ADP} \xrightarrow{\text{PK}} \text{Pyruvate} + \text{ATP}$$

Deficiency of the enzyme results in a decrease in adenosine triphosphate (ATP) production and an inability to maintain the ATPase-dependent sodium–potassium pump. Red blood cells become dehydrated and assume the shape of spherocytes or spiked cells resembling acanthocytes. These abnormal cells are removed extravascularly. As expected, the concentration of substrates proximal to the enzyme block (e.g., 2,3 bisphosphoglycerate) increases. This increase in 2,3 bisphosphoglycerate concentration shifts the oxygen dissociation curve to the right, facilitating oxygen delivery to tissue and offsetting the severity of the anemia.

35. The answer is B. *(Physiology; Pulmonary/respiratory, Cardiovascular)*
A person employed as a toll taker on the turnpike would have heavy exposure to carbon monoxide, which shifts the oxygen dissociation curve to the left. The position of the oxygen dissociation curve determines how efficiently oxygen unloads from the hemoglobin. At the tissue level, this oxygen dissolves in the capillary plasma and increases the capillary oxygen tension (Po_2), thus favoring a gradient for the diffusion of oxygen into the tissue. As long as the partial pressure in the capillary blood is greater than the partial pressure of oxygen in the tissue, oxygen diffuses into the tissue. A shift of the oxygen dissociation curve to the right is indicative of decreased affinity of hemoglobin for oxygen, which enhances the release of oxygen to the plasma. A shift of the oxygen dissociation curve to the left indicates an increased affinity of hemoglobin for oxygen with less release of oxygen from the hemoglobin into the plasma, and ultimately, to the tissue. A patient with chronic obstructive pumonary disease (COPD) would most likely have respiratory acidosis, which shifts the oxygen dissociation curve to the right. Tetralogy of Fallot produces chronic hypoxemia as a result of a right-to-left shunt in the heart. Chronic hypoxemia increases red blood cell 2,3 bisphosphoglycerate, which, in turn, shifts the oxygen dissociation curve to the right. A 103°F fever would cause a right shift of the oxygen dissociation curve. Diabetic ketoacidosis also causes the oxygen dissociation curve to shift to the right.

36. The answer is C. *(Physiology; Hematopoietic/lymphoreticular)*

Adult women have smaller iron stores than men (400 mg versus 1000 mg) as well as lower serum iron levels. Because serum ferritin is a circulating fraction that represents the iron stores in the macrophages, serum ferritin levels are also lower in women. Menstruation accounts for the difference in iron stores between men and women. Pregnancy can also diminish a woman's iron stores; women who are not taking iron supplements have a net iron deficit of 500 mg after each pregnancy.

37. The answer is C. *(Physiology; Pulmonary/respiratory)*
Most of the carbon dioxide produced in metabolism is transported as bicarbonate ions. In gas exchange at the tissue level, the carbon dioxide from tissue diffuses into the capillary blood and increases the carbon dioxide tension (P_{CO_2}). Most of the carbon dioxide diffuses into the red blood cell, where some of it combines with hemoglobin to form carbamino compounds. In addition, carbon dioxide combines with water to form carbonic acid, which dissociates into hydrogen and bicarbonate ions. The hydrogen ions displace the oxygen on the hemoglobin (Bohr effect), enabling oxygen to diffuse into the blood and out into the tissue along a diffusion gradient. The bicarbonate ions leave the erythrocyte and enter the plasma, where they serve as the primary storage form of carbon dioxide. Chloride enters the red blood cell to counterbalance the loss of the negative charge. This is called the chloride shift. When the red blood cell enters the pulmonary capillaries for gas exchange, the reverse occurs. Oxygen diffuses from the alveoli into the blood, where it increases the oxygen tension (P_{O_2}). It displaces hydrogen ions on hemoglobin; the displaced hydrogen ions combine with bicarbonate to form carbonic acid. Carbonic acid dissociates into carbon dioxide and water. The carbon dioxide exits the erythrocyte, dissolves in the plasma, and diffuses out into the alveoli, because the P_{CO_2} is higher in the pulmonary capillary than in the alveoli. Glutamine is an amino acid that is the primary carrier for ammonia.

38. The answer is B. *(Physiology; Pulmonary/respiratory)*
The patient has methemoglobinemia. In methemoglobinemic patients, the iron in the hemoglobin exists in the ferric (+3) state. Iron must be in the ferrous (+2) state in order to bind with oxygen. Methemoglobinemia lowers the oxygen saturation (S_{aO_2}), or the number of heme groups occupied by oxygen, but has no effect on gas exchange in the lung, so the arterial oxygen tension (P_{aO_2}) is normal. Administering oxygen to the patient has no effect on the S_{aO_2} because the iron is ferric rather than ferrous. Respiratory acidosis is associated with cyanosis, but is characterized by a low P_{aO_2} and a low S_{aO_2}. Patients generally respond to oxygen therapy. Although patients with intrapulmonary or right-to-left cardiac shunts do not respond to oxygen therapy (because the blood is bypassing oxygenation in the lungs), both of these conditions are characterized by a low S_{aO_2} as well as a low P_{aO_2}. A diffusion abnormality in the lungs may be associated with a low P_{aO_2} and a low S_{aO_2}, but patients should show some response to oxygen therapy.

39. The answer is D. *(Biochemistry; Multisystem processes)*
Strict vegetarians are most at risk for iron deficiency. The iron found in plants is in the ferric state. Because iron must be reduced to the ferrous state for reabsorption, its absorption is limited when compared with that of heme iron in meats, which is already in the ferrous state. Strict vegetarians are also at risk for vitamin B_{12} deficiency, because vitamin B_{12} is primarily present in meats and dairy products. However, hepatic vitamin B_{12} stores represent a supply sufficient to meet the body's needs for 6–10 years, so vitamin B_{12} deficiency would be unlikely in this patient. Folate, zinc, or vitamin C deficiencies would be unlikely in a vegetarian, regardless of whether or not she is pregnant.

40. The answer is C. *(Hematology; Skin/connective tissue)*
Haptoglobin is a protein synthesized in the liver. Haptoglobin combines with free hemoglobin in the plasma to form a complex, which is then removed from the plasma by macrophages. In this way, the body is able to retrieve the amino acids and iron associated with hemoglobin. Bilirubin is a breakdown product of hemoglobin. When macrophages remove old or deformed erythrocytes or erythrocytes coated with IgG, C3b, or both, the globin chains are separated from the heme group. The heme group is split into iron

and protoporphyrin. Iron is used for storage, either as ferritin or hemosiderin. Protoporphyrin undergoes a series of reactions to form indirect, lipid-soluble bilirubin, which is released into the blood to combine with albumin. The indirect bilirubin is taken up by the liver and conjugated into water-soluble, direct bilirubin. Ferritin and hemosiderin are storage forms of iron. Iron derives from the diet or the breakdown of erythrocytes intravascularly or extravascularly in macrophages. Ferritin is a protein–iron complex that is present in erythrocytes, macrophages, and hepatocytes. Hemosiderin consists of packets of ferritin and is primarily located in macrophages. Hematin is a Prussian blue–negative, black, crystalline product originating from the oxidation of heme from the ferrous to the ferric state.

41. The answer is C. *(Hematology; Biochemistry and molecular biology)*
Inosine, when added to stored blood, is taken up by red blood cells and converted into 2,3 bisphosphoglycerate. 2,3 Bisphosphoglycerate shifts the oxygen dissociation curve to the right, increasing the release of oxygen to tissue. Unfortunately, inosine is too toxic for general use in the blood bank, but may be used to rejuvenate blood if the cells are washed prior to transfusion. Citrate, which binds to calcium, is primarily added to blood as an anticoagulant. Dextrose improves erythrocyte survival, because erythrocytes use dextrose as an energy source for the generation of adenosine triphosphate (ATP). Phosphate and adenine are used as substrates for the synthesis of ATP. Fructose is not used in blood storage.

42. The answer is C. *(Hematology; Skin/connective tissue)*
The thalassemias are characterized by defects in globin chain synthesis. Therefore, a patient with δ β-thalassemia has a decrease in δ- and β-chain synthesis but normal synthesis of α- and γ-chains. Because hemoglobin F is composed of two α- and two γ-chains, its synthesis would increase. Hemoglobin A, which consists of two α- and two β-chains, and hemoglobin A_2, which consists of two α- and two δ-chains, would be decreased. Hemoglobin H is a tetramer of β-chains. Hemoglobin C is a lysine for glutamic acid substitution in the sixth position of the β-chain. It is not a thalassemia variant.

43. The answer is A. *(Hematology; Immune response)*
Mast cells and basophils are the primary effector cells in type I hypersensitivity reactions. When the mast cells and basophils degranulate after exposure to specific allergens, they release preformed eosinophil chemotactic factor as a primary chemical mediator. Eosinophil chemotactic factor attracts eosinophils to the area of inflammation, where they release arylsulfatase and histaminase to neutralize leukotrienes and histamine, respectively. Eosinophil major basic protein is also released and is responsible for damaging epithelial cells in the area of inflammation. Crystalline material within eosinophil granules may coalesce to form Charcot-Leyden crystals, which are commonly seen in the sputum of asthmatics. Helminths coated with IgE antibodies are killed when eosinophils hook into their Fc receptors; this is an antibody-dependent cytotoxicity (type II hypersensitivity) reaction. Interleukin-5 (IL-5) is important in the synthesis of eosinophils in the bone marrow.

44. The answer is C. *(Histology; Skin/connective tissue)*
Leukocyte alkaline phosphatase is present in the specific granules of neutrophils. It is a marker of neutrophil maturity. Staining of neutrophils in the peripheral blood for alkaline phosphatase is useful in distinguishing benign neutrophil disorders (e.g., leukemoid reactions) from chronic myelogenous leukemia (CML). One hundred neutrophils are graded on the intensity of the stain and the sum total of their individual scores is reported as the leukocyte alkaline phosphatase (LAP) score. In leukemoid reactions, which are characterized by neutrophil counts that exceed 50,000 cells/μl, the LAP score is increased. However, in CML, a neoplastic proliferation of granulocytes, the LAP score is low. Other chemicals present in the specific granules are lactoferrin, acid hydrolases, type IV collagenase, leukocyte adhesion molecules, and phospholipase A_2. Chemicals in the nonspecific (azurophilic) granules include myeloperoxidase, lysozyme, cationic proteins, elastases, and nonspecific collagenases.

45. The answer is E. *(Hematology; Immune response)*

Anti-Lewis antibodies are present in blood and body secretions and are generally not associated with hemolytic transfusion reactions or hemolytic disease of the newborn. Anti-D antibodies are the most common antibodies observed in the blood bank, but ABO incompatibility is a more common cause of hemolytic disease of the newborn. One of the major reasons for the low incidence of Rh hemolytic disease of the newborn is the practice of giving pregnant women who are negative for anti-D antibodies a prophylactic injection of Rh immune globulin. Anti-Kell antibody is a potent antibody that often produces a severe extravascular hemolytic anemia in Kell antigen–negative patients who receive Kell antigen–positive blood. Antibodies against leukocyte antigens are responsible for febrile transfusion reactions. Both anti-I and anti-I antibody are cold-reacting, IgM antibodies. Cold autoimmune hemolytic anemias associated with anti-I antibodies and anti-i antibodies may occur in *Mycoplasma pneumoniae* infections and infectious mononucleosis, respectively.

46. The answer is E. *(Hematology, Pathology; Hematopoietic/lymphoreticular)*
Sideroblastic anemias are caused by a defect in the synthesis of heme (iron + protoprophyrin) in the mitochondria. Pyridoxine (vitamin B_6) deficiency, lead poisoning, and alcoholism are common causes of these rare anemias. Because iron normally enters the mitochondria to bind with protoporphyrin, defects in the synthesis of protoporphyrin or problems with binding protoporphyrin with iron leads to an accumulation of iron in the mitochondria. These nucleated red blood cells are identified as ringed sideroblasts when bone marrow aspirates are stained with Prussian blue. Except when lead poisoning is the cause of the anemia, the ringed sideroblasts must be identified before a diagnosis of sideroblastic anemia is confirmed. In the majority of cases, a bone marrow examination is not necessary to confirm iron deficiency, pernicious anemia, anemia associated with renal failure, or the anemia of chronic inflammation. Iron deficiency and anemia of chronic inflammation are easily diagnosed by evaluating serum iron values (low in both disorders), total iron binding capacity (high in iron deficiency and low in anemia of chronic inflammation), percent saturation (low in both disorders), and serum ferritin (low in iron deficiency, normal to high in anemia of chronic inflammation). Pernicious anemia is characterized by a macrocytic anemia with low vitamin B_{12} levels. Patients with renal failure and a normocytic anemia with a low corrected reticulocyte count most commonly have reduced erythropoietin production as the cause of the anemia.

47. The answer is E. *(Genetics, Hematology; Human development and genetics)*
α globulin chain synthesis is controlled by four genes, with two genes located on each chromosome 16 (α α / α α). Bart's hemoglobin is a combination of four γ-chains. In order for a child to develop Bart's hemoglobin disease, the parents must have two gene deletions on the same chromosome (——/ α α; common in Asians), rather than one gene deletion on each of the chromosomes (—α /—α; common in blacks):

	—α	—α
—α	—α/—α	—α/—α
—α	—α/—α	—α/—α

No Bart's hemoglobin disease

	——	αα
——	——/——	——/αα
αα	——/αα	αα/αα

Bart's hemoglobin disease

The combination that results in the deletion of all four genes (— —/— —) is incompatible with life (hydrops fetalis). In α-thalassemia, there is a deletion of one or more of the four genes responsible for the synthesis of α chains. In α-thalassemia minor, there is a one- (—α/ α α) or two-gene deletion (—α /—α or— —/ α α), resulting in a mild anemia and a proportional decrease in the synthesis of hemoglobin A (two α- and two β-chains), hemoglobin A_2 (two α- and two δ-chains), and hemoglobin F (two α- and two γ-chains). In hemoglobin H disease, there is a three-gene deletion (— —/—α), resulting in a hemolytic anemia and the formation of tetramers of β-chains (hemoglobin H).

48. The answer is C. *(Pathology; Hematopoietic/lymphoreticular)*
The anemia of inflammation is the most common anemia associated with malignancy. Iron is trapped in the macrophages and is unavailable for hemoglobin synthesis. Rapidly growing tumors (e.g., leukemia, small cell carcinoma of the lung) use vitamin B_{12} and folate for DNA synthesis. However, most patients have a 6- to 10-year supply of vitamin B_{12} in the liver, and a 3- to 4-month supply of folate. Other potential causes of anemia in pregnancy include radiation, blood loss from hemorrhagic cystitis as a complication of cyclophosphamide therapy, and autoimmune hemolytic anemia (chronic lymphocytic leukemia). Blood loss in malignancy is particularly common in right-sided carcinomas of the colon. Bone marrow metastasis often results in replacement of the marrow by collagen (secondary myelofibrosis), which reduces hematopoiesis and contributes to anemia.

49. The answer is E. *(Hematology; Hematopoietic/lymphoreticular)*
Anemia is a feature of all leukemias, but microcytic anemia is an unlikely finding in patients with leukemia. Usually the anemias seen in patients with leukemia are normocytic. Anemia of inflammation is the most common anemia. Clinical features of leukemia include fatigue, infections, bleeding, massive hepatosplenomegaly, lymphadenopathy, bone pain and tenderness, and organ infiltration. Hyperuricemia is common as a result of the increased breakdown of cells and release of purines.

50. The answer is D. *(Pathology; Hematopoietic/lymphoreticular)*
The bone marrow biopsy is hypocellular. Because the patient also has pancytopenia (i.e., a low erythrocyte, white blood cell, and platelet count), the diagnosis is aplastic anemia. Drugs (e.g., chloramphenicol, phenylbutazone, chlorpromazine, streptomycin, alkylating agents) are the most common acquired cause of aplastic anemia. Other causes include Fanconi's syndrome, industrial chemicals (e.g., benzene, insecticides, inorganic arsenicals), radiation, viral infection (e.g., hepatitis C), and other anemias, such as those caused by paroxysmal nocturnal hemoglobinuria, vitamin B_{12} or folate deficiency, sickle cell anemia, or congenital spherocytosis.

51. The answer is A. *(Physiology; Cardiovascular)*
Autoregulation of flow refers to the intrinsic ability of a blood vessel to maintain a constant flow despite changes in mean arterial pressure. In the autoregulatory range (60–140 mm Hg), an increase in mean arterial pressure will result in arteriolar vasoconstriction, so that flow to the organ remains unchanged. Similarly, in the autoregulatory range, a fall in the mean arterial blood pressure leads to vasodilation. These responses do not depend on the presence of nerves.

52. The answer is C. *(Physiology; Cardiovascular)*
The extraction of oxygen by the myocardium is nearly maximal at rest (90% of the oxygen is extracted), so any additional demand for oxygen must be met by an increase in supply. In other words, oxygen delivery is flow-dependent. The sympathetic nerves do not mediate the change in oxygen delivery in response to a change in demand.

53. The answer is E. *(Physiology; Pulmonary/respiratory)*
Because a primary role of the circulations is to provide the tissues with enough oxygen to meet their metabolic needs, the vascular response to hypoxemia, in nearly all organs, is vasodilation. The lowered vascular resistance allows more blood to perfuse the organ at the same mean arterial blood pressure. The pulmonary circulation is an exception. It is the only circulation that

responds to hypoxia with vasoconstriction. The rationale behind this response is that if a site in the lung becomes hypoxic as a result of poor ventilation, an increase in vascular resistance at that site will divert blood to better ventilated regions.

54-55. The answers are: 54-D, 55-C. *(Physiology; Cardiovascular)*
When the mean arterial blood pressure falls below 40 mm Hg, the arterial baroreceptors are completely unloaded, and baroreceptor inhibition of sympathetic centers ceases. Therefore, the peripheral chemoreceptors become the main source of additional sympathetic drive to the arterioles and are primarily responsible for the maintenance of mean arterial blood pressure. The peripheral chemoreceptors are located in the carotid body at the bifurcation of the common carotid artery (afferents in cranial nerve IX) and in the aortic body at the arch of the aorta (afferents in cranial nerve X). The peripheral chemoreceptors fire in response to a low oxygen tension and a decrease in blood flow. Impulses from the peripheral chemoreceptors excite the respiratory centers in the medulla as well as the sympathetic centers in the brain stem that support the cardiovascular system.

Administration of 100% oxygen to a patient who is hemorrhaging severely may cause an additional fall in the mean arterial blood pressure because the peripheral chemoreceptors will stop firing when they sense that the oxygen tension has increased.

56. The answer is D. *(Physiology; Pulmonary/respiratory, Cardiovascular)*
A Valsalva maneuver is a forced expiration against high resistance, so that the intrapleural pressure becomes positive. This action increases the central venous pressure and decreases the gradient for venous return to the right heart, causing the jugular veins to bulge. Cardiac output and the mean arterial blood pressure fall. The fall in mean arterial blood pressure leads to a decrease in afferent activity of the nerves ascending to the brain from arterial baroreceptors and a reflex increase in sympathetic drive to the cardiovascular system. Although sympathetic drive is increased, the fall in cardiac output is so severe that blood pressure remains low.

57. The answer is A. *(Physiology; Cardiovascular)*
An increased heart rate is associated with a shortened QT interval. The duration of a ventricular action potential [reflected on the electrocardiogram (EKG) as the QT interval] is affected by the cytosolic calcium concentration. An increase in the cytosolic calcium concentration increases the membrane's conductance of potassium and shortens the duration of phase 2 of the action potential. The cytosolic calcium concentration is increased by a high heart rate and with drugs like digitalis, leading to a reduced duration of the action potential (a shorter QT interval).

58. The answer is D. *(Physiology; Cardiovascular)*
This patient is most likely experiencing atrial fibrillation. During atrial fibrillation, an excessive number of impulses are travelling in the atria in all directions, in effect cancelling one another out. Therefore, P waves, which reflect atrial depolarization, are not seen on the electrocardiogram (EKG). In atrial fibrillation, the QRS complexes are normal, but appear at irregular intervals because of the random arrival of waves of depolarization at the atrioventricular node. The overall heart rate may be as high as 150 beats per minute. P waves are seen in both first-degree heart block, which is manifested on the EKG as an increased PR interval, and second-degree heart block (although not every P wave elicits a QRS complex). In third-degree heart block, the P waves are regular but there is no relationship between the P waves and the QRS complex. In ventricular fibrillation, the QRS complex is widened and bizarre in appearance because the arrhythmia does not employ the specialized conducting system of the heart. Because the ventricles are operating independently of the atria, there is no constant relationship of the P wave to the QRS complex.

59. The answer is A. *(Physiology; Cardiovascular)*
The QRS complex would most likely be inverted in lead I in a patient with a mean electrical axis (MEA) of +150. Those circumstances that lead to right ventricular hypertrophy or right bundle branch block would result in a right axis deviation and appearance of the largest R waves in lead III. If the impulse is travelling away from the lead, the QRS complex will be inverted (in this case, lead I). If the lead is oriented

perpendicular to the direction of impulse spread, it will record no activity (in this case, lead II).

60. The answer is B. *(Genetics; Biochemistry and molecular biology)*
In normal genes, MST II endonuclease cleaves the gene on the β-chain at the same point mutation site involved in sickle cell trait and disease. Cleavage of the normal gene converts a 1.35-kb segment into two fragments, one of which is 1.15 kb long and the other of which is 0.20 kb long. However, if a point mutation has already occurred at this site, MST II endonuclease is unable to cleave the gene on the β-chain and the fragment remains 1.35 kb long. Therefore, in sickle cell trait, where the defect only involves one chromosome, the normal gene fragment splits into one 1.15-kb fragment and one 0.20-kb fragment, and the abnormal gene fragment remains whole. In sickle cell disease, the defect is on both chromosomes, so both gene fragments remain whole.

61. The answer is A. *(Physiology; Cardiovascular)*
Cerebral vascular resistance is higher at a mean arterial pressure of 140 mm Hg than at a mean arterial pressure of 60 mm Hg. The curve below illustrates the phenomenon of autoregulation and the relationship between the mean arterial pressure and cerebral blood flow.

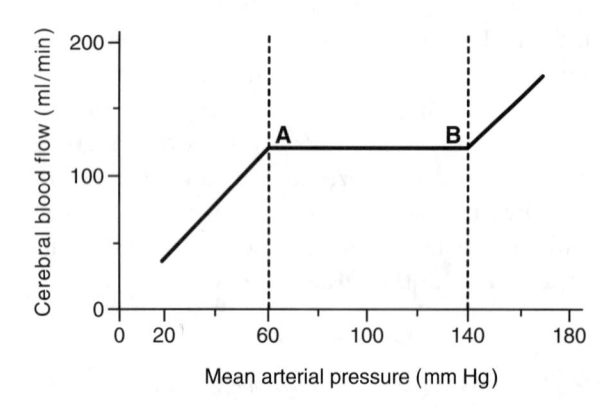

In vascular beds that are capable of autoregulation of flow, such as the cerebral, coronary, and renal circulations, an increase in the mean arterial pressure in the autoregulatory range (60 mm Hg–140 mm Hg) results in arteriolar vasoconstriction. Because flow equals the mean arterial pressure divided by resistance, in order for flow to be the same at both ends of the range, the vascular resistance to flow must be greater at the higher pressure. Outside of the autoregulatory range, increases in mean arterial pressure lead to increases in flow and decreases in the mean arterial pressure lead to decreases in flow.

62. The answer is B. *(Physiology; Cardiovascular)*
A second-degree heart block is characterized by the failure of every P wave to elicit a QRS complex on the electrocardiogram (EKG). Thus, the ventricles occasionally 'drop' a beat. In a third-degree block, the P waves and QRS complexes are dissociated from one another on the EKG. In standard electrocardiography, the paper speed is set at 25 mm/sec, so that each 1-mm box is the equivalent of 40 msec. Therefore, in this EKG, the RR interval is 800 msec, which converts to a heart rate of 75 beats per minute. Bradycardia is defined as a heart rate of less than 60 beats per minute. In atrial fibrillation, no distinct P waves are seen on the EKG. Premature ventricular contraction is associated with widening of the QRS complex because the impulse usually originates in muscle and travels through slowly conducting myocardial tissue.

63. The answer is D. *(Physiology; Cardiovascular)* The QT interval, from the onset of the QRS complex to the end of the T wave, best approximates the duration of the ventricular myocyte action potential (i.e., from depolarization to repolarization). The PP and RR intervals indicate heart rate. The PR interval includes sinoatrial nodal, atrial, and atrioventricular nodal conduction time.

64. The answer is C. *(Physiology; Cardiovascular)* This diagram illustrates how the arterial baroreceptors help mediate changes in blood pressure:

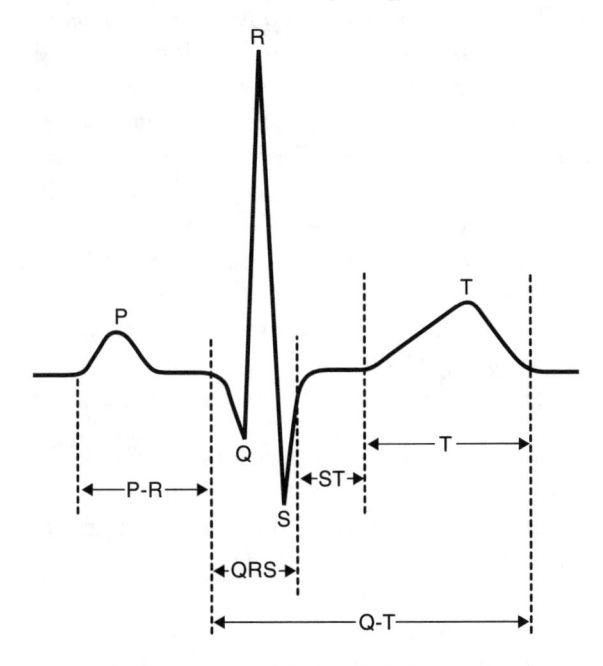

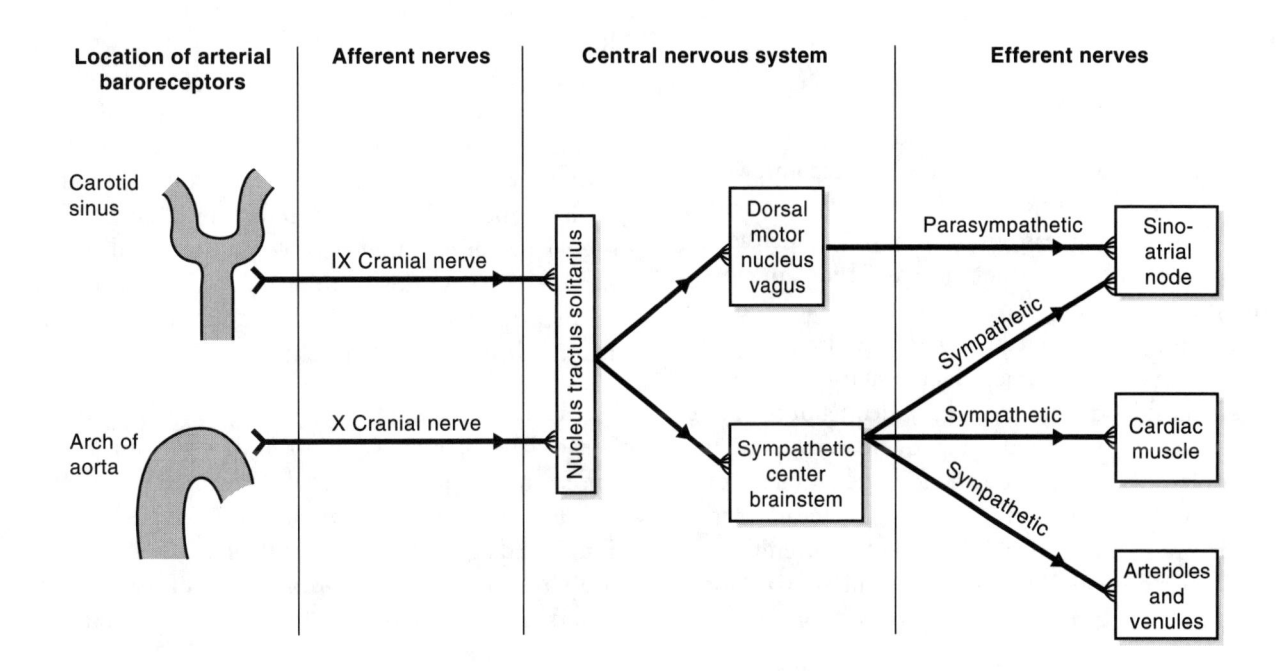

P wave = Atrial depolarization
QRS complex = Ventricular depolarization
T wave = Ventricular repolarization

The arterial baroreceptors are located in the carotid sinus and the arch of the aorta. A rise in pressure stretches them, increases their firing rate, and initiates a reflex aimed at restoring the pressure. Thus, stimulation of the arterial baroreceptors leads to a reflex decrease in heart rate, total peripheral resistance, and myocardial contractility. All of these reflex events help return the blood pressure to normal. In summary, afferent impulses ascending to the brain inhibit efferent sympathetic fibers to the heart and blood vessels and excite efferent parasympathetic (vagal) fibers to the sinoatrial node. In the case of a hemorrhage, the decrease in mean arterial blood pressure leads to the removal of inhibitory influences on sympathetic centers in the brain, resulting in a compensatory reflex increase in sympathetic drive. Therefore, the heart rate, myocardial contractility, and total peripheral resistance increase. An additional benefit of the reflex is that the increase in arteriolar resistance reduces capillary hydrostatic pressure and favors the reabsorption of interstitial fluid. Afferent impulses ascending to the brain from the baroreceptors are stimulatory to efferent parasympathetic fibers going to the heart. Thus, a fall in mean arterial blood pressure (hemorrhage) removes vagal drive to the heart.

65. The answer is D. *(Physiology; Cardiovascular)* Occlusion of the common carotid arteries lowers the blood pressure at the carotid sinus baroreceptor, which is located at the bifurcation of the artery. As a result, the baroreceptors cease firing and a relex increase in sympathetic activity ensues. Increased sympathetic activity to the heart increases the heart rate and the contractility. The increase in contractility leads to an increase in stroke volume, which results in an increase in systolic blood pressure and a widening of the pulse pressure. The increased sympathetic activity to the arterioles leads to vasoconstriction and an increase in the diastolic blood pressure. Thus, the mean arterial blood pressure increases. Injection of an arteriolar vasoconstrictor would increase the mean arterial blood pressure directly and lead to a baroreceptor-mediated reflex decrease in heart rate. Injection of a venoconstrictor would lead to an increase in venous return, which increases the end-diastolic volume (preload). The increased preload leads to an increase in stroke volume and cardiac output, causing the mean arterial blood pressure to increase. The increased mean arterial pressure would lead to a baroreceptor-mediated reflex decrease in heart rate. A venodilator would lead to a decrease in venous return, cardiac output, and mean arterial blood pressure, and a reflex increase in heart rate. Occlusion of a femoral artery will not result in any changes in blood pressure or heart rate because there are no baroreceptors in the femoral circulation.

66. The answer is D. *(Physiology; Cardiovascular)* Myocardial ischemia frequently occurs in the subendocardial muscle of patients with aortic regurgitation because the subendocardial blood flow is more affected than the epicardial blood flow by the diastolic pressure. Blood flow through the left ventricle falls during systole because of the compression of the blood vessels by the contracting muscle mass. The blood vessels most compressed by the contracting cardiac muscle are those closest to the chamber of the ventricle (i.e., those in the subendocardium). Epicardial vessels on the surface of the heart are only slightly compressed during systole. During diastole, the muscle relaxes, relieving the compression on the vessels. Therefore, the blood pressure that determines flow through the subendocardial layers of the left ventricle is the pressure during diastole. The reentry of aortic blood through the incompetent aortic valve results in a rapid fall in arterial pressure during diastole and a low diastolic pressure. Because flow through the subendocardium occurs primarily during diastole, aortic regurgitation may result in myocardial ischemia.

67. The answer is D. *(Physiology; Cardiovascular)* Aortic valve stenosis leads to a slowly rising radial pulse, a systolic murmur, a slightly reduced pulse pressure, left ventricular hypertrophy, and a left axis deviation. The diagram below places the heart in the center of a circle in which the positive ends of lead I, lead II, and lead III are at 0°, +60°, and +120°, respectively.

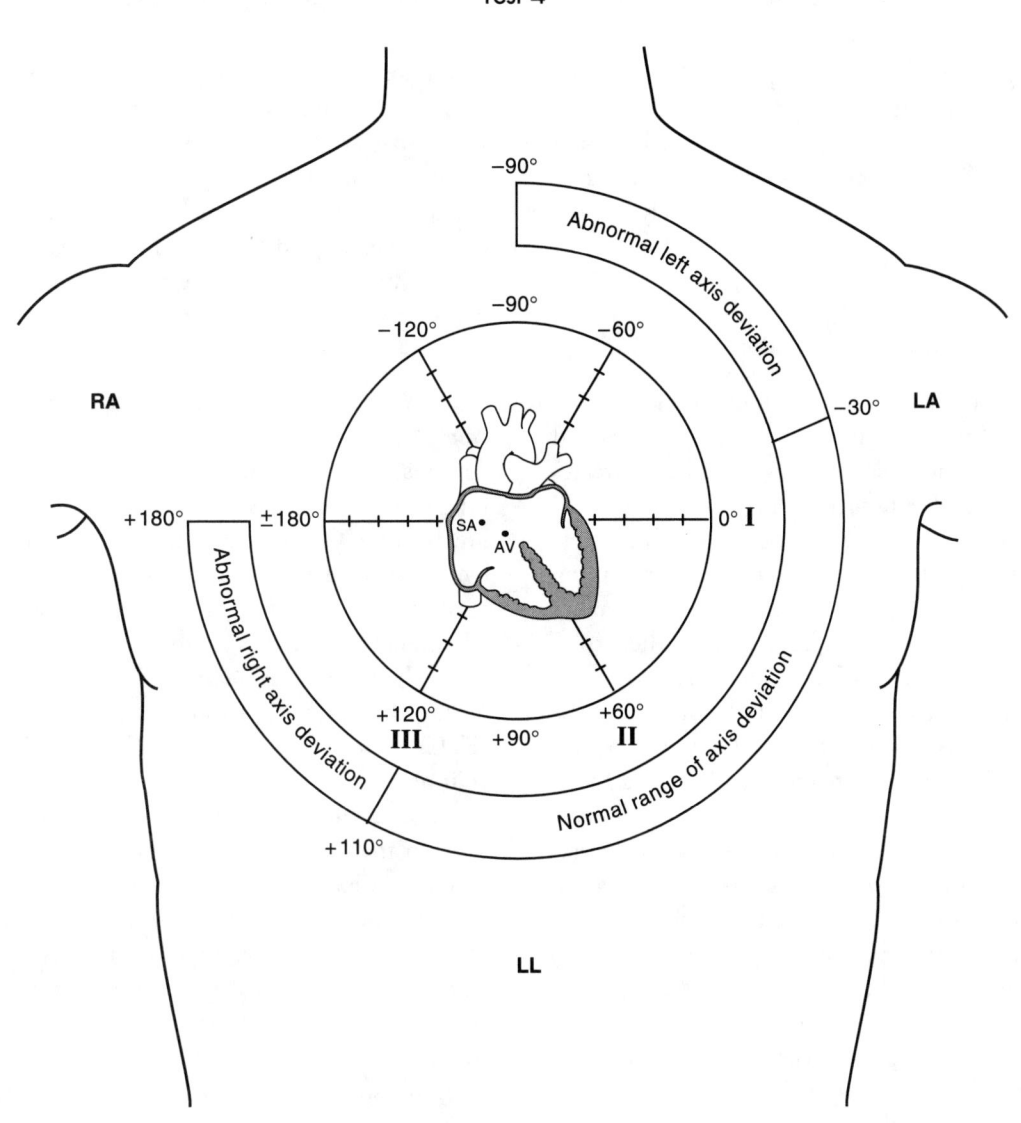

The mean electrical axis describes the average direction in which the wave of depolarization is travelling during ventricular depolarization and is normally between -30° and +110°. Ventricular depolarization is manifested on the electrocardiogram as the QRS complex. The magnitude of the recorded signal (i.e., the height of the R wave) is determined by the mass of tissue, its orientation, and the velocity of the wave of depolarization. The increased muscle mass (concentric hypertrophy) of the left ventricle associated with aortic stenosis results in the largest electrical signal on the left side of the heart. The recording electrode most closely in parallel alignment to the direction of the impulse spread (in this case, lead I) will record the largest potential.

68. The answer is E. *(Physiology; Pulmonary/respiratory)*
The increase in lung volume that results when the pressure around the lung (intrapleural pressure) becomes more negative describes the compliance of the lung. A lung with high compliance distends easily, whereas a lung with low compliance is more difficult to expand. That is, for the same decrease in intrapleural pressure, a lung that has low compliance will expand less than a normal lung. In fibrotic lung disease (curve

C), the lungs are stiffer (less compliant) than normal, whereas in emphysema, the lungs are highly compliant (curve A). As part of the aging process, lungs lose their elasticity and become more compliant. In asthma, the compliance of the lungs remains relatively normal (curve B).

69. The answer is D. *(Physiology; Pulmonary/respiratory)*
In the graph, the curves labeled *Lung A* and *Lung B* would be obtained if each of the lungs were removed from the chest and inflated. When the lungs are removed from the chest, they collapse to a very small volume. In order to inflate the lungs, air under positive pressure must be pumped through the airways. Lung B is the stiffer lung, exhibiting relatively smaller changes in lung volume than Lung A when the airway pressure is increased. Stiffness (high elasticity, low compliance) is a characteristic of individuals with fibrotic lung disease. The curve marked *Chest* would be obtained if the lungs were removed from the chest, the chest resealed, and the airway pressure altered. When the lungs are removed, the chest expands. As gas is pumped from the chest (i.e., as the airway pressure becomes more negative), the chest volume decreases. The functional residual capacity (FRC) is the volume of air in the lungs at the end of a passive expiration. It is that moment, in a respiratory cycle, when the respiratory muscles are fully relaxed and the outward recoil of the chest is exactly balanced by the inward recoil of the lungs. At this equilibrium point, the force required to expand the lungs (+20 cm H_2O in lung B) matches the force required to reduce the volume of the chest (−20 cm H_2O). Therefore, reading the graph, the FRC would be 1000 ml (1 L).

70. The answer is A. *(Physiology; Pulmonary/respiratory)*
The observed flow–volume relationship indicates that the individual has a restrictive pulmonary disease. The hallmark of pulmonary restrictive disease is a decrease in all lung volumes. Thus, the increased elastic recoil of the lungs reduces the inspiratory capacity (IC) and prevents a normal total lung capacity (TLC) from being achieved. The increased recoil also results in a residual volume (RV) that is lower than predicted and an expiratory flow rate that, at comparable lung volumes, is higher than predicted. Both the forced expiratory volume in 1second (FEV_1) and the forced vital capacity (FVC) are reduced, in a manner that results in either a normal or elevated ratio. A decreased FEV_1/FVC ratio and an increased functional residual capacity (FRC) and RV would indicate an obstructive disease (typified by a lower expiratory flow at comparable lung volumes).

71. The answer is A. *(Physiology; Pulmonary/respiratory)*
The observed relationship between expiratory flow and lung volume indicates that the subject has obstructive lung disease. Obstructive lung disease is characterized by an increase in airway resistance, which presents as a decrease in maximal expiratory flow at all lung volumes. The forced expiratory flow$_{25\%-75\%}$ ($FEF_{25\%-75\%}$), which is a measure of the rate at which gas flows during the middle portion of a forced vital capacity (FVC) maneuver, is reduced. The forced expiratory volume in 1 second (FEV_1) is reduced to a greater extent than the FVC in patients with obstructive disease, so that the FEV_1/FVC ratio is decreased. The total lung capacity (TLC) is larger than predicted and early closure of the airways during a forced expiration increases the residual volume (RV). Early closure may be caused by increased muscle tone (asthma), loss of elasticity (emphysema), or increased secretions in the airways (chronic bronchitis). The loss of elastic recoil in emphysema and the narrowed airways associated with asthma or chronic bronchitis are also responsible for the increase in the functional residual capacity (FRC). Because the intrapleural pressure becomes more negative during inspiration, causing the transmural pressure to increase and the airways to distend, maximal inspiratory flow is not as compromised as maximal expiratory flow.

72. The answer is D. *(Physiology; Pulmonary/respiratory)*
The spirogram suggests that the individual has a restrictive disease of pulmonary origin. The inspiratory capacity (IC) is reduced, as is the forced expiratory volume in 1 second (FEV_1) and the forced vital capacity (FVC). However, the FEV_1/FVC ratio is greater than predicted. Normally, the FEV_1/FVC ratio is 80%. In this patient, the ratio is 100%. Low lung

volumes coupled with a high FEV$_1$/FVC ratio usually indicates that the lung has increased elastic recoil (i.e., reduced compliance), such as occurs in patients with fibrotic lung disease. Because spirometry cannot measure the functional residual capacity (FRC) and residual volume (RV) directly, no valid judgments can be made on the basis of the data presented. However, in this condition, both the FRC and the RV are reduced. Bronchodilators do not improve pulmonary function in patients with fibrotic lung disease.

73. The answer is D. *(Physiology; Pulmonary/respiratory)*
The alveolar gas equation, given below, may be used to predict the effect of breathing various concentrations of oxygen, changing alveolar ventilation, or altering the barometric pressure on alveolar oxygen tension (PAO$_2$).

P$_{AO_2}$ = (P$_B$ − 47) × F$_{IO_2}$ − P$_{ACO_2}$/R + K, where

P$_B$ = the barometric pressure
47 = the vapor pressure of water at body temperature
F$_{IO_2}$ = the fraction of oxygen in the inspired gas
P$_{ACO_2}$ = the carbon dioxide tension in the alveolar gas
R = the ratio of the volume of carbon dioxide produced to the volume of oxygen consumed per minute, generally assumed to be 0.7
K = a constant that is relatively small and generally ignored
Thus, the P$_{AO_2}$ = 112 mm Hg:
P$_{AO_2}$ = (427 − 47) × 0.40 − 28/0.7 + K
P$_{AO_2}$ = 152 − 40, or 112 mm Hg

74. The answer is B. *(Physiology; Pulmonary/respiratory)*
Because carbon dioxide is approximately 20 times more soluble in plasma than oxygen, approximately 10 times more carbon dioxide is dissolved when the carbon dioxide tension (PCO$_2$) is 40 mm Hg, than oxygen at an oxygen tension (PO$_2$) of 100 mm Hg. As blood perfuses metabolically active tissues and the hemoglobin becomes more deoxygenated, more free amino groups capable of binding carbon dioxide (to form carbamino compound) and hydrogen ion (H$^+$) become available. This increased ability of deoxyhemoglobin to bind H$^+$ indicates that deoxyhemoglobin is a weaker acid than oxyhemoglobin. The major form in which carbon dioxide is carried in blood is as HCO$_3^-$. The shape of the carbon dioxide dissociation curve is linear; therefore, hyperventilation decreases the PCO$_2$ as well as the carbon dioxide content. The effect is significant.

75. The answer is D. *(Physiology; Pulmonary/respiratory)*
The diagram below illustrates the relationship between the carbon dioxide tension (PCO$_2$) and the carbon dioxide content of blood.

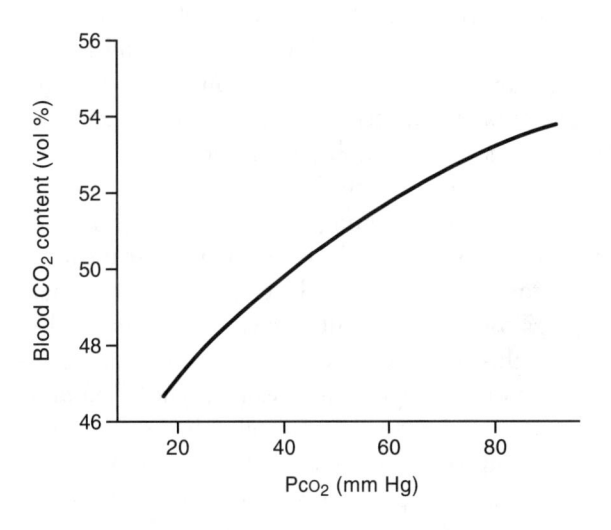

The nearly linear configuration of the carbon dioxide dissociation curve has important implications. Hyperventilation of a normal lung lowers the PCO$_2$ as well as the total carbon dioxide content. If alveolar ventilation doubles, the arterial PCO$_2$ will halve. On the other hand, the sigmoidal shape of the oxyhemoglobin dissociation curve indicates that hyperventilation of a healthy lung unit will increase the oxygen tension (PO$_2$), but will not significantly affect the total oxygen content. Thus, hyperventilation of a healthy lung unit can decrease the carbon dioxide content sufficiently, but will not increase the oxygen content enough to compensate for venous blood entering the circulation as a result of a right-to-left shunt. A right-to-left shunt always leads to hypoxemia, but not necessarily to carbon dioxide retention. In fact, depending on the degree of hyperventilation of the healthy lung units, the PCO$_2$

in the peripheral arterial blood of someone with moderate right-to-left shunting may be high, low, or normal.

76. The answer is D. *(Immunology; Immune response)*
Farmer's lung is an example of a hypersensitivity pneumonitis (extrinsic allergic alveolitis), which is an exudative lung reaction to an external antigen. Farmer's lung is characterized by both type III and type IV hypersensitivity reactions. Type III hypersensitivity reactions are immune complex—mediated reactions, in which the precipitation of immunoglobulin G (IgG) antibodies produces a localized immune complex reaction (Arthus reaction). Type IV (delayed reaction) hypersensitivity reactions involve T cells and macrophages and result in the production of noncaseating granulomas. Goodpasture's syndrome is characterized by the presence of anti–basement membrane antibodies directed against both alveolar and glomerular basement membranes, leading to type II immune destruction of these structures. Type II reactions are mediated by cytotoxic antibodies. Allergic bronchopulmonary aspergillosis is the result of both type I [immunoglobulin E (IgE)-mediated] reactions against bronchial colonization of the fungus and a localized type III immune complex reaction involving the alveoli. Antigen–antibody complexes are deposited in the alveoli, where they activate complement, which attracts neutrophils to the area. Alveolitis is the end result. Berylliosis produces an acute exudative pneumonitis or a chronic noncaseating granulomatous disease. The chronic condition is a result of a type IV hypersensitivity reaction. Histoplasmosis is a systemic fungal infection that produces granulomatous inflammation, also a type IV hypersensitivity reaction.

77. The answer is B. *(Pathology; Pulmonary/respiratory)*
The noninfectious interstitial pneumonias (fibrosing alveolitis and its variants) are a group of restrictive lung diseases of uncertain etiology. The variants begin as an alveolitis and progress to a fibrosing alveolitis. Unlike emphysema, which is characterized by dilatation of the distal airways, the noninfectious interstitial pneumonias are characterized by fibrotic alveoli and dilatation of the proximal terminal bronchioles. These changes produce a honeycomb pattern (Hamman-Rich lung). Chest radiographs demonstrate a ground-glass opacification. Bronchopneumonia, sarcoidosis, bronchioloalveolar carcinoma, and Goodpasture's syndrome are not associated with an alveolitis.

78. The answer is B. *(Pathology; Pulmonary/respiratory)*
Complications associated with respiratory distress syndrome (RDS) in newborns are those associated with the use of 100% oxygen, positive end-expiratory pressure (PEEP) therapy, and the disease itself. Oxygen-related injury (free radical injury) may result in retrolental fibroplasia and bronchopulmonary dysplasia (chronic lung disease with fibrosis). Bronchopulmonary dysplasia occurs in 20% of patients. Reduced surfactant leads to widespread atelectasis. Persistent patent ductus arteriosus and intraventricular hemorrhage are complications of RDS, but they are not oxygen-related injuries. Necrotizing enterocolitis may result from ischemia, bacterial colonization, and increased intake of protein from milk.

79. The answer is D. *(Pathology; Reproductive, Pulmonary/respiratory)*
The patient has a malignant pleural effusion secondary to metastatic serous cystadenocarcinoma of the ovary, which is the most common malignant tumor of the ovary. Surface-derived ovarian tumors are carcinoembryonic antigen (CA)-125–positive. Pleural effusions are either transudates (which involve disturbances in Starling's forces) or exudates (which involve an increase in vessel permeability that leads to protein loss and emigration of leukocytes). Congestive heart failure is the most common cause of transudative pleural effusions. Causes of exudative pleural effusions include pneumonia, pulmonary infarction, and malignancy, which is the most common cause in older patients. The most common malignancy associated with a pleural effusion is peripherally located primary lung cancer, usually an adenocarcinoma. Metastatic breast cancer and malignant lymphoma are also relatively common causes of pleural effusion. Metastasis from the ovaries, gastrointestinal tract, and thyroid are less common causes. A bloody pleural fluid is most commonly associated with trauma (iatrogenic or otherwise), malignancy, or a pulmonary infarction. Most malignant

effusions are symptomatic; patients present with dyspnea, chest pain, and cough. Pleural fluid cytology, pleural biopsy, or both confirm the diagnosis.

80. The answer is A. *(Pathology, Physiology; Pulmonary/respiratory)*
Pleural effusions associated with congestive heart failure are transudative and are most likely to contain less than 3 g/dl of protein and very few inflammatory cells. Transudative effusions are caused by an increase in hydrostatic pressure or a decrease in oncotic pressure. Exudative effusions are caused by an increase in vessel permeability. The pleural fluid protein content is usually greater than 3 g/dl and numerous cells are present. Causes of exudative pleural effusions include pneumonia, pulmonary infarction, acute pancreatitis, and malignancy.

81. The answer is D. *(Pathology; Pulmonary/respiratory)*
More than 90% of bronchial adenomas are bronchial carcinoids. Bronchial carcinoids are derived from Kulchitsky cells, which are of neural crest origin and contain neurosecretory granules. Bronchial carcinoids are not related to smoking, unlike small cell carcinomas of the lung, which are also neurosecretory tumors. Grossly, bronchial carcinoids project into the lumen like an "iceberg" and readily bleed when biopsied. Approximately 40% metastasize locally to hilar nodes and 5%–10% metastasize to distant sites, like the liver. Bronchial carcinoids are rare causes of the carcinoid syndrome. Metastasis is not essential for the development of carcinoid syndrome because the serotonin is released directly into the systemic circulation. Bronchioloalveolar carcinoma, squamous carcinoma, large cell undifferentiated carcinoma, and scar carcinoma do not contain neurosecretory granules.

82. The answer is E. *(Pathology; Pulmonary/respiratory)*
Hypercalcemia is related to the ectopic secretion of a parathyroid hormone–like peptide and can be induced by a primary squamous cell carcinoma. Hypercalcemia is not a manifestation of local invasion. Hoarseness is associated with invasion of the recurrent laryngeal nerve. Malignant pericardial effusions are most commonly the result of direct spread from a primary lung cancer. Endogenous lipoid pneumonia is most commonly secondary to bronchial obstruction by a primary lung cancer (squamous cell carcinoma or small cell carcinoma). Dysphagia for solids is caused by compression or invasion of the esophagus by cancer.

83. The answer is C. *(Microbiology; Pulmonary/respiratory)*
Pulmonary cryptococcosis can be caused by the inhalation of *Cryptococcus neoformans*, which grows abundantly in pigeon excretions. The birds are reservoirs for, but are not infected by, the fungus. Most cases of cryptococcosis are opportunistic infections in immunocompromised patients. In tissue, the yeast forms produce a granulomatous reaction if host immunity is intact, but no inflammatory reaction occurs if the host is immunocompromised. This is particularly true in the central nervous system (CNS). *Candida albicans*, *Coccidioides immitis*, *Chlamydia trachomatis*, and *Histoplasma capsulatum* are not associated with pigeon excreta.

84. The answer is E. *(Pathology; Pulmonary/respiratory)*
Silicosis is a type of pneumoconiosis characterized by inhalation of insoluble silica particles, such as sand or quartz. The silica particles are ingested by alveolar macrophages, which rupture, releasing interleukin-1 (IL-1), proteases, free radicals, and fibrogenic factors that stimulate fibrogenesis. The pathology in the lung is characterized by extensive nodulation caused by concentric areas of fibrosis that contain polarizable silica particles. Farmer's lung is caused by the inhalation of thermophilic actinomycetes in moldy hay. Silo filler's disease is related to the inhalation of nitrous oxide fumes originating from fermenting corn. Bagassosis is caused by exposure to moldy sugar cane, while byssinosis is associated with exposure to cotton, hemp, or linen.

85. The answer is B. *(Pathology; Reproductive, Pulmonary/respiratory)*
This patient most likely has metastasis from a choriocarcinoma originating in his testicles. β-Human chorionic gonadotropin (β-hCG) is released from the syncytiotrophoblast portion of the tumor. The gynecomastia most likely results from stimulation of

breast tissue by the β-hCG, which is a luteinizing hormone (LH) analogue. Leukemias or multiple myeloma originating from the bone marrow are not associated with gynecomastia. Kidney tumors, especially renal adenocarcinoma, can ectopically secrete gonadotropins; however, renal tumors do not usually occur in patients this young. Islet cell tumors of the pancreas do not ectopically secrete β-hCG or gonadotropins. Most anterior pituitary tumors are benign. Primary gonadotropin-secreting tumors are rare.

86. The answer is A. *(Pharmacology; Pulmonary/respiratory)*
Bleomycin and amiodarone are associated with pulmonary fibrosis. Of the diseases listed, only progressive systemic sclerosis is associated with pulmonary fibrosis. Progressive systemic sclerosis produces a honeycomb lung and subpleural fibrosis. Pulmonary hypertension, adenocarcinoma of the lung, and cor pulmonale are possible sequelae. Other drugs implicated in producing interstitial fibrosis include nitrofurantoin, cyclophosphamide, and methysergide. Fibrosis is not a prominent feature of bronchial asthma. Aspergillosis is associated with asthma, fungus balls, and an invasive pneumonia with hemorrhagic infarctions. Wegener's granulomatosis produces a necrotizing granulomatous vasculitis involving the upper airways, lungs, and kidneys. Pulmonary alveolar proteinosis is an unusual disease that is characterized by a protein-rich alveolar fluid that is rich in surfactant. Pulmonary alveolar proteinosis does not produce fibrosis and is not an interstitial lung disease.

87. The answer is A. *(Pharmacology)*
Mefloquine is a fluorinated quinoline derivative that is currently effective in the prophylaxis and treatment of malaria caused by *Plasmodium falciparum*. Mefloquine is also used to prevent malaria caused by *P. vivax*. Doxycycline is an alternative agent for the prophylaxis and treatment of chloroquine-resistant malaria. Another, less suitable regimen for the prophylaxis of chloroquine-resistant malaria is a combination of chloroquine, proguanil, pyrimethamine, and sulfadoxine.

88. The answer is D. *(Pharmacology; Gastrointestinal)*
Only primaquine is effective in eradicating the persistent hepatic infection that occurs in *Plasmodium vivax* and *Plasmodium ovale* malaria. Primaquine is combined with a blood schizonticide (e.g., chloroquine) in order to treat acute infections and may be taken with chloroquine at the end of a period of potential exposure in order to provide complete prophylaxis against these species. Because primaquine may cause hemolysis in patients with glucose-6-phosphate dehydrogenase deficiency, patients should be screened or monitored for this deficiency.

89. The answer is C. *(Pharmacology; Hematopoietic/lymphoreticular)*
MOPP, the combination regimen of meclorethamine, vincristine (Oncovin), prednisone, and procarbazine, is often alternated with the ABVD regimen [doxorubicin (Adriamycin), bleomycin, vincristine, and dacarbazine] in the treatment of patients with Hodgkin's disease. These regimens are based on the use of drugs with different mechanisms of action and different toxicities, so that the cancer cells are attacked at different sites of action, and the collective toxicity is not directed toward one organ system. Mechlorethamine is a nitrogen mustard alkylating agent, vincristine is a mitotic spindle poison, prednisone is a glucocorticoid with lymphotoxic effects, and procarbazine has multiple actions directed against nucleic acids and proteins. Vincristine and prednisone produce less bone marrow suppression than alkylating agents and antimetabolites.

90. The answer is B. *(Pharmacology; Pharmacodynamic and pharmacokinetic processes)*
Although didanosine and zalcitabine produce less bone marrow suppression than zidovudine in some patients, they are associated with a significant risk of sensory and motor neuropathy than can be quite painful. Dosage reduction may partially alleviate this condition while still providing some antiviral activity. Didanosine has been associated with fatal pancreatitis as well.

91. The answer is B. *(Pharmacology; Pharmacodynamic and pharmacokinetic processes)*
Human immunodeficiency virus (HIV) binds to a surface molecule on CD4 lymphocytes and enters the cell, where viral RNA serves as a template for both

positive and negative strands of complementary DNA. Reverse transcriptase (RNA-dependent DNA polymerase) transcribes the viral RNA. The double-stranded complementary DNA is then spliced into the host cell genome by a virus-encoded integrase. The integrated complementary DNA is then transcribed and translated into viral proteins. The didoxynucleosides, including zidovudine, selectively inhibit reverse transcriptase. Host cell DNA polymerase is more than 100 times less sensitive than the HIV polymerase to zidovudine.

92. The answer is D. *(Pharmacology; Pharmacodynamic and pharmacokinetic processes)*
Cyclosporine and other antibiotic immunosuppressants are less likely to cause bone marrow suppression than the cytotoxic agents (e.g., methotrexate, mercaptopurine, cyclophosphamide). Azathioprine is converted to mercaptopurine in the body.

93. The answer is D. *(Pharmacology; Pharmacodynamic and pharmacokinetic processes)*
The steady-state level of digoxin is primarily dependent on renal clearance of the drug and the dosing rate. Digoxin is eliminated primarily by glomerular filtration and digoxin clearance is proportional to creatinine clearance, which declines significantly in renal disease and requires a dosage reduction to avoid toxicity. Probenecid, which inhibits renal tubular secretion, has no significant effect on digoxin clearance. Hepatic disease or inhibition of hepatic drug metabolizing enzymes by cimetidine also has relatively little effect on digoxin clearance and serum concentrations. Administration of gastric acids and cholestyramine may reduce digoxin absorption and lower serum levels; therefore, these drugs should be administered several hours before or after administration of digoxin.

94. The answer is B. *(Pharmacology; Pharmacodynamic and pharmacokinetic processes)*
The time required to establish a new steady-state plasma drug concentration when maintenance doses are changed is approximately four half-lives, the same amount of time required to establish the initial steady-state level. Because the half-life of digoxin is approximately 1.6 days, the time required to reach a new steady state is approximately 7 days. Digitoxin has a much longer half-life of approximately 7 days and the time required to reach steady state with digitoxin is approximately 4 weeks. The time required to reach steady state is independent of dose or dosage interval and will not change unless drug clearance or volume of distribution are altered, because these two variables determine half-life.

95. The answer is E. *(Gross anatomy; Cardiovascular)*
The right coronary artery lies in the coronary, or atrioventricular, sulcus. This artery is a direct branch of the ascending aorta, supplies the right atrium and ventricle, and sends a branch to the sinoatrial node. The posterior interventricular artery is a branch of the right coronary artery and lies in the posterior interventricular sulcus. This branch supplies the right and left ventricles and the posterior portion of the interventricular septum.

96. The answer is C. *(Gross anatomy; Pulmonary/respiratory)*
The inferior limit of the lungs is at the sixth, eighth, and tenth ribs at the midclavicular, midaxillary, and paravertebral lines, respectively. The inferior limit of the parietal pleura is two rib levels lower at each location (i.e., at the eighth, tenth, and twelfth ribs). The cupola of the pleura rises above the first rib into the root of the neck. These surface landmarks are used to determine the site for thoracentesis. For example, a pleural tap is performed in the eleventh intercostal space in the paravertebral line, because this level is below the lung but within the pleural space. In the midaxillary line, the same procedure is performed in the ninth intercostal space.

97. The answer is A. *(Gross anatomy; Cardiovascular)*
When the ductus venosus closes, the bypass of blood around the liver ends, and blood must pass through the liver. The adult remnant of the ductus venosus is the ligamentum venosum. The increased venous return from the lungs increases the blood pressure in the left atrium, and the reduced blood flow from the umbilical vein reduces blood pressure in the right atrium. This change in pressure differential causes the foramen ovale to close. The ductus arteriosus closes

in response to elevated oxygen tension and reduced circulating prostaglandins.

98. The answer is A. *(Pathology; Cardiovascular)*
An infection is least likely to be primarily involved in the pathogenesis of a berry aneurysm in the central nervous system. The majority of these aneurysms are caused by a congenital defect that is present at birth, in which the internal elastic lamina and muscle wall are absent. In the presence of hypertension, an aneurysm with the potential to rupture and produce a subarachnoid bleed develops.

Osler nodes, which are painful nodules on the pads of the fingers and toes, are seen in patients with infective endocarditis. They are an immune complex vasculitis, secondary to bacterial antigen–antibody complex deposition.

The "spots" in Rocky Mountain spotted fever are caused by a direct invasion of the endothelial cells by *Rickettsia rickettsii*. The vessels rupture and produce the spots, which first appear in the periphery of the hands and feet and extend towards the trunk.

Both myocarditis and pericarditis are most commonly caused by a viral infection, usually coxsackievirus B.

99. The answer is A. *(Physiology; Cardiovascular, Pulmonary/respiratory)*
Well-trained athletes have a slower resting heart rate, less increase in heart rate during exercise, and a greater stroke volume during exercise than a person who is not conditioned. The increase in stroke volume either during exercise or at rest is caused by increased cardiac contractility associated with left ventricular hypertrophy, not an increase in preload. Cardiac hypertrophy allows more blood to be ejected out of the left ventricle (greater ejection fraction) during systole than in a non-conditioned person. For example, a well-trained athlete most likely has a resting stroke volume of 100 ml that increases to greater than 150 ml during exercise, whereas a non-conditioned person most likely has a resting stroke volume of 80 ml and an average stroke volume of 120 ml after exercise.

100. The answer is E. *(Physiology, Pathology; Cardiovascular)*
Cardiac output typically ranges from 4 to 7 liters/min. To correct for differences in body size, cardiac output is often expressed as the cardiac index:

$$\text{Cardiac index} = \text{cardiac output}/m^2 \text{of body surface area}$$

Cardiac output = heart rate × stroke volume. Therefore, an increase in heart rate, stroke volume, or both increases cardiac output. Reducing the preload in the left ventricle, or the amount of blood in the left ventricle during diastole, decreases the force of contraction, which is based on the length–tension relationship in cardiac muscle. Less filling decreases the stroke volume and reduces cardiac output.

101. The answer is B. *(Pathology; Cardiovascular)*
The patient has evidence of both left-sided heart failure (wet inspiratory rales) and right-sided heart failure (neck vein distention and heptatojugular reflux caused by liver congestion). The left-sided heart failure is caused by mitral regurgitation, which produces a pansystolic murmur radiating into the axilla. The patient has carditis and polyarthritis, two of the five major Jones criteria used for diagnosis of acute rheumatic fever.

Rheumatic fever is a systemic disease caused by an immunologic hypersensitivity reaction (e.g., cytoxic antibodies, immune complexes) that primarily, but not exclusively, affects the heart. It is most commonly associated with group A β-hemolytic streptococcal infections of the pharynx (e.g., exudative tonsillitis). Because it is an immunologic disease, a positive blood culture for group A streptococcus is not likely. The pathogenesis is believed to be associated with the host developing antibodies against the M protein (virulence factor) of certain types of streptococcus, which have cross-reactivity (antigen mimicry) against the heart and other tissues and causes fibrinoid necrosis and healing by scar formation. Immunologic damage to the valves, most commonly the mitral valve, is the most serious complication of rheumatic fever. Nonembolic, sterile, warty-appearing vegetations are noted along the lines of closure of the involved valve. Initially, the mitral valve disease is one of insufficiency (i.e., regurgitation) because the warty-appearing vegetations line up on the line of closure of the valve. However, mitral stenosis is the most common chronic valvular lesion in long-standing rheumatic fever.

The diagnosis of acute rheumatic fever requires the use of the Jones criteria. The major criteria area migratory, asymmetric polyarthritis (75% of cases); carditis (35% of cases); chorea (10% of cases); subcutaneous nodules (10% of cases), which are pea-size, nontender nodules with a granulomatous reaction around a center of fibrinoid necrosis that develop on the extensor tendons; and erythema marginatum (10% of cases), which are skin lesions with a circular ring of erythema around normal skin. Patients commonly have elevation of their antistreptolysin O titers.

102. The answer is B. *(Pathology, Physiology; Cardiovascular)*
In normal fetal circulation, unoxygenated fetal blood is carried by two umbilical arteries to the placenta for oxygen transfer. The high affinity of hemoglobin F for oxygen facilitates this transfer. The oxygenated blood enters a single umbilical vein. A portion of this blood drains into the liver, and the other portion drains into the inferior vena cava through the ductus venosus. Blood empties into the right atrium, where two thirds of it is shunted through a patent foramen ovale, into the left atrium, and out to the systemic circulation. The remaining one third enters into the right ventricle and out the pulmonary artery. Blood entering the pulmonary artery encounters increased pulmonary resistance from medial smooth muscle proliferation of the vessels, thus it is shunted out through the ductus arteriosus, which connects with the aorta. The ductus arteriosus is kept open by hypoxemia and prostaglandin E_2, which is synthesized in the placenta. At birth, a newborn's first breath increases the partial pressure of oxygen (PaO_2), which lowers the pulmonary artery resistance and increases pulmonary blood flow and return of blood to the left atrium, functionally closing the foramen ovale that connects the two atrial chambers by 2 weeks of age. The ductus arteriosus undergoes muscle spasm because the vasodilator effect of prostaglandins is reduced after the placenta is delivered. The presence of hypoxemia or acidosis, both potent vasoconstrictors, enhances pulmonary vascular resistance and keeps the ductus arteriosus open. Indomethacin, a potent nonsteroidal anti-inflammatory agent, may be used to close the ductus, because it inhibits prostaglandin E_2.

103. The answer is B. *(Pathology; Cardiovascular)*
Constrictive pericarditis occurs when the pericardium and visceral, parietal, or both layers, become noncompliant. The stiffness of the pericardium is often enhanced by dystrophic calcification, which is visible on a routine chest radiograph. In cardiac tamponade, ventricular filling is impeded throughout early and late diastole. In constrictive pericarditis, ventricular filling is normal in early diastole but reduced abruptly in later diastole when the diastolic volume reaches the confines of the thickened pericardial shell, producing a sound known as the pericardial (diastolic) knock.

The mean pressures in all chambers of the heart and the pulmonic and systemic veins are equally elevated. In other words, the ventricles are not completely filled, thus the stroke volume and cardiac output are reduced. Because the heart is not completely filled, there is a backup of venous pressure. On inspiration, the typical neck vein collapse associated with decreased intrathoracic pressure drawing blood into the right heart is reversed. Neck vein distention on inspiration is called Kussmaul sign. Because less blood enters the right heart on inspiration, less blood leaves the left heart, and the amplitude of the pulse and the blood pressure decrease (pulsus paradoxicus).

104-105. The answers are: 104-C, 105-B. *(Pathology, Physiology; Cardiovascular)*
The following diagram shows all of the key events in the cardiac cycle.

312　Body Systems Review I: Hematopoietic/Lymphoreticular, Respiratory, Cardiovascular

the TV). The *a* wave would be increased in mitral stenosis. The *B section* begins systole with closure of the MV and TV (S1 heart sound; MV before TV) followed by isovolumic contraction of the ventricles against the closed aortic (AV) and pulmonic (PV) valves (not depicted). In *section C*, when the AV and PV open, there is rapid ventricular ejection of blood into the aorta and pulmonary arteries (not depicted). In *section D*, ventricular ejection slows down and ends systole with closure of the AV and PV. In aortic stenosis, the aortic pressure between the beginning of *section C* to the end of *section D* would not increase to the same degree as does the left ventricular (LV) pressure because the valve has a problem opening and the stroke volume is decreased. *Section E* marks the beginning of diastole with isovolumic relaxation of the ventricles followed by opening of the MV and TV. During *D* and *E*, there is filling of the atria when the MV and TV are closed, which produces the *v* wave in the left atrium and JVP. The *v* wave is accentuated in mitral and tricuspid insufficiency as blood enters the atria through the incompetent valves. In *section F*, there is rapid ventricular filling (correlates with the y descent in the left atrial pressure) followed by *section G*, where ventricular filling slows down (portion of diastole most affected by heart rate). In cardiac tamponade, one would not expect to see a *v* descent in the jugular venous pulse wave in *F*, because the ventricles do not fill properly in both early and late diastole owing to pressure exerted on these chambers by blood in the pericardial sac. Between *E* and *A*, there would be a greater increase in LV volume in aortic insufficiency owing to regurgitation of blood back into the ventricle during diastole.

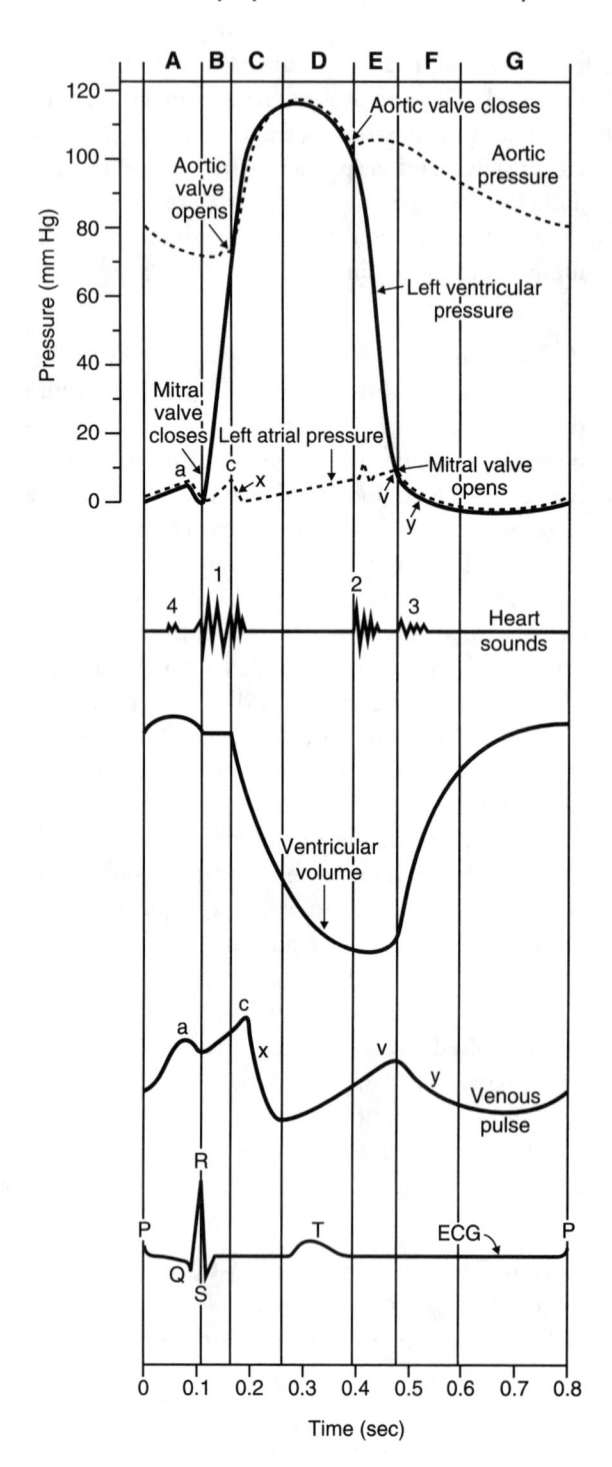

In *section A*, the final events in diastole (beginning of *E* to the end of *A*) are occurring while the mitral valve (MV) and tricuspid valve (TV) are still opened. The *a* wave in the jugular venous pulse (JVP) represents atrial contraction in late diastole (not closure of

106. The answer is D. *(Physiology, Pathology; Cardiovascular, Pulmonary/respiratory)*
The ventricular function curve shows the effect of preload (i.e., left ventricular end-diastolic volume) on stroke volume using Frank-Starling mechanism of increasing the force of contraction by increasing the length of the cardiac muscle. The ejection fraction is calculated by dividing the stroke volume by the left ventricular end-diastolic volume. Therefore, for *point A*, 60/90 = 0.67; for *point B*, 80/120 = 0.67; and for *point C*, 100/150 = 0.67. The ejection fraction for each point on the curve remains the same, indicating

that Frank-Starling mechanisms are operating rather than an increase in contractility, which would have increased both the stroke volume and the ejection fraction.

107-108. The answers are: 107-A, 108-A. *(Hematology; Biochemistry and molecular biology)*
The following diagram shows the steps in heme synthesis. Porphyrins, which are precursors for heme synthesis, are also involved in oxidative or oxygen-transferring functions. Enzyme defects in the heme synthesis pathway account for the clinical porphyrias.

$$\begin{array}{c} \text{Succinyl CoA + Glycine} \\ \downarrow \textit{δ-Aminolevulinic acid synthase} \\ \text{δ-Aminolevulinic acid} \\ \downarrow \textit{δ-Aminolevulinic acid dehydrase} \\ \text{Porphobilinogen} \\ \downarrow \textit{Uroporphyrinogen I synthetase} \\ \text{Uroporphyrinogen III} \\ \downarrow \textit{Uroporphyrinogen III decarboxylase} \\ \text{Coproporphyrinogen III} \\ \downarrow \\ \text{Protoporphyrin IX + Iron} \\ \downarrow \textit{Ferrochelatase} \\ \text{Heme + Globin chains } (\alpha, \beta, \delta, \gamma) \rightarrow \text{Hemoglobin} \end{array}$$

Succinyl coenzyme A (succinyl CoA), which combines with glycine to initiate porphyrin synthesis, is not a water-soluble vitamin. Lead denatures two enzymes that play a role in heme synthesis, mainly δ-aminolevulinic acid dehydrase and ferrochelatase. Inhibition of δ-aminolevulinic acid dehydrase and ferrochelatase causes a build-up of δ-aminolevulinic acid and protoporphyrin IX, respectively, behind the enzyme blocks. Therefore, the free erythrocyte protoporphyrin IX concentration is increased in patients with lead poisoning. It is also increased in patients with iron deficiency and anemia of chronic inflammation, because in both anemias, there is insufficient iron available to combine with the protoporphyrin. Heme inhibits δ-aminolevulinic acid synthase via a negative-feedback loop; therefore δ-aminolevulinic acid synthase is the rate-limiting reaction in heme synthesis. The availability of globin chains is reduced in patients with thalassemia syndromes.

Lead does not inhibit δ-aminolevulinic acid synthase. The two most common porphyrias in the United States are acute intermittent porphyria and porphyria cutanea tarda. Acute intermittent porphyria has two basic defects: increased activity of δ-aminolevulinic acid synthase and decreased activity of uroporphyrinogen synthetase. The net effect of the two defects in acute intermittent porphyria is an excessive quantity of δ-aminolevulinic acid and porphobilinogen. When drugs are metabolized in the liver by the cytochrome P-450 system, heme is utilized in the process. Because δ-aminolevulinic acid synthase activity goes unchecked as a result of the loss of negative feedback, certain drugs can precipitate an attack of acute intermittent porphyria. Porphyria cutanea tarda is caused by decreased activity of uroporphobilinogen decarboxylase. The net result of the enzyme defect in porphyria cutanea tarda is an increased excretion of uroporphyrin, a slight increase in formation of coproporphyrins, and normal porphobilinogen levels.

109. The answer is D. *(Immunology; Immune response)*
Scratch (intradermal) testing, not radioimmunosorbent and radioallergosorbent testing (RIST and RAST), is used to detect the late phase reaction. RIST and RAST are used in the work-up of type I hypersensitivity disorders involving immunoglobulin E (IgE). RIST measures the IgE concentration in serum. RAST detects patient IgE antibodies against specific allergens (e.g., ragweed, dust). When two subjacent IgE antibodies on the mast cell membrane are bridged by a specific allergen, the mast cells degranulate, causing the release of preformed primary mediators from the granules. These mediators include histamine, serotonin, neutral proteases, and chemotactic factors for neutrophils and eosinophils. Histamine and serotonin are biogenic amines. They produce vasodilation and increased vessel permeability and the classic wheal and flare reaction in the skin. This release reaction also triggers the de novo synthesis of prostaglandins and leukotrienes over the next few hours. These chemical mediators are released into the tissue and produce a wheal and flare reaction as well (late phase reaction). During the late phase reaction, prostaglandins and leukotrienes, which potentiate the initial allergic response, are released. Other stimuli for degranulation of mast cells are the anaphylatoxins C3a and C5a, warm water, and certain drugs (e.g., codeine, morphine). Interleukins 3 and 4 are important in mast cell synthesis in the bone marrow.

110. The answer is C. *(Hematology; Skin/connective tissue)*
Platelets have ABO antigens but lack Rh antigens. Platelets have a peripheral zone, a sol–gel zone, and an organelle zone. The peripheral zone is composed of an extramembranous glycocalyx that surrounds a plasma membrane. An open canalicular system lies underneath the plasma membrane. The sol–gel zone has microtubules and microfilaments and a dense tubular system that contains adenine nucleotides and calcium ions. Thrombosthenin, a contractile protein, is also present in this zone. The organelle zone is composed of dense bodies, alpha granules, mitochondria, lysosomes, and endoplasmic reticulum. Alpha granules release fibrinogen and lysosomal enzymes, while dense bodies release adenosine diphosphate (ADP), coagulation factors (e.g., factor VIII), serotonin, histamine, catecholamines, platelet factor 3, and platelet factor 4 (a heparin-like factor). Thromboxane synthetase converts prostaglandin H_2 into thromboxane A_2, a potent platelet aggregator and vasoconstrictor. Platelets have human leukocyte antigens (HLA) antigens and specific platelet antigen-1 (PLA-1) antigens. Patients who are negative for the PLA-1 antigens may develop anti–PLA-1 antibodies if they are exposed to this antigen in blood products. Re-exposure to platelets containing the antigen may result in severe thrombocytopenia. Platelets have glycoprotein IIb (GP IIb) and glycoprotein IIIa (GP IIIa) for binding fibrinogen, which is important in the formation of temporary hemostatic plugs. Platelets also have glycoprotein Ib (GP Ib) receptors for binding von Willebrand's factor.

111. The answer is E. *(Histology; Cardiovascular)*
Endothelial cells are not positive for S100 antigen. This antigen is primarily found in tissue derived from the neural crest. Endothelial cells play a pivotal role in the hemostatic process. They synthesize tissue plasminogen activator, von Willebrand's factor, prostacyclin (PGI_2), and nitric oxide. Tissue plasminogen activator initiates the fibrinolytic system. Von Willebrand's factor is necessary for platelet adhesion when vessels are injured. PGI_2 inhibits platelet aggregation and produces vasodilatation. Nitric oxide is a potent vasodilator that is released when endothelial cells are injured. On electron microscopy, endothelial cells contain Weibel-Palade bodies. This finding is useful for identifying a tumor as vascular in origin.

112. The answer is D. *(Histology; Skin/connective tissue)*
Monocytes release interleukin-1 (IL-1), which enhances CD4 T helper cell activity. Helper T cells release interleukin-2 (IL-2). Monocytes normally live in the peripheral blood for 24 hours and then emigrate into tissue to become wandering or fixed macrophages in the mononuclear phagocyte system (MPS). As a primary effector cell in chronic inflammation, they are important in the phagocytosis and killing of microbial agents, the removal of debris, the processing of antigens for presentation to B and T lymphocytes, and the enhancement of host responses via secretory products [e.g., IL-1, tumor necrosis factor (TNF), and B and T cell growth factors]. Monocytes kill microbial agents by releasing lysosomal enzymes, generating free radicals, releasing chemotactic factors, and activating the oxygen-dependent myeloperoxidase system. They also play a prominent role in the atherosclerotic process by releasing free radicals that contribute to the formation of oxidized low-density lipoprotein, supplying growth factors for smooth muscle hyperplasia, and participating in the formation of foam cells by phagocytizing (scavenging) low-density lipoprotein. Monocytes harbor human immunodeficiency virus (HIV); this is the primary mechanism by which the virus is introduced to the central nervous system (CNS).

113. The answer is A. *(Microbiology; Microbial biology and infection)*
There is no risk for developing hepatitis or any infectious disease with injection of hepatitis B immune globulin. Cryoprecipitate, packed red blood cells, platelet concentrates, and fresh frozen plasma all pose a risk for transmitting hepatitis. Hepatitis C virus (HCV) is the most common cause of post-transfusion hepatitis. Although blood is screened for HCV, there is still a 1/3300 risk per unit of blood for contracting the virus. This incidence should decrease with the introduction of assays that detect anti-HCV antibodies within 4 weeks of exposure.

114. The answer is C. *(Physiology; Cardiovascular)*
The decrease in blood pressure and increase in heart rate after sitting up indicates the presence of significant hypovolemia. The drop in stroke volume from hypovolemia is counterbalanced by a compensatory increase in heart rate in order to increase the cardiac output. Splenic rupture is the most likely cause of the hypovolemia, and is consistent with the tenderness over the rib. Because normal saline was not infused prior to collecting blood for a hemoglobin and hematocrit, the loss of whole blood did not reduce the hemoglobin and hematocrit concentration because of the proportional loss of plasma and red blood cells. Therefore, the hemoglobin and hematocrit do not accurately reflect the hematologic status of this patient. However, if the hemoglobin and hematocrit had been obtained after infusion of normal saline, the normal saline would begin replacing the lost plasma and would reveal the red blood cell deficit. A key point to remember in acute blood loss is that the plasma volume is restored first by fluid entering the vascular compartment from the interstitial space. Eventually, the bone marrow begins repairing the erythrocyte deficit, but this can take a few hours to a few days. Because the plasma volume deficit is restored almost immediately, the erythrocyte deficit becomes apparent, as reflected by a drop in the hemoglobin and hematocrit.

115. The answer is C. *(Pathology; Hematopoietic/lymphoreticular)*
Excessive exposure to ultraviolet light predisposes to primary skin malignancies (e.g., basal cell carcinoma, squamous cell carcinoma, malignant melanoma), but it does not predispose to leukemia. Factors that predispose to leukemia include chromosomal abnormalities (e.g., trisomy 21 in Down's syndrome), chromosome instability syndromes, which are characterized by an increased number of mutations (e.g., Fanconi's syndrome, ataxia telangiectasia, Bloom's syndrome), oncogenic viruses, chemicals (e.g., benzene, alkylating agents), and immunodeficiency syndromes (e.g., Wiskott-Aldrich syndrome).

116. The answer is A. *(Physiology; Cardiovascular)*
The pulmonary and cutaneous circulations do not autoregulate flow. Autoregulation does occur in the renal, cerebral, and coronary circulations.

117. The answer is B. *(Physiology; Cardiovascular)*
A decreased plasma potassium concentration would not predispose to the development of a reentrant signal. Under normal conditions, continued spread of a cardiac action potential ceases when the action potential reaches refractory tissue. In abnormal situations, an impulse may arrive at tissue that has returned to an excitable state. Therefore, conditions that predispose to reentry include lengthening of the conduction path (as can occur with cardiac dilatation), slowing of the velocity of conduction (as can occur with myocardial ischemia, hyperkalemia, or bundle branch blockade), or shortening of the refractory period (as occurs with epinephrine or excessively rapid pacing of the heart).

118. The answer is E. *(Physiology; Pulmonary/respiratory)*
The flowmeter tracing depicts the relationship between the rate at which gas is flowing during a forced vital capacity (FVC) maneuver and lung volume. An FVC maneuver begins at the total lung capacity (TLC, point A) and ends at the residual volume (RV, point D). At the onset of the FVC, peak expiratory flow depends on the force with which the respiratory muscles contract and is said to be 'effort-dependent.' The distance between points A and B represents the peak expiratory flow. As expiration proceeds, the flow rate decreases progressively as the lung volume decreases. The flow at these lower lungs volumes is maximal, and any additional effort will not increase the rate of flow. In this range (points C to D), expiratory flow rates are said to be "effort-independent." A TLC of 9 liters and an RV of 3 liters would be consistent with obstructive, not restrictive, lung disease, because air is trapped behind the area of obstruction. In restrictive lung disease, the reduced compliance of the lungs decreases the TLC and RV.

119. The answer is A. *(Physiology; Pulmonary/respiratory)*
The inside of an alveolus is lined with a watery fluid that forms an air–liquid interface with alveolar gas. The attraction between water molecules at the surface results in a tendency for the surface to get smaller (i.e., the attraction gives the surface the property known as surface tension). An attempt to expand the surface

requires work to overcome these surface forces. At tidal volumes taken just above the functional residual capacity (FRC), most of the work of inspiration is expended to overcome the surface tension that exists at the gas–liquid interface of each alveolus. Any substance that can lower the surface tension will make the surface easier to expand. Surfactant, produced by alveolar type II cells, is composed mainly of dipalmitoylphosphatidylcholine (DPPC). Surfactant lowers the surface tension, thus reducing the elastic recoil of the lungs. In its absence, alveoli would be smaller and harder to expand, leading to a reduced FRC. Surfactant has the unique property of being able to reduce the surface tension to a greater extent at low lung volumes than at high lung volumes, which promotes lung stability. In its absence, small alveoli are prone to collapse, followed sequentially by larger and larger alveoli, a condition known as progressive atelectasis. Premature infants are born with insufficient quantities of surfactant, and as a result suffer from respiratory distress syndrome (RDS). Their lungs have reduced compliance, and, because of the high surface tension, the alveoli are collapsed. Lung maturation, and therefore surfactant secretion, is inhibited by the presence of excessive amounts of insulin, making maternal diabetes a risk factor for RDS.

120. The answer is E. *(Physiology; Pulmonary/respiratory)*
Increased lung compliance, not reduced lung compliance, occurs as part of the normal aging process as elasticity in the lungs decreases. Sarcoidosis, a restrictive lung disease, reduces compliance, as do conditions that involve mechanical restriction of the lung (e.g., thoracic kyphoscoliosis, mesothelioma). Conditions that increase pulmonary venous pressure (e.g., mitral stenosis), lead to increased fluid in the interstitial space (e.g., bronchial asthma), or involve loss of surfactant [e.g., respiratory distress syndrome (RDS)] also reduce compliance.

121. The answer is C. *(Pharmacology; Hematopoietic/lymphoreticular)*
Chlorambucil is a slow-acting, orally administered alkylating agent that is useful in the treatment of chronic lymphocytic leukemia. Vincristine and prednisone are often used for inducing remission in patients with acute lymphocytic leukemia. Methotrexate, mercaptopurine, and cyclophosphamide are used for remission maintenance.

122. The answer is D. *(Pharmacology; Hematopoietic/lymphoreticular)*
Hormone antagonists such as tamoxifen are least likely to cause acute myelogenous leukemia in a patient who has been treated for another malignancy. Several of the nitrogen mustards (e.g., mechlorethamine, cyclophosphamide) and procarbazine have been shown to cause a secondary leukemia in a small percentage of patients who were first treated for another malignancy. Moderate to high dosages of ionizing radiation can also predispose to acute myelogenous leukemia.

123. The answer is B. *(Pharmacology; Hematopoietic/lymphoreticular)*
Clinical trials have demonstrated that zidovudine suppresses human immunodeficiency virus (HIV) replication and reduces patient morbidity and mortality. However, severe anemia, granulocytopenia, and thrombocytopenia may occur due to bone marrow suppression resulting from zidovudine's inhibition of DNA synthesis in human cells.

124. The answer is D. *(Pharmacology; Pharmacodynamic and pharmacokinetic processes)*
Unlike cyclosporine, azathioprine, prednisone, and cyclophosphamide, levamisole stimulates (rather than inhibits) the maturation and proliferation of T cells in patients with impaired immune function, and it is used as adjunctive therapy with fluorouracil and other agents in treating colorectal cancer. Cyclosporine inhibits the early stages of T-cell differentiation and blocks the synthesis of factors that stimulate the growth of T cells. Prednisone, cyclophosphamide, and azathioprine inhibit the proliferation of lymphocytes; these agents have a greater effect on T cells than on B cells.

125. The answer is D. *(Pharmacology; Pharmacodynamic and pharmacokinetic processes)*
Digoxin is administered to patients with low-output congestive heart failure to improve tissue perfusion and oxygenation. It does so by increasing cardiac contractility, which, in turn, increases the stroke volume and cardiac output at any given ventricular filling

pressure. The ejection fraction is also increased as a result of increased cardiac contractility. Venous pressure usually decreases following the administration of digoxin, secondary to a reduction in blood volume and a shift of blood volume from the venous to the arterial circulation as a result of increased cardiac output, diuresis, and the withdrawal of compensatory angiotensin-induced aldosterone augmentation of blood volume.

126. The answer is C. *(Pharmacology; Pharmacodynamic and pharmacokinetic processes)*
Diuretics such as furosemide are useful in the treatment of patients with congestive heart failure because of their ability to counteract edema by increasing urine output, but they do not increase cardiac output. Amrinone, a bipyridine phosphodiesterase inhibitor, and digoxin are positive inotropic agents that increase cardiac contractility, stroke volume, and cardiac output. Captopril, an angiotensin-converting enzyme (ACE) inhibitor, reduces angiotensin levels, thereby decreasing peripheral resistance and afterload and increasing cardiac output. Nitroprusside sodium is a vasodilator that decreases afterload and increases cardiac output.

127. The answer is E. *(Pharmacology; Pharmacodynamic and pharmacokinetic processes)*
Cardiac glycosides such as digoxin are believed to act by inhibiting sodium–potassium–adenosine triphosphatase (ATPase), thereby increasing intracellular sodium and leading to a secondary increase in intracellular calcium. The augmented intracellular calcium levels serve to increase cardiac contractility, resulting in an increased peak systolic tension. Although the rate of tension development increases, the time to reach peak tension remains unchanged and the duration of the cardiac systole is not increased by digitalis.

128. The answer is D. *(Pharmacology; Pharmacodynamic and pharmacokinetic processes)*
Digoxin augments vagal tone and thereby slows atrioventricular conduction. Slowing of atrioventricular conduction is reflected on the electrocardiogram (EKG) as prolongation of the PR interval. The atrioventricular refractory period is also increased. Digoxin accelerates ventricular repolarization and the action potential duration and refractory period are decreased, resulting in a shortened QT interval. In contrast, digoxin does not slow ventricular conduction velocity; therefore, the QRS interval is not prolonged.

129. The answer is C. *(Pharmacology; Pharmacodynamic and pharmacokinetic processes)*
Toxic concentrations of cardiac glycosides increase automaticity by producing afterpotentials (afterdepolarizations). Because excessively elevated intracellular calcium is the mechanism behind the arrhythmias, calcium administration is not an effective treatment for digitalis toxicity. Digitalis toxicity may result from diuretic-induced hypokalemia that augments digoxin inhibition of sodium–potassium–adenosine triphosphatase (ATPase); potassium administration may be indicated if hypokalemia is present. Lidocaine is an effective antiarrhythmic agent for treating ventricular arrhythmias caused by digitalis. Magnesium salts may also be useful, partly because magnesium ions are physiologic antagonists of many calcium-induced effects on excitable tissue. If severe toxicity is present, digoxin antibody should be administered to rapidly reduce serum digoxin levels and reverse toxic effects.

130-132. The answers are: 130-A, 131-F, 132-H. *(Pathology, Physiology; Cardiovascular)*
Venoconstriction increases venous return to the right heart, which increases the preload, or the volume of blood that fills the ventricles in diastole. Recall the Frank-Starling length–tension relationship: the more it fills, the more it spills.

Restricting salt and water intake in a patient with congestive heart failure reduces blood volume, which reduces preload in the ventricles.

Increasing contractility, or the force of contraction, increases the stroke volume by increasing the ejection fraction without affecting the preload or afterload.

133-134. The answers are: 133-D, 134-A. *(Pathology; Tissue biology and associated response to disease)*
Nonspecific esterase stains are useful for identifying the monocytes of acute monocytic leukemia. Acute monocytic leukemia frequently infiltrates the gums.

In acute lymphoblastic leukemia, the periodic acid–Schiff (PAS) stain reveals chunks of PAS-positive material in the cytoplasm. The PAS stain also reveals bizarre erythroblasts in Diguglielmo's erythroleukemia (normally, red blood cells are PAS-negative). In the Sezary syndrome, the PAS stain reveals neoplastic T cells with a cleft in the nucleus.

Specific esterase stains are useful in differentiating acute myelogenous leukemias from lymphoid and monocytic leukemias. This stain is also useful in identifying immature granulocytes in soft tissue collections of leukemic cells. Tartrate-resistant acid phosphatase stain is specific for hairy cell leukemia, a B cell leukemia. Platelet peroxidase stains that are visible by electron microscopy are useful for identifying acute megakaryocytic leukemia. The leukocyte alkaline phosphatase stain is useful in differentiating chronic myelogenous leukemia (CML) from other disorders associated with increased granulocyte counts (leukemoid reactions). In CML, the granulocytes do not stain for alkaline phosphatase, whereas leukemoid reactions result in cells that stain positive for alkaline phosphatase.

135-137. The answers are: 135-C, 136-B, 137-A. *(Physiology; Pulmonary/respiratory, Cardiovascular)*
The total oxygen content of arterial blood is determined by the amount of oxygen in solution plus the oxygen that is chemically combined with hemoglobin. The oxygen in solution is responsible for the arterial oxygen tension (PaO_2). The PaO_2 is determined by alveolar ventilation and pulmonary capillary blood flow.

The PaO_2 determines the percent saturation of hemoglobin with oxygen.

The hemoglobin concentration does not determine the PaO_2; therefore, anemic individuals have a normal PaO_2 and a normal percent saturation of hemoglobin, but a lower total oxygen content.

138-140. The answers are: 138-D, 139-B, 140-C. *(Pathology; Pulmonary/respiratory)*
Neurogenic tumors are most commonly located in the posterior mediastinum (D). These tumors account for approximately 20% of all mediastinal tumors, regardless of location. In adults, neurogenic tumors are usually benign (e.g., ganglioneuroma) and asymptomatic, whereas in children, they are usually malignant (e.g., neuroblastoma) and symptomatic. Erosion of the spinal neural foramen is a useful diagnostic sign.

Thymomas are associated with myasthenia gravis and pure red blood cell aplasia. These tumors are most commonly located in the anterior mediastinum (B). Most thymomas (approximately 70%) are benign. Invasiveness, rather than histologic appearance, is the best way of distinguishing malignant tumors.

Pericardial cysts are the most common cystic lesion in the mediastinum and are most frequently located in the middle mediastinum (C). Bronchogenic cysts also occur in this compartment.

141-146. The answers are: 141-H, 142-B, 143-D, 144-E, 145-A, 146-F. *(Pharmacology; Hematopoietic/lymphoreticular)*
Anemia associated with aplastic bone marrow is best treated with erythropoietin and granulocyte colony stimulating factors (G-CSFs, such as filgrastim) or granulocyte-macrophage CSFs (GM-CSFs, such as sargramostim). Filgrastim and sargramostin are synthetic hematopoietic growth factors produced by recombinant DNA technology, and are indicated for bone marrow disorders including aplastic anemia, myeloproliferative disorders, multiple myeloma, and chemotherapy-induced leukopenia.

Vitamin B_{12} deficiency causes a megaloblastic anemia with the delayed onset of neurologic signs such as paresthesias (pernicious anemia); therefore, the disorder is best treated with cyanocobalamin (vitamin B_{12}).

Folic acid deficiency, such as may occur in malnourished patients, chronic alcoholics, and patients receiving drugs such as phenytoin, causes a megaloblastic anemia and is treated with oral folic acid supplementation. Folic acid will also correct the hematologic deficiency in pernicious anemia, but will not correct the neurologic deficit.

Ferrous sulfate and other oral iron preparations are suitable for all but the most severe form of iron deficiency anemia, which is a microcytic and hypochromic anemia. Ferrous sulfate would be indicated for a patient with a decreased mean corpuscular volume and hemoglobin concentration.

Patients with end-stage renal disease lack erythropoietin, which must be replaced. Erythropoietin is also useful in treating anemias associated with AIDS, cancer, and chronic inflammation.

147-151. The answers are: 147-E, 148-A, 149-B, 150-D, 151-C. *(Pharmacology; Pharmacodynamic and pharmacokinetic processes)*

Aldesleukin is a recombinant interleukin-2 (IL-2), a lymphokine that promotes differentiation and activity of killer lymphocytes. It is used in treating renal cell carcinoma.

Prednisone is preferred for the treatment of several autoimmune disorders, including acute glomerulonephritis, thrombocytopenic purpura, and hemolytic anemia. Prednisone is also used in the treatment of systemic lupus erythematosus (SLE) and rheumatoid arthritis.

Cyclosporine is the primary agent used to prevent rejection of transplanted organs.

Muromonab-CD3 is a monoclonal antibody to the CD3 antigen on the surface of T cells; therefore, it blocks the cytotoxic effect of T cells. It is used to counteract renal allograph rejection.

Levamisole is an immunostimulant that promotes maturation and proliferation of lymphocytes and is useful as an adjunct to chemotherapy in colorectal cancer.

152-162. The answers are: 152-K, 153-C, 154-H, 155-A, 156-D, 157-E, 158-I, 159-A, 160-G, 161-F, 162-J. *(Gross anatomy; Pulmonary/respiratory)*

In a patient who is sitting or standing up, a foreign body or aspirated material would most commonly localize in the posterobasal segment of the right lower lobe. Aspirated material or foreign bodies can lead to a lung abscess with a mixed aerobic and anaerobic flora.

Panacinar emphysema most commonly involves the respiratory unit (i.e., the respiratory bronchioles, alveolar ducts, and alveoli). α_1-Antitrypsin deficiency is the most common cause of this type of emphysema. Absence of α_1-antitrypsin allows the elastases produced by neutrophils to destroy the elastic tissue that supports these respiratory units.

In a patient who is lying on his right side, a foreign body or aspirated material would most likely localize to the posterior segment of the right upper lobe.

Small cell carcinomas of the lung are centrally located bronchogenic carcinomas that are derived from Kulchitsky cells of neural crest origin. There is a high association between the incidence of small cell lung carcinoma and cigarette smoking. Small cell carcinomas are highly malignant and aggressive tumors. Surgery is not usually a treatment option because disease is often disseminated at the time of presentation.

The most common primary site of malignancy for the Pancoast tumor and for patients who present with hoarseness is the apex of the lung. Invasion of the brachial plexus (usually T1–T2) and the cervical sympathetic ganglion by a primary lung cancer is referred to as Pancoast syndrome. Destruction of the T1–T2 sensory and motor nerves produces unilateral arm pain, medial paresthesias, and wasting of the small muscles of the hand. Destruction of the cervical sympathetic ganglion produces Horner's syndrome, which is characterized by the triad of ipsilateral lid lag, meiosis (pinpoint pupil), and anhidrosis (lack of sweating). The recurrent laryngeal nerve can also be destroyed, resulting in hoarseness.

The periphery of the lung is the most common site for adenocarcinoma (the most common type of primary lung cancer), scar cancer (commonly associated with adenocarcinoma), Ghon foci (i.e., subpleural areas of caseous necrosis in primary tuberculosis), and pulmonary infarction.

In a patient who is lying on her back, aspiration of a foreign body or aspirated material would most commonly localize in the superior segment of the right lower lobe.

Like small cell carcinomas of the lung, primary squamous cell carcinomas have a strong association with cigarette smoking. Primary squamous cell carcinomas are most commonly centrally located in the first few segments of the bronchus. In this location, they commonly produce cough, hemoptysis, and obstruction.

The hilar lymph nodes are the most common site for metastasis of primary lung cancers. The hilar lymph nodes are involved in 93%–96% of patients.

In a patient who is lying on his left side, aspiration of a foreign body or aspirated material most commonly localizes in the lingula of the left lung.

A consolidation (e.g., lobar pneumonia) that could obliterate a border of the heart on a postero-anterior radiograph would most commonly be located in the right middle lobe. The four "bumps" making up the left heart contour on a normal postero-anterior chest film are, from top (superior) to bottom (inferior), the aortic knob, the main pulmonary artery, the left atrial appendage, and the left ventricle. The right heart contour has three potential "bumps"—from top to bottom, the superior vena cava, the ascending portion of the aorta, and the right atrium. The right ventricle lies anteriorly and forms the anterior cardiac border on a lateral film. The posterior heart border on a lateral film is made up of the superiorly located left atrium and the more inferiorly located left ventricle.